THEORY CHANGE, ANCIENT AXIOMATICS, AND GALILEO'S METHODOLOGY

VOLUME I

SYNTHESE LIBRARY

STUDIES IN EPISTEMOLOGY,

LOGIC, METHODOLOGY, AND PHILOSOPHY OF SCIENCE

VOLUME 145

THEORY CHANGE, ANCIENT AXIOMATICS, AND GALILEO'S METHODOLOGY

Proceedings of the 1978 Pisa Conference on the History and Philosophy of Science

VOLUME I

Edited by

JAAKKO HINTIKKA, DAVID GRUENDER

Dept. of Philosophy, Florida State University, Tallahassee

and

EVANDRO AGAZZI

Dept. of Philosophy, University of Genoa, Italy

D. REIDEL PUBLISHING COMPANY

DORDRECHT : HOLLAND / BOSTON : U.S.A.

LONDON : ENGLAND

Library of Congress Cataloging in Publication Data

Pisa Conference on the History and Philosophy of Science, 1978.
Theory change, ancient axiomatics, and Galileo's methodology.

(Its Proceedings of the 1978 Pisa Conference on the History and Philosophy of Science; v. 1) (Synthese library; v. 145)
Includes index.
1. Science—Philosophy—Congresses. 2. Science—Methodology—Congresses. 3. Change—Congresses. 4. Axioms—Congresses. I. Hintikka, Kaarlo Jaakko Juhani, 1929– II. Gruender, C. David, 1927– III. Agazzi, Evandro, 1934– IV. Title.
Q124.6.P57 1978, vol. 1 [Q174] 509s [501]
ISBN-13: 978-94-009-9047-0 e-ISBN-13: 978-94-009-9045-6
DOI: 10.1007/978-94-009-9045-6

Published by D. Reidel Publishing Company,
P.O. Box 17, 3300 AA Dordrecht, Holland.

Sold and distributed in the U.S.A. and Canada
by Kluwer Boston Inc.,
190 Old Derby Street, Hingham, MA 02043, U.S.A.

In all other countries, sold and distributed
by Kluwer Academic Publishers Group,
P.O. Box 322, AH Dordrecht, Holland.

D. Reidel Publishing Company is a member of the Kluwer Group.

TABLE OF CONTENTS

SECTION III: THE PHILOSOPHICAL PRESUPPOSITIONS AND SHIFTING INTERPRETATIONS OF GALILEO

TABLE OF CONTENTS OF COMPANION VOLUME

SECTION VI: WHAT CAN THE HISTORY AND PHILOSOPHY OF SCIENCE AND PHILOSOPHY DO FOR EACH OTHER? A ROUND-TABLE DISCUSSION

PREFACE

The two volumes to which this is a preface consist of the Proceedings of the Second International Conference on History and Philosophy of Science. The Conference was organized by the Joint Commission of the International Union of History and Philosophy of Science (IUHPS) under the auspices of the IUHPS, the Italian Society for Logic and Philosophy of Science, and the Domus Galilaeana of Pisa, headed by Professor Vincenzo Cappelletti. Domus Galilaeana also served as the host institution, with some help from the University of Pisa. The Conference took place in Pisa, Italy, on September 4–8, 1978.

The editors of these two volumes of the Proceedings of the Pisa Conference acknowledge with gratitude the help by the different sponsoring organizations, and in the first place that by both Divisions of the IUHPS, which made the Conference possible. A special recognition is due to Professor Evandro Agazzi, President of the Italian Society for Logic and Philosophy of Science, who was coopted as an additional member of the Organizing Committee. This committee was otherwise identical with the Joint Commission, whose members were initially John Murdoch, John North, Arpad Szabó, Robert Butts, Jaakko Hintikka, and Vadim Sadovsky. Later, Erwin Hiebert and Lubos Novy were appointed as additional members.

The announced program of the Conference is included in each of the two volumes. A number of changes as compared with this planned schedule are worth mentioning. A reception in honor of the Conference was given by the city of Pisa on September 6, and is hereby gratefully acknowledged. Several of the expected participants were unable to come to Pisa. They include Erwin Hiebert, A. T. Grigorian, Wesley C. Salmon, M. Clavelin, Karel Berka, I. A. Akchurin, B. M. Kedrov, V. S. Kirsanov, and S. R. Mikulinsky, (Arpad Szabó participated only in the first two and a half days of the meeting.) Of the absentees, Hiebert, Salmon, Clavelin, Kirsanov (jointly with Markova), and Berka have nevertheless contributed to the proceedings of the Conference. As will be seen from these volumes, interesting contributions were also made by originally unscheduled participants. L. Jonathan Cohen's contribution to the Pisa Conference has meanwhile appeared in a modified form in the *Journal of the History of Ideas* **41** (1980), 219–231. The modifications have rendered redundant the interesting comments which Robert Butts contributed to the Conference. This is not the case with Ian Hacking's remarks in our Vol. II below, which now should be referred to Cohen's paper, published as indicated above, even though Professor Hacking, through no fault of his own, wrote his contribution on the basis of what Dr. Cohen said in Pisa.

As is seen from the program below, the Conference was organized in six sections. Papers from Sections I–III are published as Volume I, and those from Sections IV–VI as Volume II of these Proceedings.

The papers published in these two volumes speak for themselves. They are not primarily intended to present the last word on any one subject. Rather, their purpose is not just to contribute to a discussion of a worthwhile subject which is of interest to both historians of science and philosophers of science, but also to exemplify the actual co-operation of representatives of these two sister disciplines. The co-operation of those two fields of study has in fact long been one of the aims of Synthese Library. Hence it was not hard to find a natural publication forum for these proceedings, especially as the output of its predecessor conference in 1964 has been partly published in the same series. It is not for the managing editor of Synthese Library to judge the papers appearing here in the light of this objective of the Conference. It is in any case amply obvious that the Conference produced extremely lively exchanges and several unexpected combinations of ideas.

Another aim of the organizers was to have a variety of approaches and a variety of geographical areas represented at the Conference. Whether this aim was realized or not as fully as it could have been, we are in any case pleased to be able to include in this volume several contributions by participants from Eastern Europe which will allow the readers of Synthese Library an interesting glimpse of the work done there in the field of the Conference. Again, the aims of the organizers and the policies of Synthese Library converge.

There could not have been a more felicitous location for a joint meeting of philosophers and historians of science than Domus Galiaeana. The organizers are most grateful to Professor Cappelletti for this privilege. It was symbolically appropriate that the first volume of the *Carteggio dei Discepoli di Galileo* was officially presented to the public as a part of the Conference. In a wider sense, we are all disciples of Galileo, even though these Proceedings appear in Synthese Library and not in the *Carteggio.*

We are thankful to D. Reidel Publishing Company for including these two proceedings volumes in one of their series. Much of the detailed editorial work was done by Ms. Jayne Moneysmith, to whom our warm thanks are hence due. In the compilation of the indexes, she was assisted by Mr. Jim Garrison and Ms. Susan Leffler, whom we would also like to thank.

On behalf of the editors and organizers,

JAAKKO HINTIKKA

PROGRAM OF THE SECOND INTERNATIONAL CONFERENCE ON THE HISTORY AND PHILOSOPHY OF SCIENCE

Pisa, Domus Galilaeana: 4–8 September 1978

OPENING ADDRESS: Vicenzo Cappelletti

TOPIC I:	The Structure of Theory Change
FOCUS:	The Sneed–Stegmüller "Rational Reconstruction" of Kuhn's Views
CHAIRMAN:	Lorenz Krüger
PAPERS BY:	Ilkka Niiniluoto (Finland) – Zev Bechler (Israel)
COMMENTATORS:	Joseph Sneed (Holland) – Vadim Sadovsky (USSR) – Paolo Rossi Monti (Italy)

TOPIC II:	The Early History of the Axiomatic Method
FOCUS:	Szabo's Theory of the Origin of the Axiomatic Method
CHAIRMAN:	John Murdoch
PAPERS BY:	Wilbur Knorr (USA) – Patrick Suppes (USA) – M.V. Popovich (USSR)
COMMENTATORS:	Árpád Szabó (Hungary) – Karel Berka (Czechoslovakia)

TOPIC III:	Philosophical Presuppositions and Shifting Interpretations of Galileo
FOCUS:	The Shift in the Interpretation of Galileo from Koyré to Drake
CHAIRMAN:	Vincenzo Cappelletti
PAPERS BY:	David Gruender (USA) – W. Wisan (USA) – V.S. Kirsanov (USSR) – L.A. Markova (USSR)
COMMENTATORS:	M. Clavelin (France) – Carlo Maccagni (Italy) – A.C. Crombie (GB) – N. Jardine (GB) – B. Kouznetsov (USSR)

TOPIC IV:	Probability Theory and Probabilistic Thinking in the Seventeenth to Nineteenth Centuries
FOCUS:	Hacking's View on the Development of Probability Concepts
CHAIRMAN:	Robert Butts
PAPERS BY:	Ivo Schneider (BRD) – Anne Fagot (France)
COMMENTATORS:	Ian Hacking (Canada and USA) – L. Jonathan Cohen (GB) – A.T. Grigorian (USSR) – V.N. Kostjuk (USSR) – Lubos Novy (Czechoslovakia)

TOPIC V:	Thermodynamics and Physical Reality since 1850
FOCUS:	The Mach–Boltzmann Contrast
CHAIRMAN:	Patrick Suppes
PAPERS BY:	Erwin Hiebert (USA) – Lorenz Krüger (BRD) – Ulises Moulines (Mexico)
COMMENTATORS:	John North (GB and Holland) – Nancy Cartwright (USA) – I.A. Akchurin (USSR) – Kurt M. Pedersen (Denmark) – E. Bellone (Italy)

TOPIC VI:	Round Table Discussion: WHAT CAN THE HISTORY AND PHILOSOPHY OF SCIENCE DO FOR EACH OTHER?
CHAIRMAN:	Evandro Agazzi
	John Murdoch (USA) – Robert Butts (Canada) – Erwin Hiebert (USA) – B.M. Kedrov (USSR) – S.R. Mikulinsky (USSR) – Evandro Agazzi (Italy) – A.T. Grigorian (USSR)

The first volume of the *Carteggio dei Discepoli di Galileo* will be presented during the course of the meeting (Thursday 7 September, 6 p.m.)

DIRECTORS	SECRETARIAT
Jaakko Hintikka	Guido Cimino
Evandro Agazzi	Marcello Tricarico
Vincenzo Cappelletti	

SECTION I

THE STRUCTURE OF THEORY CHANGE

ILKKA NIINILUOTO

THE GROWTH OF THEORIES: COMMENTS ON THE STRUCTURALIST APPROACH

I. INTRODUCTION: THE BACKGROUND AND THE RISE OF THE SNEED-STEGMÜLLER-PROGRAMME

The nature and function of scientific theories is perhaps the most central problem within the philosophy of science. To study the basic features of theories and their role in scientific inquiry, a philosopher should be able to construe them as entities which he can handle and investigate. It is therefore no wonder that questions concerning the most appropriate ways of reconstructing scientific theories have a crucial importance in discussions about the proper method which philosophers of science should follow.

In this century, perhaps most philosophers of science have employed the *logistic method* for the analysis of the structure of scientific theories. This programme was initiated in Gottlob Frege's *Begriffschrift* (1879), where Frege defined a "formula language" by means of which he attempted to "provide a more detailed analysis of the concepts of arithmetic and a deeper foundation for its theorems". Frege was also confident that his ideography could be successfully applied to the study of the foundations of the differential and integral calculus, to geometry, and eventually "to the pure theory of motion and then to mechanics and physics" (see Frege, 1967, pp. 7–8). The new symbolic logic of Frege, Peano, and Russell provided philosophers with a tool for formalizing the informal axiomatic method which had been successfully applied in mathematics and physics. David Hilbert coined the term 'metamathematics' for the proof-theoretical study of mathematical theories, where theories are construed as *sets of sentences in a formal* (usually first-order) *language.* In the 1920's, the same approach was extended from mathematical to empirical theories by the logical positivists who treated theories as consisting of a set of axioms in an uninterpreted theoretical language (pure theory) together with a set of sentences which, perhaps partially, define or interpret theoretical terms by means of observational terms (correspondence rules). Emended with some insights and

J. Hintikka, D. Gruender, and E. Agazzi (eds.), Pisa Conference Proceedings, Vol. I, 3–47.

concepts from Alfred Tarski's logical semantics, this 'partial interpretation view' was further developed and modified by Rudolf Carnap, Carl G. Hempel, Ernest Nagel, and others; in its established form it is usually referred to as the 'Received View' or the 'Standard Conception' of scientific theories.[1]

The Received View was heavily attacked from many angles during the 1960's. One source of its troubles is the questionable assumption that theories contain, in their object language, a class of syntactic entities (correspondence rules) which serve special semantic *and* methodological functions (interpretation of theoretical terms). Another is the theoretical-observational dichotomy which has convincingly been shown to be untenable – at least as a general semantic division. Thirdly, it has been argued that the 'semantical empiricism' underlying the partial interpretation view does not do justice to the antecedent meaning of terms occurring in scientific theories. Fourthly, in the logistic reconstruction theories are treated as 'Finished Research Reports' (to use N.R. Hanson's words), not as historically developing and changing creations of scientific communities. The latter conception of theories was, however, strongly suggested by the work of such historians and philosophers as Thomas Kuhn, Stephen Toulmin, Paul Feyerabend, and Imre Lakatos. Finally, some philosophers argued that the first-order formalizations of physical theories are either too clumsy to be useful or else restricted to relatively uninteresting simple situations, while some others entirely denied the value of the formalization and the axiomatization of theories.

At the end of the 1960's, it was quite generally agreed that there is something wrong with the Received View. Several different attitudes towards the treatment of theories prevailed among philosophers of science at that time. Let us mention some of the most important ones.

(1) Many philosophers continued to work within the Carnap–Reichenbach–Hempel–Scheffler–Braithwaite–Nagel tradition, but only cautiously and with growing scepticism. Wolfgang Stegmüller's survey *Theorie und Erfahrung* (1970) is an excellent document of this attitude. Two serious blows to this approach were given by Hempel's rejection of the Received View in 1969 (see Hempel, 1970) and by Carnap's death in 1970.

(2) Some philosophers suggested that, apart from a few elementary results, the available logical techniques have played "a surprisingly

small role" in the methodological literature dealing with scientific theories (Hintikka and Tuomela, 1970, p. 298). Among these available but unused logical tools one can mention the logical theory of definition (cf. Rantala, 1977) and Tarski's model theory. Since the 1950's, the latter has been applied with very impressive results to the study of mathematical theories, especially of algebraic theories and of set theory. Model-theoretic methods have later been used also in the philosophy of science, most notably by Polish philosophers – in particular, by Marian Przełecki in *The Logic of Empirical Theories* (1969).[2] Raimo Tuomela's *Theoretical Concepts* (1973) applies these strong logical tools to the syntactic and semantic analysis of first-order theories; in his 'realist' view of theories Tuomela also tries to take into account the criticism which has been directed against the Received View, thus separating the logistic approach from the typical philosophical views which were combined with it by the logical empiricists.

(3) In his article 'Deterministic Theories' (1961), Richard Montague gave an explicit first-order axiomatization to Newton's mechanics. As he had to build all the required mathematics into the postulates, his axiomatization looked formidably complex. Some philosophers have suggested that, in axiomatizing specific physical theories, one should not imitate the metamathematicians (i.e., the logistic approach) but rather the mathematicians or the physicists themselves. Inspired by von Neumann's treatment of quantum mechanics, Evert Beth outlined a 'semantic' approach to physical theories (see Beth, 1961) which has been further developed by Bas van Fraassen (1970); related proposals have been made, independently, by Erhard Scheibe (1973) and Frederick Suppe (1972, 1974). Another approach, which follows the practice of mathematicians to define theories directly by means of their models (e.g., "a triple $\langle S, \mathcal{A}, P\rangle$ is a probability space if and only if...."), was started with the axiomatization of Classical Particle Mechanics by J. McKinsey, A. Sugar, and P. Suppes in 1953 and with the axiomatization of the Rigid Body Mechanics by Ernest Adams in 1955. This approach is characterized by the slogan: *To axiomatize a theory is to define a set-theoretical predicate*; it has been defended and applied by Patrick Suppes in a number of works and articles.[3]

(4) Several philosophers have searched for conceptual units which are more comprehensive and more flexible than the notion of an axiomatized theory. Of these attempts, which are mainly inspired by a

desire to be able to discuss scientific change, one can mention W.V. Quine's 'holistic' network model for theories (cf. Hesse, 1974), Kuhn's (1962) 'paradigms', Toulmin's (1972) 'conceptual systems', Lakatos's (1970) 'research programmes', Dudley Shapere's 'scientific domains' (Shapere, 1974), and Larry Laudan's 'research traditions' (Laudan, 1977). Most of these concepts, especially Kuhn's, contain pragmatic notions which refer to the community of investigators who support a scientific theory; they also emphasize the role of fundamental theories as deep commitments to comprehensive world views.

(5) Finally there is the view (sometimes combined with (4)) which is sceptical over the possibility of analysing the axiomatic structure of theories. Thus, Peter Achinstein suggests that theories can be characterized only by first defining the pragmatic concept '*A* has a theory *T*' and then saying that theories are, roughly speaking, those entities held by some people at some time (see Achinstein, 1968, Ch. 4).

The set-theoretic approach of Suppes and Adams has made rapid progress in this decade. Joseph Sneed studied its foundations during 1966–70 and published his results in *The Logical Structure of Mathematical Physics* (1971). In this technical work, Sneed proposed a new theory-relative notion of theoretical functions; he defined theories of mathematical physics as pairs consisting of a mathematical structure ('core') and its putative applications.[4] He further introduced the notion of 'constraint' (which essentially binds together some of the applications of a theory) and proposed an 'emended Ramsey view' for expressing the empirical content of theories. In the final chapter, 'The Dynamics of Theories', Sneed discussed some aspects of the development of theories – how people come to have them and how they cease to have them.

The first reactions to Sneed's work concerned his notion of Ramsey-eliminability.[5] At the end of 1972, his book was discussed and compared to the model-theoretic approach by the Polish philosophers (cf. Przełecki, 1974, and Wojcicki, 1974b). At the same time, Stegmüller read Sneed's book – hoping to find in it something interesting about the Ramsey view of theories. What he discovered there was quite different: first, a new conception of scientific theories, viz. a *non-statement* or *structuralist view* (i.e., theory is a mathematical structure together with a set of applications) as opposed to the traditional *statement view* (i.e., theory is a set of statements), and, secondly, a sketch of the dynamics of theories which, he thought, for the first time made possible the reconstruction of Kuhn's conception of the development of science.[6] In 1973, Stegmüller published a sequel *Theorienstruktur und Theoriendynamik* to his *Theorie und Erfahrung*; it presented Sneed's formalism by using a simplified and more perspicious notation together with a more detailed discussion of the development of theories by using this formalism. Stegmüller also made a number of general comments concerning Kuhn's, Popper's, Lakatos's, and Feyerabend's views on theory-change. He has later clarified, extended, and defended in several papers this new programme for studying the structure and the growth of scientific theories.[7]

Stegmüller's enthusiasm brought Sneed to Munich in 1974–75 and to work on physical theories again. Together with Wolfgang Balzer, Sneed started to revise, generalize and simplify his formalism. The results have been published in Sneed (1976) and Balzer & Sneed (1977).[8] Other collaborators of this programme have done technical work on the Sneedian approach in thermodynamics (Moulines, 1975b), on the approximative application of theories (Moulines, 1976), and on the concept of reduction (Mayr, 1976).

One of the most impressive achievements of the set-theoretical approach has been Kuhn's very favorable response to it. Stegmüller sent in early 1974 his book on theory dynamics to Kuhn. On September 1975, Sneed, Stegmüller, and Kuhn presented papers in a Symposium on Theory Change in the London, Ontario, Congress on Logic, Methodology, and Philosophy of Science.[9] Even though Kuhn made the reservation that the Sneed formalism "currently does virtually nothing to clarify the nature of revolutionary change", he added that "to a far greater extent and also far more naturally than any previous mode of formalization, Sneed's lends itself to the reconstruction of theory dynamics" (Kuhn, 1976, p. 184). He also said that "Stegmüller, approaching my work through Sneed's, has understood it better than any other philosopher who has made more than passing reference to it" (*ibid.*, p. 179).

Even though Sneed is not the first who has proposed formal models for scientific change,[10] his set-theoretic treatment of theories is the first one which has gained any support from a leading exponent of the historical approach. Against those who are opposed to all kinds of formalization, Kuhn's comments suggest that a sufficiently rich formal description of the structure of theories which at the same time includes a reference to a number of pragmatic factors may be a valuable tool in the analysis of theory dynamics. The admission that such formal models may be relevant from the viewpoint of the historians gives a serious challenge to logically minded philosophers of science to work in this direction. (Note that the more structure one puts in the description of a theory, the more possibilities one has for distinguishing between different types of theory change.) It also indicates that the diverse approaches within the philosophy of science can not only complement but also co-operate with each other.

How adequately does Sneed's set-theoretical formalism represent the structure of scientific theories and the patterns of their growth? To what direction can it be further developed? Is there an essential difference between the set-theoretic and the model-theoretic formalizations? Can the Sneedian approach be extended from mathematical physics to other fields of science? To what extent can one reconstruct actual examples from the history of science by means of it? What new insights does this formalism give about theory dynamics? To what extent does it allow for the formalization of Kuhn's conception of scientific change? Does it favor, in some systematic sense, Kuhnian ideas in contrast with such rival accounts as Popper's or Lakatos's? These are examples of questions which are raised by the work of Sneed, Stegmüller, and their collaborators. One cannot attempt to answer in detail to all these questions in a single paper; therefore this paper contains only a preliminary survey of them.

2. THE STRUCTURALIST CONCEPTION OF THEORIES

In this section, I shall outline the elements of the structuralist notion of a scientific theory. At the same time, I shall make some technical and philosophical comments which seem to me to give good reasons for modifying and complementing some of the basic definitions.

(a) *Core*

A non-empty class of $m+k$-tuples of the form $\langle n_1, \ldots, n_m, t_1, \ldots, t_k \rangle$, $m > 0$, $k \geq 0$, is called a $m+k$-matrix, if $n_1, \ldots, n_m, t_1, \ldots, t_k$ are sets, relations, or functions. Here $n_1, \ldots, n_m$ are the non-theoretical components and $t_1, \ldots, t_k$ are the theoretical components of the matrix. (For this distinction, see below.) Let M_p be a $m+k$-matrix, and let M_{pp} be corresponding $m+0$-matrix which is obtained from M_p by the restriction function, i.e. by $r: M_p \to M_{pp}$ such that $r(\langle n_1, \ldots, n_m, t_1, \ldots, t_k\rangle) = \langle n_1, \ldots, n_m\rangle$. If $M \subseteq M_p$, then the quadruple $F = \langle M_p, M_{pp}, r, M\rangle$ is a *frame*. Here M_p is the class of the *potential models* of F, M_{pp} is the class of the *partial potential models* of F, and M is the class of the *proper models* of F.

A *core* $K = \langle F, C\rangle$ is obtained from a frame $F = \langle M_p, M_{pp}, r, M\rangle$ by adding to it a *constraint* C for M_p, i.e., a class $C \subseteq \text{Pot}(M_p)$ such that (i) $\emptyset \notin C$, (ii) if $x \in M_p$, then $\{x\} \in C$, and (iii) if X, $Y \in \text{Pot}(M)$, $X, Y \neq \emptyset$, $X \in C$, and $Y \subseteq X$, then $Y \in C$.[12]

Comments: The intuitive idea is that M_p is "the set of all possible models for the *full* conceptual apparatus of a theory" including theoretical components, M_{pp} is the set of the corresponding non-theoretical models, M is the set of those potential models which are not excluded by the fundamental law of the frame, and finally C is a set of restrictions which rule out certain combinations of components in different potential models (Balzer and Sneed, 1977, p. 198). However, these ideas are only partly formalized by the given definitions – which thus seem to be too general. In the first place, the components of an element of a matrix are allowed to be any sets, relations or functions whatsoever.[13] For example, one element of a matrix may contain only binary relations as components, while another element of the same matrix may contain only 5-place relations. Secondly, the definition allows M_p to be any $m+k$-matrix of any non-empty size. For example, M_p might contain only one element. These unintended cases can be eliminated by requiring that an $m+k$-matrix is a collection of many-sorted[14] structures of a given *type*, where a type is essentially a finite set of functions which indicate what sort of set-theoretical entities can be chosen as the different components, and by assuming that M_p is the class of *all* structures of the given type (cf. Rantala, 1978a). With these stipulations (which we shall always assume below), the representation of the 'fundamental law' by means of a set M of models in M_p becomes identical to the standard 'semantic' treatment of propositions as classes of models,[15] which has the virtue that one can speak of propositions independently of their various linguistic formulations.

Let us denote the types of M_p and M_{pp} by τ_p and τ_{pp}, respectively. Then τ_p and τ_{pp} are finite set-theoretical entities, while M_p and M_{pp} are proper classes. From the set-theoretical viewpoint, it therefore would be much more convenient to operate with types than with the corresponding proper classes. Moreover, when we later say that a person holds a theory, it might seen more realistic to assume that, in using a core, this person is operating with the finite types rather than with the proper classes. We could thus define cores as quadruples $\langle \tau_{pp}, \tau_p, M, C\rangle$, where τ_p is a $m+k$-type, τ_{pp} is a $m+0$-type which is a restriction of τ_p, M is a class of structures of type τ_p and C is a constraint for structures of type τ_p. However, here M and C are still quite complicated set-theoretical entities.[16] This is one reason – but, as we shall see later, not the only reason – for replacing M and C with their finite formulations.

Let us say that a pair of first-order languages $\langle L_0, L\rangle$, where L_0 is of type τ_{pp}, L is of type τ_p, and L_0 is a sublanguage of L, is a *language-pair for core* $K = \langle M_{pp}, M_p, M, C\rangle$. Here L_0 is a non-theoretical language and L is the full theoretical language; the terms belonging to L but not to L_0 constitute the *theoretical vocabulary* of L. All the ordinary model-theoretical notions can now be defined in the usual way. If a sentence σ of L is true in a structure $z \in M_p$, we shall write $z \models \sigma$ or $z \in \text{Mod}(\sigma)$. If Σ is a set of sentences in L, then $\text{Mod}(\Sigma) = \bigcap_{\sigma\in\Sigma} \text{Mod}(\sigma)$. If there is a set of sentences Σ in L such that $M = \text{Mod}(\Sigma)$, then we shall say that Σ is a *formulation of the fundamental law M.*[17]

The formulation of the constraint C is not possible in a language L for core $K = \langle M_{pp}, M_p, M, C\rangle$ (cf. Przełecki 1974). To do this, we need a language L^* which is interpreted on the domain M_p and which contains the vocabulary of set theory (cf. Harris, 1978). As L^* speaks of the models of the language L, it is essentially a semantic metalanguage for L. If f is a function symbol in L and if z is a variable in L^* ranging over the elements of M_p, then f^z is a function symbol in L^* which denotes the interpretation of f in z. One can now *formulate constraints in L**: for example,

$$\text{Dom}(z_1) = \text{Dom}(z_2)$$

says that models z_1 and z_2 have the same domain, and

$$\forall z_1 \forall z_2 (f^{z_1} \restriction_{\text{Dom}(z_1)\cap\text{Dom}(z_2)} = f^{z_2} \restriction_{\text{Dom}(z_1)\cap\text{Dom}(z_2)})$$

says that function f has the same value for the same object in all models where that object appears.[18] In this way, the constraint C corresponds to a set of sentences in language L^*.

A Sneedian core $K = \langle M_{pp}, M_p, M, C\rangle$ can thus be replaced, if one wishes, by a seven-tuple $\langle \tau_{pp}, \tau_p, L_0, L, L^*, \Sigma, \Gamma\rangle$, where τ_{pp} and τ_p are types, τ_{pp} is a restriction of τ_p, L_0 is a first-order language of type τ_{pp}, L is a (first-order) language of type τ_p, L^* is a set-theoretical metalanguage for L, Σ is a formulation of M in L, and Γ is a formulation of C in L^*.

(*b*) *K-theoreticity*

The notion of a core presupposes a distinction between non-theoretical and theoretical components. For Sneed, this distinction is relative to a theory and to the existing expositions of their applications. There are two difficulties in giving a precise definition of the notion of T-theoreticity.[19] First, in the pure 'non-statement view' of theories which tries to avoid all linguistic or syntactic entities it becomes difficult to say *what* is T-theoretical, for the various models of a physical theory contain only specific functions. Moreover, it would obviously be impossible to use Sneed's criterion for showing separately of all the theoretical components in all the elements of the proper class M_p that they are T-theoretical. Therefore, we should like to say that there is some single entity which is T-theoretical and which is interpreted in the different models by the specific functions. The most natural way of doing this is to introduce a language for a core (see above), but this step already goes beyond the pure non-statement view of theories (cf. Stegmüller's difficulties on this matter in Stegmüller, 1976a, pp. 41–45.) Secondly, Sneed and Stegmüller define the notion of T-theoreticity before they have given a general definition of theories, and it turns out that their later definition of a theory (or theory-element) does not coincide with those entities T which are presupposed in the definition of T-theoreticity (cf. Harris, 1978) – the notion of an 'application' of a theory is different in these cases. The following reconstruction of the criterion tries to avoid these difficulties.

Let $E_t(K)$ be the set of the *existing theoretical applications* of a core K at time t. This means that (i) $E_t(K) \subseteq M_p$, (ii) for each $z \in E_t(K)$ there is an exposition of z, (iii) each existing exposition for

z gives such a description of z that it can be rationally believed that $z \in M$, (iv) $E_t(K) \in C$. In other words, $E_t(K)$ contains those structures in M_p which have successfully been shown (at time t) to be models of the core K, and $E_t(K)$ itself satisfies the constraints of K.

Let P be a term which belongs to the theoretical vocabulary of a language L for core K. Let P^z be the interpretation of P in model $z \in M_p$; here P^z is either a n-place relation or a n-place (real valued) function. Let K' be a core which may be identical or non-identical with K. Then P is *measured in a K'-dependent way in z at time t* iff there are objects $a_1, \ldots, a_n$ in the domain of P^z such that in each exposition of z the methods of establishing the truth or falsity of $\langle a_1, \ldots, a_n \rangle \in P^z$ or $P^z(a_1, \ldots, a_n) = x$ $(x \in \mathbf{R})$ presuppose that $E_t(K') \neq \emptyset$. Further, P is *K'-theoretical at time t* iff P is measured in a K'-dependent way at time t in each theoretical application $z \in E_t(K)$. In particular, if K' is the same as K, P is K-theoretical at time t.

One way of modifying these definitions is to replace the talk of the 'expositions' (i.e., textbooks, lecture notes, etc.) by the talk of the *actions* (measurement, determination of truth-values) of typical scientists within certain scientific communities (see Tuomela, 1973, 1978b), Instead of saying that there is an exposition for $z \in M_p$, we could say that the scientists at time t are able to 'handle' the structure z by means of their shared knowledge and abilities. In this case, we may relativize the set $E_t(K)$ to a certain *scientific community SC*, and denote it by $E_{SC,t}(K)$. In fact, this modification would bring the structuralist conception closer to the Kuhnian way of viewing science, since Kuhn has repeatedly emphasized the importance of the research activity of the scientists as compared to the expositions which record the scientific achievements (see, for example, Kuhn, 1962, p. 1).

Stegmüller has argued that the notion of presupposition used above has to be analysed in the sense that 'A presupposes B' means the same as 'B is a logical consequence of A' (see Stegmüller, 1976a, pp. 45, 55). This interpretation is extremely strong, and one may doubt whether there exist, in this sense, any K'-theoretical terms in scientific theories. To see this, let us mention some typical examples of theoretical terms in science. Radio astronomy studies radio sources (supernovas, quasars, pulsars) by means of radio telescopes. In using a radio telescope, the astronomers assume that it is functioning in accordance with the theory of electromagnetic radiation. Thus, the term 'radio source' is theoretical relative to the theory of electromagnetism. Similarly, many terms in astronomy are theoretical relative to optics. For example, in using telescopes to detect double stars we assume that optics can be succesfully applied to these instruments.[20] However, it is sufficient to presuppose (in Stegmüller's strong sense) that *something like our optics* holds of the telescope – in spite of some changes in our optics we might still be able to argue that we detect double stars (rather than something created by the instrument). The same remark applies also to the case where the scientists measure theoretical functions: an application of classical particle mechanics CPM to a projectible problem may involve the belief that CPM itself applies *at least approximately* (see Sneed, 1971, p. 32) to the systems which are used to measure the mass function. Again, we can only infer that some theory which is approximately equal to *CPM* is applicable.

For these reasons, I shall assume that the notion of 'presupposition' is interpreted in a pragmatic sense, so that it refers to the activities which the scientists in a community *SC* actually are capable of performing in a given situation. I shall leave it as an open problem whether Stegmüller's strong interpretation of 'presupposition' could be maintained by using in the above definition of K-theoreticity some suitable notion of existing theoretical *approximative* applications (cf. Moulines, 1976).

Let P be a term which occurs in a language for core K, and assume that P is K-theoretical. According to Sneed, mass and force in *CPM* are examples of such theoretical terms. Let us say that P is *strongly K-theoretical* iff for each theoretical application $z \in E_t(K)$ the measurement of P in z at time t presupposes not only that $E_t(K) \neq \emptyset$ but also that $z \in E_t(K)$. Do scientific theories contain any strongly theoretical terms? The answer seems to be positive: for example, if a medical theory is only able to describe the symptoms of a desease, then the only method of diagnosing that a person has this disease is to argue from the symptoms by assuming the truth of the theory in this situation. More generally, many dispositional terms in scientific theories seem to be strongly theoretical – at least so long as the scientists have not yet been able to discover an independent way of determining whether the base of the disposition is actualized in some situation or not.

(c) *Possible applications*

A core K can be applied to a non-theoretical structure $z \in M_{pp}$ if z can be expanded to a structure in M_p which satisfies both the fundamental laws of K and the constraint C for M_p. This idea is made precise in the following definitions. Let $\bar{r}: \mathrm{Pot}(M_p) \to \mathrm{Pot}(M_{pp})$ be the function defined by

$$\bar{r}(X) = \{r(z) \mid z \in X\}, \quad \text{for } X \subseteq M_p.$$

Then the class of the *sets of the possible non-theoretical applications* of core K is defined by

$$(1) \qquad A(K) = \{\bar{r}(Z) \mid Z \in \mathrm{Pot}(M) \cap C\}.$$

Thus, $A(K) \subseteq \mathrm{Pot}(M_{pp}) - \{\emptyset\}$ (cf. Fig. 1).

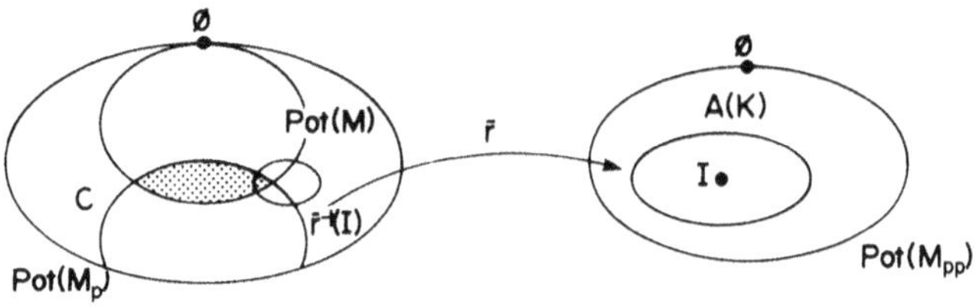

Fig. 1.

In addition to the non-theoretical applications, we can define the class of the *sets of the possible theoretical applications* of core K simply by

$$(2) \qquad AT(K) = \mathrm{Pot}(M) \cap C.$$

Then, by (1), $A(K) = \bar{r}(AT(K))$.

Classes $A(K)$ and $AT(K)$ are objectively determined as soon as core K is given. However, it should be remembered that the non-theoretical/theoretical distinction within K is relative to the set $E_{SC,t}(K)$ of the existing theoretical applications of K.[21]

(d) *Intended applications and theory-elements*

Sneed (1976) defines 'theory-elements' as pairs $\langle K, I\rangle$, where K is a core and $I \subseteq M_{pp}$ is a non-empty set of the 'intended applications' of K. Following Moulines, a pragmatic characterization of theory-elements can be given as follows.

Let SC be a scientific community and t a point (or an interval) of time. Let $I_{SC,t}(K)$ be the set of the non-theoretical structures in M_p to which SC intends to apply K at time t. Then $I_{SC,t}(K)$ is the set of the *intended non-theoretical applications* of K relative to SC and t. (If no confusion is to be expected, we shall sometimes refer to this set simply by 'I'.) The pair $\langle K, I_{SC,t}(K)\rangle$ is a *theory-element* relative to SC and t.

Some of the applications of K in $I_{SC,t}(K)$ have been sufficiently confirmed by the community SC before time t. If $C_{SC,t}(K)$ is the set of these *confirmed* (non-theoretical) *applications* or K at time t, then $C_{SC,t}(K) \subseteq I_{SC,t}(K)$ and the community SC believes that $C_{SC,t}(K) \in A(K)$. The community claims something more, however, namely that all the intended applications of K are possible applications of K. In general, the *claim* of a theory-element $\langle K, I\rangle$ is simply

(3) $\quad I \in A(K).$

Note that the claim (3) is vacuous if $M = M_p$ and $C = \text{Pot}(M) - \{\emptyset\}$, i.e., if core K does not contain any non-trivial laws or constraints.

Formula (3) implicitly contains existential quantified variables ranging over theoretical components. Therefore, it may be regarded as a modification of the ordinary Ramsey-sentence for (finitely axiomatizable) first-order theories (cf. Sneed, 1971). This suggests that we might regard (3) as an expression of the *non-K-theoretical content* of theory-element $\langle K, I\rangle$. It should be remembered, however, that the non-K-theoretical terms may be K'-theoretical for some other cores K' at time t.

Does (3) express the whole claim that the supporters of K in SC wish to make? Sneed and Stegmüller seem to think so, since for them the intended applications of a core K are always members of the class M_{pp} of partial potential models. But why could we not have 'intended applications' in class M_p as well? A scientific realist might indeed argue that it is precisely the K-theoretical terms which are the most interesting elements connected with a core K: theories are attempts to obtain 'deep' theoretical knowledge of the reality, and one should not assume that the theoretical terms refer only to fictitious entities in the various applications of a theory. A putative description of a model which includes the theoretical relations and functions is more informative than, and hence preferable to, a purely non-theoretical description. Moreover, we defined above the notion of K-theoreticity by referring to the set of the existing theoretical applications of K at time t, i.e., to set $E_{SC,t}(K) \subseteq M_p$. The definition of this set explicitly involves the assumption that the community SC believes that $E_{SC,t}(K) \in AT(K)$ (cf. (2)). (Another natural requirement is the following: $\bar{r}(E_{SC,t}(K)) \subseteq C_{SC,t}(K)$.)

These remarks suggest that it is natural to define a set of the *intended theoretical applications* of core K relative to SC and t. Let us denote this set by $J_{SC,t}(K)$ or J. The triple $\langle K, I_{SC,t}(K), J_{SC,t}(K)\rangle$ is a *full theory-element* relative to SC and t, and the *theoretical claim* of this theory-element is

(4) $\quad J_{SC,t}(K) \in AT(K).$

It is required that the set $J_{SC,t}(K)$ satisfies the conditions

(5) $\quad E_{SC,t}(K) \subseteq J_{SC,t}(K)$

and

(6) $\quad \bar{r}(J_{SC,t}(K)) = I_{SC,t}(K).$

In other words, J is an element in the inverse image $\bar{r}^{-1}(I)$ of I. The claim (3) is equivalent to $\bar{r}^{-1}(I) \cap \mathrm{Pot}(M) \cap C \neq \emptyset$, while (4) says that J which by (6) is chosen as a member of $\bar{r}^{-1}(I)$ is also a member of $\mathrm{Pot}(M) \cap C$. It thus follows immediately from (6) that the theoretical claim (4) entails the claim (3), i.e., the full theoretical content of a full theory-element is at least strong as its non-theoretical content. On the other hand, it need not be the case that (3) entails (4).

Against this suggestion it might be objected that the claim (4) could not possibly have more 'empirical' or 'testable' content than (3). The reason for this objection is the peculiar character of the K-theoretical terms, i.e., the difficulty which Sneed calls 'the problem of theoretical terms' (see Sneed, 1971, p. 38; Stegmüller, 1976a, Chapters 3 and 4). To make the statement (4) we have to pick up a set J in $\bar{r}^{-1}(I)$. But how can we ever know that any such set J belongs to $AT(K)$? No matter whether J is defined extensionally (i.e., by enumerating its members) or intensionally (i.e., by the means of a set-theoretical property), it seems that, to settle this question, one should be able to determine the values of the theoretical components of the members of J. As these components are K-theoretical, this determination is possible only on the condition that K has some successful theoretical applications. This condition presupposes again that K has successful applications, and so on. It therefore seems that we have ended up either with a vicious circle or with an infinite regress (cf. Stegmüller, 1976a, p. 58). Note also that if the theoretical components are strongly K-theoretical, then the choice of their values in an application $z \in M_p$ guarantees that z is a model of K. But if these values are chosen in this way, then the claim that $z \in M$ seems to become true *a priori*. For these reasons, Stegmüller has argued that the Ramsey–Sneed solution is "absolutely compulsory" (Stegmüller, 1978a, p. 44), i.e., that it is the only known way of solving the problem of theoretical terms.

As an answer to these objections, I shall present six different arguments to defend the theoretical claims and their testability. Let us first note that in making the distinction between non-K-theoretical and K-theoretical terms it was already assumed that K has or at least may have some successful theoretical applications, viz. the members of $E_{SC,t}(K)$. Now there are two possibilities: either the claim that these applications are successful is based on a viciously circular argument or it is not. In the former case, it seems that the whole idea of separating K-theoretical and non-K-theoretical terms was after all groundless. In the latter case, it seems that one can go on to extend the set $E(K)$ of the existing theoretical applications of K and thus justify claims of the form (4) without bothering too much about the 'problem of theoretical terms'.[22]

Secondly, it should be noted that already the problem concerning the testability of the claim $I \in A(K)$, which involves only non-K-theoretical terms, may be problematic. To see this, assume that the values of all the non-K'-theoretical functions (or relations) in a core K' are known. Then, by the definition of K'-theoreticity, all the methods for measuring the K'-theoretical functions presuppose that K' has successful applications. This does not imply – even if there were expositions of K' – that there in fact are any methods for measuring these functions. Moreover, in those cases where K' can be used for making such measurements, there need not be any guarantee that this can be done in a unique or exact way: the K'-theoretical functions may happen to be definable or identifiable in K' by the non-K'-theoretical functions only in some weak sense.[23] Let K now be another core which contains terms which are non-K-theoretical but K'-theoretical. In this case, which is not untypical,[24] there are special problems already in ascertaining the values of these non-K-theoretical functions – contrary to what Stegmüller claims (Stegmüller, 1976a, p. 57). To call partial possible models 'observable facts' (cf. Stegmüller, 1976a, p. 59; Balzer and Sneed, 1977, p. 199) is therefore very misleading – it may give the wrong impression that the distinction between K-theoretical and non-K-theoretical functions is a revived version of the old theoretical – observational dichotomy.

Thirdly, it was seen above that the measurement of the K-theoretical components in an application $z \in M_p$ of K normally presupposes only that something like K has a successful application z'. If we used K itself to fill in the theoretical components of z', an infinite regress or a vicious circle would be started. However, it may be sufficient to apply some $K' \neq K$ to z'; and I can see no reason to exclude the possibility that P is not measured in a K'-dependent way in z'. In this case, the vicious circle resulting from the K-theoreticity of the theoretical components of z is blocked. (We may of course have troubles which are similar to those mentioned in the second point above, but they do not show that there is some difference in the testability of theoretical and non-theoretical claims.) Another way of putting this argument is the following: we may have good reasons (perhaps from other areas of theoretical science) to claim that *some* theory in a larger class $\mathcal{K}$ of theories (where $K \in \mathcal{K}$) can be applied to z'. If now all the members of $\mathcal{K}$ give essentially the same description of z', then we have good – and non-circular – grounds for deciding whether K can be successfully applied to z or not.

Fourthly, it seems that there are examples involving strongly theoretical terms in which the testing (i.e., confirmation or disconfirmation) of a theoretical claim is possible. Assume that a psychologist proposes a theory that certain persons who behave in certain ways in certain situations suffer from a mental disorder, which he calls P-fobia, and that this disorder has been caused by some injury in the earlier history of the person. Write Px for 'x has P-fobia', O_1x for 'x has been injured earlier', and $O_2x, \ldots, O_nx$ for various predicates expressing different behavioral patterns of x. Then a formulation of the theory of P-fobia can be given by the conjunction σ of the following two sentences

(7) $$(x)(Px \supset O_1x)$$
$$(x)(O_2x \supset (Px \supset O_3x \& \ldots \& O_nx)).$$

Let us call these sentences σ_1 and σ_2. The potential models are now structures of the form $z = \langle D_z, O_1^z, \ldots, O_n^z, P^z\rangle$, where D_z is a domain of individuals, the partial potential models are of the form $r(z) = \langle D_z, O_1^z, \ldots, O_n^z\rangle$, and the class M of proper models is simply Mod(σ). As the constraint C we may assume a principle to the effect that an individual b has P-fobia in one of the models of the theory if and only if it has P-fobia in all such models. This defines the core K of the P-fobia theory. Here P is strongly K-theoretical and $O_1, \ldots, O_n$ are non-K-theoretical predicates.

What are the intended applications of K? In the Sneedian approach, it is assumed that "what the theory is about" is a class I of *partial* potential structures (cf. Balzer and Sneed, 1977, p. 199). In this case, there does not seem to be any natural choice for such a class I. Moreover, it is more natural (at least from the realist viewpoint) to say that the theory of P-fobia is a theory *about P-fobia* rather than about observable behavior. In other words, this theory should be applicable to all cases which are instances of P-fobia – and these applications are theoretical rather than non-theoretical. Thus, we can say that the set $J(K)$ of the intended theoretical applications of K is the class of all actual and potential victims of *P-fobia*, i.e., the set of all individuals who now suffer, or have suffered, or will suffer, or would suffer, from this disease. Set-theoretically, $J(K)$ can be defined as the set of all structures of the following form

$$z = \langle\{a_z\}, O_1^z, \ldots, O_n^z, P^z\rangle$$

where a_z is a human being, O_i^z is $\{a_z\}$ or $\emptyset$ depending on whether a_z has or does not have the property expressed by $O_i(i = 1, \ldots, n)$, and $a_z \in P^z$ (that is, the individual a_z has P-fobia in z). It follows immediately that $J(K)$ satisfies the constraint C.

How can we now test the claim that $J(K) \in AT(K)$, or equivalently $J \subseteq M = \text{Mod}(\sigma)$? Note first that if z is a fully specified model in M_p then the claim $z \in M$ is a priori true or false – it is just an exercise in logic to determine which one of these alternatives is valid. In the testing process, we rather start from an incompletely known model consisting of an individual with its so far unknown properties. Choose an individual b which satisfies the condition $O_2(b)$ and assume that it also shows the symptoms $O_3, \ldots, O_k (k \leq n)$. Together with the assumption that σ_2 holds of b, the evidence $O_2(b), \ldots, O_k(b)$ now gives us *inductive* reasons for believing that $P(b)$, i.e., that $\{b\}$ is an intended application of K. This belief can be further strengthened by deriving the claims $O_{k+1}(b), \ldots, O_n(b)$ from σ_2 and $P(b)$ and by showing that these claims are true. If some of these predictions turn out to be false, then it follows from σ_2 that after all b does not have P-fobia. In both cases, the law σ_2 will be true of b simply for the reason that we have presupposed it in determining the extension of P in the domain $\{b\}$.

Thus, assuming that, on condition $O_2(b)$, b has been shown to have all the symptoms $O_3, \ldots, O_n$, we have by σ_2 strong support for the claim that $P(b)$. If we now find that $O_1(b)$ is also true, then, by assuming σ_1, the support for $P(b)$ will be even still stronger. Hence, if σ_1 and σ_2 are assumed, we may have very strong evidence for claiming that $\{b\}$ with its real properties is an intended application of K. On the other hand, if we fail in our attempts to show that $O_1(b)$, then we can blame either the assumption $P(b)$ or σ_1. If we keep hold on σ, then the claim that $\{b\}$ is an intended application of K· will be disconfirmed. If we wish to keep the claim that $P(b)$, then our failure in verifying $O_1(b)$ disconfirms σ_1.

As we usually have no final guarantee of the truth of such theoretical statements as $P(b)$ above, the intended theoretical applications in $J(K)$ cannot simply be enumerated, but they have to be characterized by means of an intensional theoretical description. In testing the theoretical claim $J(K) \subseteq M$ there are thus two possible situations. First, we try *to show that a structure to which σ applies is an intended model*. In this case, we argue as follows: Here is a structure z which is a model of M. I believe that z is in fact the structure $z^* \in J$. Therefore, I believe that $z^* \in M$. Secondly, we try *to determine whether the law σ applies to a structure which has been identified, with some degree of certainty, as an intended model*. In this case, we use one part of the theory to appraise another part. In these ways, both the confirmation and the disconfirmation of the theoretical claim $J(K) \subseteq M$ is possible.

Fifthly, in practice theoretical claims are often tested by means of non-theoretical structures which contain individuals or properties that are analogous with but not identical to those that the theory is supposed to speak about.[25] These 'testing models' are obtained from the full potential models by 'operationalization', i.e., by replacing the theoretical components by their operationalized counterparts. The effect of this operationalization can be described as a mapping of the original core K to another core K' such that the K-theoretical components will be mapped to non-K'-theoretical components. More precisely let us say that o is an *operationalization* of K by K' iff (i) $o: M_p \to M'_p$, (ii) $o(\langle n_1, \ldots, n_m, t_1, \ldots, t_k \rangle) = \langle n_1, \ldots, n_m, t'_1, \ldots, t'_k \rangle$, where $n_1, \ldots, n_m, t'_1, \ldots, t'_k$ are all non-K'-theoretical. The claim

$$(8) \qquad \bar{o}(J(K)) \in \bar{o}(AT(K)),$$

which does not involve any K'-theoretical components, is an *operationalized version* of the theoretical claim (4), i.e., of $J(K) \in AT(K)$. The verification of (8) gives support to (4) to the extent that the values of o 'validly' measure the original theoretical terms.[26]

Sixthly, the status of the assumption that a term is K-theoretical is historically relative and can change when new applications of the core K are found. Even if a theoretical claim of the form (4) may *now* seem to be untestable, it may still have factual content which becomes testable when our knowledge grows. This seems to happen to many dispositional terms which in their 'infancy' are strongly K-theoretical (cf. P-fobia in the above example) but may later become non-K-theoretical. This change may be due to the discovery of new instruments or to the invention of new theories (possibly introducing new theoretical terms) relative to which the measurement of these terms become possible.

(e) *Theoretization and specialization*

The above description of cores K is incomplete for the reason that the different models in M may satisfy some *special laws* and *special constraints.* The formalisms of Sneed (1971) and Stegmüller (1976a) contain devices for defining *expanded cores* which take into account these restrictions. Following a suggestion of Balzer, Sneed (1976) construes these 'expanded cores' as cores which are specializations of the original core. Sets of cores, structured by the specialization relation, define 'core-nets'; together with nets of intended applications these core-nets constitute sets of theory-elements or 'theory-nets'.

Balzer and Sneed (1977) consider three different basic relations between theory-elements: theoretization, specialization, and reduction; they also claim that all the other interesting relations can be defined by means of them.

Let $T' = \langle K', I' \rangle$ be a theory-element which is obtained from another theory-element $T = \langle K, I \rangle$ by adding new theoretical components to the members of M_p. Assume that an element x' of M'_p satisfies the fundamental laws of T' (*i.e.*, $x \in M'$) only if the corresponding element $x = \theta(x')$ of M_p satisfies the fundamental law of T(i.e., $x \in M$). (Here $\theta: M'_p \to M_p$ is the restriction function.) Then T' is obtained from T by *theoretization*, which is denoted by $T'\tau T$.

This definition is ambiguous, since it does not tell whether some of the K-theoretical terms turn out to be non-K'-theoretical or not. The definition D6 in Balzer and Sneed (1977), p. 200, indicates that *no* such changes occur: M'_p and M_p are supposed to be $k+n$-and $k+m$-matrices $(m \leq n)$, which means that $M'_{pp} = M_{pp}$. On the other hand, the authors suggest that one could require that $M'_{pp} \subseteq M$ (*ibid.*, p. 201), which would mean that *all* K-theoretical terms become non-K'-theoretical. In the general case, we should allow that of m such possible changes $j (0 \leq j \leq m)$ will be realized.

A core K' is a specialization of another core K if $M'_p = M_p$, $M'_{pp} = M_{pp}$, $r' = r$, $M' \subseteq M$, and $C' \subseteq C$, i.e., if K' has a stronger fundamental law than K or more restrictive constraints than K. A theory-element T' is a *specialization* of another theory-element T if core K' is a specialization of core K and $I' \subseteq I$. This is denoted by $T'\sigma T$. The condition $T'\sigma T$ guarantees that $A(K') \subseteq A(K)$ and that $I' \in A(K)$ whenever $I \in A(K)$.

Let $\langle L_0, L \rangle$ be a language-pair for core K and $\langle L'_0, L' \rangle$ a language-pair for core K'. If $T'\sigma T$, then L_0 and L'_0, and similarly L and L', are of the same type. But if $T'\tau T$, then the type of L'_0 (resp. L') is an extension of the type of L_0 (resp. L). Specialization does not involve any conceptual change, while theoretization is connected with conceptual enrichment. The logical theory of the definability and the eliminability of new concepts is thus applicable in the study of the theoretization relation τ.

The notions of theoretization and specialization can be generalized to full theory-elements. Thus, let us say that $\langle K', I', J' \rangle$ is a theoretization of $\langle K, I, J \rangle$ iff (i) $\langle K', I' \rangle \tau \langle K, I \rangle$ and (ii) $z \in AT(K')$

entails $\theta(z) \in AT(K)$ for all $z \in \text{Pot}(M'_p)$. It follows from these conditions that the theoretical claim $J' \in AT(K')$ entails the claims $J \in AT(K)$ and $I \in \underline{A}(K)$. Moreover, if θ' is the restriction function θ': $M'_{pp} \to M_{pp}$, then $\bar{\bar{\theta}}'(I') \in A(K)$.

Let $\langle K', I', J' \rangle$ and $\langle K, I, J \rangle$ be full theory-elements such that $\langle K', I' \rangle$ is a specialization of $\langle K, I \rangle$. The definition of σ guarantees that $AT(K') \subseteq AT(K)$. We shall say that $\langle K', I', J' \rangle$ is a specialization of $\langle K, I, J \rangle$ iff (i) $\langle K', I' \rangle \sigma \langle K, I \rangle$ and (ii) $J' = J \cap M'$. Condition (ii) guarantees that the set of the intended theoretical application remains as large as possible in the step from K to K'.

For the notions of reduction ρ and strong reduction $\bar{\rho}$ between theory-elements, and for the corresponding relations of equivalence $\equiv_\rho$ and $\equiv_{\bar{\rho}}$, see Section 3(b) below.

(f) *Theory-nets*

Let X be a non-empty finite set, and let $\leq$ be a reflexive and transitive relation on X'. Define a relation $\sim$ on X by

$$x \sim y \quad \text{iff} \quad x \leq y \text{ and } y \leq x.$$

Then $N = \langle X, \leq, \sim \rangle$ is a *net*. The *basic set* of a net N is defined by

$$B(N) = \{z \in X \mid z \leq y \to z \sim y\}.$$

A net may contain several basic elements. A net N is *connected*, if for each two basic elements $x, y \in B(N)$ there is a $z \in X$ such that $z \leq x$ and $z \leq y$. A net is thus trivially connected if $B(N)$ contains only one element. A net N' is an *expansion* of another net $N (N \sqsubseteq N')$ iff $X \subseteq X'$ and $\leq$ is the restriction of $\leq'$ to $X \times X$. Here N' may contain elements which are 'between' some elements of N; moreover, an element $z \in B(N)$ need not belong to $B(N')$. These possibilities are excluded if N is required to be an *initial part* of N' in the following sense: $y \nleq x$ for all $x \in N$ and $y \in N'$.

A net $N = \langle X, \leq, \sim \rangle$ is a *theory-net* iff X is a finite set of theory-elements and $I' = I$ implies $K' = K$ for all $\langle K, I \rangle$ and $\langle K', I' \rangle$ in X. To each theory-net there naturally corresponds an associated *core-net* N^* and an *application-net* N^+. If $N = \langle X, \leq, \sim \rangle$ is a theory-net, then the claim of N is the following

(9) $\quad I \in A(K) \quad$ for all $\quad \langle K, I \rangle \in X$

Obviously (9) reduces to (2) if X has only one element. Among the most important special cases of theory-nets are *σ-theory-nets* of the form $N = \langle X, \sigma, =\rangle$, i.e., such that $\leq$ is the specialization relation σ and, hence, $\sim$ is the identity (Sneed, 1976). In this case, $\langle X, \sigma\rangle$ is a partially ordered set of theory-elements. Other examples are of the form $N = \langle X, \tau, \equiv_\tau\rangle$, where τ is the theoretization relation and $x \equiv_\tau y$ iff $x\tau y$ and $y\tau x$, and of the form $N = \langle X, \rho, \equiv_\rho\rangle$ and $N = \langle X, \bar{\rho}, \equiv_{\bar{\rho}}\rangle$ (see Balzer and Sneed, 1977, pp. 210–211).

Following Moulines, a theory-net N is said to be a theory-net relative a scientific community SC and time t iff each $\langle K, I\rangle$ in N is a theory-element relative to SC and t. This pragmatic conception of theory-nets can be used to represent the growth of scientific theories (see Section 3a).

A net $N = \langle X, \leq, \sim\rangle$ is a *full theory-net* iff X is a finite set of full theory-elements such that $J' = J$ entails $I' = I$, and $I' = I$ entails $K' = K$, for all $\langle K, I, J\rangle$ and $\langle K', I', J'\rangle$ in X. The *theoretical claim* of N is

(10) $J \in AT(K)$ for all $\langle K, I, J\rangle \in X$;

this claim (10) entails the claim (9). To a full theory-net N there corresponds a net $N^{\oplus}$ of the theoretical applications J.

(g) *Remarks on the structuralist concept of a theory*

After this extensive survey of the basic concepts of the structuralist conception of a theory, we can now make some methodological and philosophical comments on its general nature.

Let us first note that the structuralist conception is very general: it seems that it can be applied within all fields of empirical science and equally well in mathematics. To distinguish mathematical theories and empirical theories from each other further conditions are needed. Thus, in his definition of 'theories of mathematical physics' Sneed requires that the set I of intended applications should consist of 'physical systems' which are 'of the same kind' and which have 'connected' domains (see Sneed, 1971, pp. 250–260; cf. Stegmüller, 1976a, pp. 163–165).[28] These more or less intuitive requirements are not very clear – here, again, it seems that a more precise definition could be given by talking about the vocabulary associated with a theory. Moreover, it is conceivable that two theories in different fields of science are concerned with precisely the same nontheoretical structures and differ from each other only by virtue of their specific theoretical vocabulary. Thus, we might say that a theory-element $\langle K, I, J\rangle$ (or a theory-net) belongs to the field X if and only if the formulation of the fundamental law (and special laws) and constraints is given by using the theoretical vocabulary of field X.

It was seen, in subsection (a) above, that to a core $K = \langle M_{pp}, M_p, r, M, C\rangle$ one can correlate two languagees L and L^* and the formulations Σ and Γ of M and C in L and L^*, respectively. Hence, to a theory-element $\langle K, I\rangle$ (or $\langle K, I, J\rangle$) one can correlate a pair (or a triple) which involves the formulations – rather than the set-theoretical representations – of the laws and the constraints. It is further interesting to ask whether it is possible to give formulations to the claims $I \in A(K)$ and $J \in AT(K)$, too, in L^*.

The theoretical claim $J \in AT(K)$ can be formulated in L^* by the following sentence:

$$(11) \qquad \forall z \in J(z \models \Sigma) \;\&\; \Gamma^J,$$

where Γ^J is the relativization of Γ with respect to J.[27] Similarly, the claim $I \in A(K)$ can be formulated by

$$(12) \qquad \exists y(y \subseteq M_p \;\&\; \bar{r}(y) = I \;\&\; \forall z \in y(z \models \Sigma) \;\&\; \Gamma^y).$$

Here (11) is a *formulation of the theoretical claim* of the full theory-element $\langle K, I, J\rangle$.[29] Alternatively, we might say, as Harris (1978) essentially suggests, that the set of all the (logical and mathematical) consequences of (11) in L^* is an L^*-theory *based on* Σ and Γ. This proposal gives *a dual-level statement view* of scientific theories: a theory consists of first-order axioms Σ together with some metalinguistic postulates Γ.

Given a full theory-element $\langle K, I, J\rangle$, one can find an L^*-theory corresponding to it, and *vice versa*. It follows that the structuralist conception of theories is essentially equivalent to a suitably emended statement view of theories. It seems, therefore, that Stegmüller, at least in (1976a), has overstated the importance of the question whether a theory consists of statements or of set-theoretical structures.

But a more important difference between the *old* statement view and the Sneedian concept of a theory seems to underlie the notion of an 'intended model' or 'intended application'. According to Stegmüller (1976a), pp. 70–71, the former view is based upon a "fiction" about one "cosmic application" of a theory, while the latter assumes that there are several intended applications for a theory. The importance of this difference can be emphasized by noting that the notion of constraint is dispensable if a core has only one intended application – in this case the structuralist conception of a theory would be essentially equivalent to the old one-level statement view.

In motivating the multi-application approach, there are three different grounds which should be kept separate. One of them could be called 'Wittgensteinian', the second 'instrumentalist', and the third 'pragmatist'.

What could it mean that a theory has one "cosmic application"? The answer is given by Hempel:

> one source of misunderstanding is the view ... that a hypothesis of the simple form 'Every P is a Q', such as 'All sodium salts burn yellow', asserts something about a certain limited class of objects only, namely, the class of all P's. This idea involves a confusion of logical and practical considerations: Our interest in the hypothesis may be focussed upon its applicability to that particular class of objects, but the hypothesis nevertheless asserts something about, and indeed imposes restrictions upon, *all* objects .... (Hempel, 1965, p. 18.)

In other words, the law 'Every P is a Q' asserts about *all* objects x a conditional statement that *if* x is P *then* it is also a Q. But another way of viewing this law is to say that in it the concept Q is applied to certain *intended cases*, namely, to all P's. (More generally, one could say that in this law the concept 'If P then Q' is applied to some class which includes all P's.) According to this assumption, the range of a generalization is not the whole universe of discourse but rather a pragmatically relativized class of 'intended' cases. If the different P's could be enumerated by $a_1, \ldots, a_n$, then the generalization 'Every P is a Q' could, on this account, be replaced by the conjunction $Q(a_1) \& \ldots \& Q(a_n)$.

By deliberate inclusion of pragmatic factors within the notion of law, the latter alternative involves a "confusion of logical and practical considerations". This is, in fact, the key idea of the Sneed–Stegmüller approach to theories. Instead of formulating Newton's theory by the *statement*

$$(x)\ (x \text{ is a physical system} \supset x \text{ satisfies Newton's laws}),$$

which applies to the whole universe, they represent it by the *pair* ⟨Newton's laws, Intended applications⟩. This is analogous to replacing the generalization 'Every P is a Q' by the pair $\langle Q, I \rangle$, where I the class of P's. The main virtue of this approach seems to be based on the fact that it can be used in those situations where the class of the intended applications is an 'open class', that is, a class which is defined by the *method of paradigmatic examples*, i.e., by giving a list of some typical cases and admitting into this class other cases which are sufficiently similar to these exemplars (cf. Stegmüller, 1976a, pp. 170–185). As this sort 'openness' of concepts was emphasized by Wittgenstein, one can say that the need for the pragmatic 'non-statement' approach may arise from the 'Wittgensteinian' nature of concept formation in science.

Another motivation for the assumption of multiple intended applications of theories is based upon the idea of theories as *tools*. Reference to theories of pure mathematics is illustrative here. In mathematics, there are two basically different types of theory construction. One of them starts from a particular mathematical structure $\mathfrak{A}$ and tries to find a theory which gives as complete axiomatization as possible for the class $Th(\mathfrak{A})$ of all true sentences about this structure (e.g. axiomatic theories of number theory). Another starts from a set of abstract mathematical assumptions and tries to find as many applications as possible where these assumptions are true (e.g. algebraic theories like group theory).[30] In the former case, the theory has *one intended application* (e.g. the set N of natural numbers), while in the latter case the theory has *several intended applications*. The criteria for appraising these sorts of theories are also different: in the former case, the number of obtained truths about the given structure, and, in the latter case, the number of interesting theoretical and practical applications.[31]

In the Sneedian approach, theories of mathematical physics are thought to be more like group theory than like number theory: such theories contain a mathematical core which is a "tool" or an "instrument" that is hoped to be useful in dealing with certain "chunks of experience" (Sneed, 1971, p. 306). A theoretician creates mathematical cores (Newton's laws, Maxwell's equations, hyperbolic geometry, etc.), and after some initial successes in applying this core to some phenomena attempts will be made to enlarge the scope of these applications to other phenomena.

One cannot deny that there is an instrumentalist flavor in this conception of theories. A realist might indeed argue that *scientific theories should be regarded as analogous to number theory rather than to group theory*: the ultimate aim of theorizing in science is to find the truths about one particular structure, viz. the actual world.[32] Sneed regards his view as "a kind of compromise between an instrumentalist view of theories and a descriptive, or statement view" (Sneed, 1971, p. 306). Moulines says that it would be "unjust to call Sneed's view an instrumentalist one" (Moulines, 1975a, p. 429), and Stegmüller points out that, according to the structuralist conception, "theories are not instruments for creating fictitious pictures of reality, but tools for making empirical claims about reality" (Stegmüller, 1976, p. 163). However, in formulating the 'empirical claim' of a theory-element $T = \langle K, I \rangle$, no function symbols for the K-theoretical components but only existentially quantified function variables are needed. This is one of the reasons why Przełecki (1974) and Tuomela (1978) have concluded that the Sneed–Stegmüller approach better suits an instrumentalist than a full-blown realist.

One difficulty in classifying the position of Sneed and Stegmüller[33] is the fact that their approach does not contain any rules against using the K-theoretical functions – which are *not* used in formulating the 'empirical claim' of $\langle K, I \rangle$ – in the formulation of the 'empirical claim' of *other* theory-elements $\langle K', I' \rangle$, even if their use in the latter connection need not be more testable than in the former. On the basis of this fact, Sneed might rebut the charge of instrumentalism and claim that he takes the theoretical function as seriously as the nontheoretical ones. Pursuing this realist line, I have suggested above that one should add *theoretical claims* of theory-elements to the structuralist formalism. Stegmüller's inclinations are more decidedly instrumentalist, however, which can be best seen in his treatment of theory-dynamics and of the concept of scientific progress (cf. Section 3 below).

Third reason for not accepting the idea that theories have only one intended application is a practical one: the whole universe in its all variety is never the object of one single inquiry. All inquiries concern only some aspects of some parts of the universe, where the focus depends upon the problems and the corresponding 'demands of information'[34] which have given rise to the inquiry. But while inquiries are in this sense 'local', theories are not supposed to be restricted in their applicability to one single research problem. The theoretical importance of these points have been emphasized especially by the pragmatist philosophers of science.

By saying above that one should keep separate the 'Wittgensteinian', 'instrumentalist', and 'pragmatist' motivations for the multi-application approach I mean the following: one can appreciate both the method of paradigmatic examples in the formation of 'open' concepts and the pragmatist emphasis on the local aims of inquiries without committing oneself to an instrumentalist view of theories. This is a philosophical thesis which cannot be proved by any simple argument, but in 3(c) below I try to outline some of my reasons for endorsing it.

3. THE DYNAMICS OF THEORIES

According to Stegmüller, the structuralist conception of theories has five important advantages over the statement view:

> (1) With it a concept corresponding to Kuhn's notion of '*normal science*' can be introduced in an unforced way. (2) With it a *concept of progress* can be introduced which also covers the revolutionary cases. (3) The phenomenon of the immunity of theories to 'recalcitrant experience' can be made clear and understandable. (4) It permits an elegant simplification of what Lakatos intended his *theory of research programmes* to achieve. (5) It removes the danger – and this is perhaps the main advantage – of falling into a rationality monism and, thus, into the *rationality rut* of assuming there could be but *one single* source of scientific rationality. (Stegmüller, 1977, p. 272.)

All these points are related to the dynamic aspect of scientific theorizing.

This section is a commentary on Sneed's and Stegmüller's treatment of theory-dynamics. I shall discuss, in this order, (a) the structuralist explication of Kuhnian normal science, (b) the characterization of scientific revolutions, and (c) the concept of scientific progress.

(a) *Normal science*

Two different strategies can be used in the explication of normal science. First, one defines what is meant by saying that *a person p holds a theory*; then persons who hold the same theory can be said to *belong to the same normal scientific tradition.* This gives an explication of 'Kuhnian' normal science if the persons are required to hold the same *Kuhn-theory* – that is, essentially, a theory in which the class of the intended applications is defined by the method of paradigmatic examples.[35] In the other approach, one employs theory-nets which are relativized to a scientific community *SC* and time t; certain temporal sequences of such nets (relative to the same *SC*) are then characterized as *Kuhnian theory evolution.*[36]

Before stating any formal definitions, let us note that in both of these strategies one has to presuppose the notion of a scientific community as a primitive concept (cf. Section 2(b) above). Therefore, I shall not assume any sharp difference between them.

According to Balzer and Sneed, a quadruple $\langle K_0, I_p, N_p, \mathfrak{N}\rangle$ is a *Kuhn-theory* if and only if (i) $\langle K_0, I_p\rangle$ is a theory-element, (ii) N_p is a σ-theory-net with $B(N_p) = \langle K_0, I_p\rangle$, and (iii) $\mathfrak{N}$ is the set of all σ-theory-nets N such that $N_p \sqsubseteq N$ and $B(N) = \langle K_0, I_0\rangle$ for some $I_0 \supseteq I_p$.

Here K_0 is a core which defines the basic conceptual structure of the theory. I_p is the set of the *paradigmatic applications* of core K_0, i.e., the first successful applications of K_0 by the founder(s) of the theory. N_p is the *paradigmatic σ-theory-net* which indicates how core K_0 is applied to I_p. A *σ-theory-net* N is *admissible* relative to K_0, I_p, and N_p if and only if N is an expansion of N_p and the (unique) basic element of N contains the core K_0 and a set I_0 of intended applications which is an enlargement of I_p. Thus, $\mathfrak{N}$ is the set of all admissible σ-theory-nets relative to K_0, I_p, and N_p.

The set $\mathfrak{N}$ contains all the potential further developments of the theory. The state of the actual development of the theory at time t is a net $N^t \in \mathfrak{N}$. The basic element $\langle K_0, I_0^t\rangle$ of N^t contains all up to t successful (non-theoretical) applications of K_0, not the hoped-for applications. The *historical development* of the Kuhn-theory $\langle K_0, I_p, N_p, \mathfrak{N}\rangle$ can thus be described as a sequence $N^0, N^1, \ldots, N^t, \ldots$, where $N^0 = N_p$ and $N^t \in \mathfrak{N}$ for all t. A step from N^t to N^{t+1} in this development is *cumulative* if N^{t+1} is a proper expansion of N^t. This can happen in three different cases: (i) the core-nets $(N^t)^*$ and $(N^{t+1})^*$ are identical but some of the elements I_i^t of the application-net $(N^t)^*$ are replaced by their proper supersets I^{t+1} in $(N^{t+1})^+$, (ii) $I_0^t = I_0^{t+1}$ but $(N^{t+1})^*$ is a proper expansion of $(N^t)^*$ (i.e., the range of applications remains the same, but new special laws are added), (iii) both the application-net and core-net grow as in (i) and (ii).

Sneed and Balzer suggest that Kuhn-theories correspond precisely to the products of scientific communities during periods of Kuhnian normal science. The historical development of a Kuhn-theory may be only partly cumulative (in the technical sense), but in any case the scientists who are working with this theory at time t are committed to "the use of a certain set of basic concepts" (i.e., core K_0 remains constant) which are "applied to a certain kind of phenomena" (i.e., the nets in $\mathfrak{N}$ are uniquely based). There is also a historical continuity between these scientists and the creator of the theory: they accept the initial paradigm applications of K_0 ($I_p \subseteq I_0^t$) and the paradigmatic theory-net $N_p(N_p \sqsubseteq N^t)$. If they change any of the elements K_0, I_p, or N_p, they have abandoned to work under the original 'paradigm' or 'disciplinary matrix'[37] and, therefore, do not belong to the same tradition of normal science any more.

In the definition of Kuhn-theories $\langle K_0, I_p, N_p, \mathfrak{N}\rangle$, the status of class $\mathfrak{N}$ is problematic. In the first place, it is a huge class: as Kuhn-theories are supposed to be products of scientific communities, one cannot restrict the admissible theory-nets N e.g. to those which have a true non-theoretical claim. Secondly, $\mathfrak{N}$ seems to be superfluous in the sense that it is definable by means of the three other elements (i.e., K_0, N_p, and I_p) of a Kuhn-theory. For these reasons, one might suggest that $\mathfrak{N}$ should be replaced by a class $\mathfrak{M} \subseteq \mathfrak{N}$ which contains those admissible σ-theory-nets whose claims can be reasonably be expected to be successful. Class $\mathfrak{M}$ thus introduces a *heuristics* to a Kuhn-theory. Let us call the members of $\mathfrak{M}$ *promising* theory-nets. Then we can require that persons who *hold* the Kuhn-theory $\langle K_0, I_P, N_p, \mathfrak{M}\rangle$ at time t agree about the confirmed applications I_0^t of K_0 and the specializations of K_0 in N^t and about the class $\mathfrak{M}^t \subseteq \mathfrak{M}$ of the promising expansions of N_p at t but they may disagree in their *claims* that certain expansions of N^t are (or will be found to be) applicable to sets including I_0^t. This separation of the confirmed and intended applications at t shows that there may exist 'subschools' within a normal scientific tradition. Moreover, a person may hold the same theory at different times even if his beliefs change.[38]

The historical development of Kuhn-theories describes a pattern of the growth of scientific theories. But does it correspond to what Kuhn has called 'normal science'? And does it represent anything which takes place within the actual development of science?[39] Of these two questions, I shall concentrate on the first one.

Let us recall that, for Kuhn, normal science is research carried out by a scientific community which is bounded together by a shared 'disciplinary matrix', i.e., a constellation of group commitments which may include such elements as 'symbolic generalizations', beliefs in particular 'models', and values. Further, they include 'exemplars' or "concrete problem-solutions that students encounter from the start of their scientific education, whether in laboratories, on examinations, or at the ends of chapters in science texts" and "at least some of the technical problem-solutions found in the periodical literature..." (Kuhn, 1970, p. 187). Scientists within a community learn to do their job – to use the symbolic generalizations of a matrix – through the study of exemplars, and the existence of a shared matrix explains the "relative fulness of their professional communication and the relative unaminity of their professional judgments" (*ibid.*, p, 182). Through the matrix, the scientific community acquires "firm answers to questions like the following: What are the fundamental entities of which the universe is composed? How do these interact with each other and with the senses? What questions may legitimately be asked about such entities and what techniques employed in seeking solutions?" (*Ibid.*, pp. 4–5.) As the matrix provides a criterion for choosing problems which can be assumed to have solutions, normal science can be described as an activity of 'puzzle-solving'. On the other hand, as the possession of a matrix involves a commitment to a *Weltanschauung*, the replacement of one matrix by another – a scientific revolution – can be characterized as a 'change of world view'.

In a Kuhn-theory $\langle K_0, I_p, N_p, \mathfrak{M}\rangle$, the elements which correspond to Kuhn's exemplars and disciplinary matrix are I_p and $\langle K_0, I_p, N_p\rangle$, respectively. As a first objection to these identifications, one could point out that it would seem more natural to take the exemplars to be possible models in M_p rather than partial possible models in M_{pp}. In other words, (at least some of) the exemplars should include also the K_0-theoretical functions – for example, the mass and force functions in the case of Newtonian mechanics (cf. Kuhn's examples of exemplars in Kuhn, 1970, pp. 187–190). For this reason, the Kuhn-theory $\langle K_0, I_p, N_p, \mathfrak{M}\rangle$ could be replaced by a quadruple $\langle K_0, J_p, N_p, \mathfrak{M}\rangle$, where $J_p \subseteq M_p$ is the class of the paradigmatic theoretical applications and, further, N_p and the members of $\mathfrak{M}$ are *full* theory-nets (cf. Section 2(f)). This modification of the notion of Kuhn-theory will be assumed below.

As a second objection, let us note that Kuhn's discussion seems to allow for the possibility that the set of exemplars grows over time. Thus, if J_p^t is the set of paradigmatic applications at time t, we require that $J_p \subseteq J_p^t \subseteq J_0^t$. In the same way, net N_p can grow, so that we require $N_p \sqsubseteq N_p^t \sqsubseteq N$ for all N in $\mathfrak{M}$. With this modification, normal science becomes more dynamic than in the Sneedian reconstruction.

A third objection goes in the same direction as the second: the requirement that core K_0 remains constant – or at best is replaced with its specializations – throughout the whole historical development of the theory seems too strict. This requirement excludes from normal science all forms of conceptual change – no matter how trivial they are. For example, one cannot introduce new concepts by means of explicit definitions or make for convenience inessential modifications in the mathematical formalism without breaking the bounds of normal science. It also seems that even more 'radical' conceptual enrichment should be allowed within normal scientific development. Even though it may be reasonable to exclude theoretization (i.e., the introduction of new theoretical concepts) from normal science, one should not rule out the possibility of expanding models with new non-theoretical components. For example, it seems reasonable to expect that at least some periods in the development of modern molecular genetics could be described as normal science. The phenotypes and thus the relevant features of the organisms vary from one application to another. In extending the genetic theory to account for new kinds of hereditary phenomena one therefore may be forced to use in an essential way new non-theoretical concepts.[40]

Continuing this objection, it also seems too strict to require that the fundamental assumptions associated with core K_0 remain entirely unchanged: if the notion of normal science is to be used in the historiography and the sociology of science, we should allow some small changes in the basic laws and constraints of K_0. Arguing for this kind of flexibility, Laudan assumes in his characterization of 'research traditions' (Newtonians, Cartesians, etc.) that at each time there is a set of core assumptions which define a tradition – but this core may change through time (Laudan, 1977, p. 99). We could perhaps go farther and treat even the temporal slices of research traditions as defined by cluster concepts; instead of giving necessary and sufficient conditions for the membership among the Newtonians at time t, we could say that a person supports the Newtonian tradition at t if and only if he satisfies a sufficient number of criteria which are 'semantically relevant' to this classification.

As a fourth objection, one may argue that Kuhn-theories are not able to serve all functions of Kuhn's disciplinary matrixes. It should be noted, however, that nothing told above prevents the formulation of the fundamental law of K_0 to be an existential claim or a statement involving value concepts, so that core K_0 may express ontological commitments and axiological assumptions. But, on the other hand, it is not quite clear to what extent core K_0 can be made to reflect various methodological assumptions concerning proper methods and techniques in 'puzzle-solving'. Perhaps the most crucial question in this direction is the following: does the notion of Kuhn-theory account for the problem generating and problem determining character of disciplinary matrixes?[41] The answer seems to be negative: even though the core K_0 generates questions of the form 'Does K_0 apply to model z?', the Sneedian notion of Kuhn-theory does not explain how such questions turn out to be (or involve) 'puzzles' in Kuhn's sense. The requirement that the set I_0^t of successful applications of K_0 at time t includes as a subset the set I_p of paradigmatic applications is not enough. Following Kuhn's ideas, one should require that the new applications are *similar* to the paradigmatic ones. One should assume that there is a – more or less articulated – relation $\mathfrak{R}$ of similarity or resemblance by means of which proposed applications can be compared to the exemplars.[42] In a formal level, this requirement has been taken into account in my proposal to include the class $\mathfrak{M}$ in Kuhn-theories: one could define the class $\mathfrak{M}^t$ of promising applications at time t as the set of all possible models which are in the relation $\mathfrak{R}$ to some accepted exemplar in I_p^t at time t (and have not yet been tested and ruled out as models of the core K_0).

As a fifth objection, we can point out that a person p may hold, in Sneed's and Stegmüller's sense, two different Kuhn-theories at the same time – even when these theories concern the same non-theoretical structures but conflict with each other in their theoretical part. It is quite possible to believe that the partial possible models can be filled in by theoretical functions in essentially different ways. Instrumentalists have traditionally argued that a scientist can use for his purposes different and conflicting theories.[43] Kuhn explicitly denies such a freedom in the case of 'paradigms': they have a dominant position in scientific communities, and a paradigm change is like a gestalt switch which involves a scientific revolution. In this respect, Sneedian Kuhn-theories and Kuhn's paradigms are not analogous. On the other hand, Kuhn-theories modified so that they involve full theory-nets with intended theoretical applications are more like Kuhnian paradigms – in the sense that a person cannot hold at the same time two such Kuhn-theories which conflict with other.

For Stegmüller, one of the main virtues of the structuralist explication of normal science is the fact that it explains why it is not irrational to assume theories to be immune against refutation by recalcitrant experience. First, a theory need not be 'immunized' against refutation, because "it is the sort of entity of which 'falsified' cannot sensibly be predicated" (Stegmüller, 1976a, p. 13). Secondly, no finite number of unsuccessful attempts to find applications for a core can conclusively show that there are no such applications (cf. *ibid.*, p. 198). Thirdly, if it turns out that a core is not applicable to an intended model *z* (which is not one of the paradigm applications) "the theoretician can decide to drop this element from the set of intended applications" (ibid., p. 174). Because of this immunity, Popper's claim that those working within normal science are uncritical dogmatists is unjustified.

Two comments on these points are in order. First, as Feyerabend (1977) points out, one need not adopt the structuralist view of theories to account for the relative immunity of theories against refutation: this feature has been built e.g. in the 'methodology of research programmes' of Lakatos.[44] Secondly, it is not clear to which extent the structuralist analysis of the situation really supports Kuhn's claim that in normal science negative test results give discredit to the scientist rather than to the theory (cf. Stegmüller, 1976a, p. 13). Kuhn's main ground for this thesis is the assumption that 'paradigms' define legitimate problems and guarantee the existence of their solutions. This is quite different from the idea that a theory can be saved from refutation by making adjustments in its range of applications. Moreover, by dropping an element from the set of the intended applications we in fact change the theory – recall that these applications are a part of theory – and in this way give discredit to the theory rather to the scientist.

As a concluding remark in this subsection, I should like to point out that an explication of normal science is incomplete as long as it does not contain *a theory of belief formation in scientific communities* –

that is, an account of how the members of a community are influenced by the opinions of other members and how they are able to reach a consensus. An interesting theory towards this direction has recently been proposed by Keith Lehrer (1977) who shows that the members of a community bound together by a chain of 'positive mutual respect' can combine their different non-dogmatic degrees of belief to 'consensus probabilities'. The members are even allowed to hold different meaning postulates (i.e., statements which they believe in degree one), so that this account also clarifies the question to what extent communication across different 'paradigms' is possible.

(b) *Scientific revolutions*

Elaborating on Sneed (1971), pp. 297–306, Stegmüller distinguishes between two kinds of scientific revolutions: the transition of 'pretheory' to theory and the 'dislodgment' of one theory by another (Stegmüller, 1976a, pp. 202–219). Stegmüller has emphasized that his notion of 'theory-dislodgment' is not meant as an *explicate* of Kuhn's concept of scientific revolution (Stegmüller, 1978a, p. 72). He has nevertheless argued what revolutions would be like: the dislodged theory should be reducible to the dislodging theory (Stegmüller, 1976a, p. 216).

By scientific revolutions Kuhn means "those non-cumulative developmental episodes in which an older paradigm is replaced in whole or in part by an incompatible new one" (Kuhn, 1970, p. 91). We have already observed that the Sneedian notion of Kuhn-theory does not adequately explain why such paradigm change should be treated as a 'change in world view' and that changes from one Kuhn-theory to another cannot always be called 'revolutionary'.

The emergence of primary theories is obviously a special case of theoretization (see Section 2(e)), and presumably some examples of revolutionary theory-change could be analysed by means of this relation. The main tool which Sneed and Stegmüller use in treating theory-change is the structuralist notion of reduction: a theory-element $T' = \langle K', I' \rangle$ *reduces* to another theory-element $T = \langle K, I \rangle$ if and only if there is a many-one relation R from M'_{pp} to M_{pp} such that the intended applications of T' are correlated with intended applications of T and what T' says about these applications is entailed by what T says about the corresponding applications. A stronger concept of reduction is obtained by requiring the existence of a reducing relation R from M_p' to M_p. These concepts can then be generalized to cover reduction between theory-nets and reduction between Kuhn-theories.[45] They can also be used in illuminating the sense in which rival theories may be said to be 'incommensurable' (Balzer, 1978).

The reduction relation can in principle obtain between theory-elements which employ entirely different conceptual frameworks. Reduction thus defines a 'translation' between these conceptual systems (cf. Balzer and Sneed, 1977, p. 204) – but only on the purely semantical level of potential models or partial potential models. In fact, if L and L' are theoretical languages for cores K and K', respectively, and if T' reduces to T, then L and L' need not be of the same type at all.

I cannot here discuss problems which are involved in the attempt to use the reduction relations in a (partial) characterization of scientific revolutions.[46] Instead, I make only the general remark that perhaps one should reject the *dichotomy* between normal science and scientific revolutions and replace it by an evaluation of the *degree of radicality* of theory-change. Such evaluation should take into account both the difference between the conceptual frameworks of the theories and the distance between their postulate sets.[47]

(c) *Scientific progress*

According to Stegmüller, normal scientific *progress* can be defined as cumulative development of Kuhn-theories (see Stegmüller, 1976a, p. 220). The emergence of primary theories is always progressive (*ibid.*, p. 212), while theory-dislodgment is progressive if it satisfies – at least approximatively – the reducibility condition (Stegmüller, 1978a, pp. 60–62). Thus, in progressive scientific revolutions "the displaced theory can be partially and approximatively imbedded into the supplanting theory" (*ibid.*, p. 62).

An important feature of this account of scientific progress is the possibility of *progress branching*, both in normal science (set I_0^t can be extended in two different ways either to $I_0^t \cup \{z\}$ or to $I_0^t \cup \{z'\}$) and in revolutionary theory-dislodgment (a theory can be approximately imbeddable in two non-equivalent theories). (See Stegmüller, 1977, pp. 275–277, 284–285.) In such cases, the scientist has to decide on the basis of *value judgments* which way to proceed (*ibid.*, pp. 276, 284).

Stegmüller takes the existence of progress branching to be in sharp contradiction with all *teleological* conceptions of scientific progress (*ibid.*, p. 284), and he thinks that Kuhn is correct in rejecting "a teleological myth of progress" (cf. Kuhn, 1970, pp. 170–173, 206–207), viz. "all variants of a teleological metaphysics explaining progress in terms of 'coming closer and closer to the truth'" (Stegmüller, 1978a, p. 59).

This account of scientific progress is in close agreement with the view that the main virtue of a scientific theory is its *problem-solving ability* (cf. Kuhn, 1970, p. 169, and Laudan, 1977). This view admits an instrumentalist and a realist interpretation, where the former denies, and the latter claims, that the notion of truth is essentially involved in the notion of problem. In giving an account of scientific progress in terms of the problem-solving ability of theories, one can follow Kuhn and *try* to avoid the use of such "metaphors" as "true constitution of nature" (Stegmüller, 1978a, p. 59). But, in spite of his rejection of "teleological metaphysics", Stegmüller's characterization of scientific progress can be claimed to *involve as a special case Popper's problem of verisimilitude.*

To see this, let us see how Stegmüller's definition of progress can be motivated. The class of problems which a Kuhn-theory has solved at time t is represented by the set I_0^t (or, alternatively, by the set J_0^t). A step from $T = \langle K_0, I_0^t \rangle$ to $T' = \langle K_0, I_0^{t+1} \rangle$, where $I_0^t \subseteq I_0^{t+1}$, is progressive, since T' solves more problems than T. A step from $T = \langle K_0, I_0^t \rangle$ to $T' = \langle K_1, I_0^t \rangle$, where K_1 is a specialization of K_0, is progressive, since T' gives a better solution to the same problems as T. Finally, a step from $T = \langle K_0, I_0^t \rangle$ to $T' = \langle K_1, I_1^t \rangle$, where T reduces to T', is progressive, since T' at least solves, at least equally well as T, the same problems as T.[48]

In this account, it is assumed that a 'problem' $z \in M_{pp}$ (alternatively: $z \in M_p$) has been solved by K_0 at time t if and only if the claim that K_0 applies to z is well-confirmed at t (or accepted by the scientific community SC at t). But a well-confirmed claim may be false. It is thus possible that a step from $\langle K, I \rangle$ to $\langle K', I' \rangle$ which exhibits normal scientific progress in Stegmüller's sense leads from a *true* claim $I \in A(K)$ to a *false* claim $I' \in A(K')$ (or from a false claim to another false claim).

In particular, suppose that $I_0^t = \{z\}$. Then one can make normal scientific progress by proposing stronger and stronger laws concerning z, i.e., by specializing the core K_0, and finding empirical confirmation to these laws. But, again, a step from a law σ_1 to law σ_2 may lead us from truth to falsity – or from falsity to another falsity.

Is it reasonable to speak of progress in the cases mentioned above? A positive answer could perhaps be defended by asking: What else could we mean by progress than the introduction of new well-confirmed applications and laws? What other grounds there could be for making claims about progress than the accepted empirical and theoretical evidence? But suppose that at time t we propose to replace a law σ_1 with a stronger law σ_2 and at time $t+1$ we find new evidence which gives us good reasons for rejecting σ_2. Then, instead of still claiming that the step from σ_1 to σ_2 was progressive, it is much more natural to say only that at time t it *seemed* that this step is progressive, while at time $t+1$ it seems otherwise. In other words, we are here employing the notion 'the step from T to T' seems progressive on evidence e'. This seems to presuppose, however, that it is meaningful to distinguish between *real* and *apparent* progress – even though it were the case that we can never know for sure which steps in theory-change are 'really' progressive. Namely, a natural analysis for the claim

(11) The step from T to T' seems progressive on e

can be given by

(12) The claim 'The step from T to T' is progressive' is well-confirmed by e.

In Niiniluoto (1978c), I have shown how the notion of truthlikeness or verisimilitude can be used in defining a non-epistemic notion of 'progress' and its epistemic counterpart 'seems progressive on evidence e'.[49] This account does not involve any kind of "teleological metaphysics": it does not presuppose that in making claims about the progressiveness of scientific developments the scientists already possess the knowledge of the true constitution of the reality, and it does not contain any theses about the future development of science. Moreover, it does not exclude the possibility of 'progress branching' – on the contrary, given a law σ, there are normally several other non-equivalent laws which have a greater degree of truthlikeness than σ.[50] It does not follow, however, that the choice between these alternatives should be made on the basis of value judgments.

One way of utilizing this sort of 'realist' theory of progress in connection with the structuralist theory conception is to define the *value* (or degree of adequecy) *of the solution* that K gives to a 'problem' $z \in M_p$ by the (estimated) degree of truthlikeness of (the formulation of) the fundamental and special laws of K relative to z.[51] The value of a theory would then depend on the class of problems which it claims to solve and on the value of the given solutions.

University of Helsinki

NOTES

[1] For a classical exposition of the logistic method, see Church (1956), pp. 1–68. For an excellent account of the development of the Received View and of its problems, see Suppe (1974). See also Tuomela (1973).

[2] See also the articles of Przełecki and Wojcicki in Przełecki and Wojcicki (1977).

[3] See Suppes (1957), Ch. 12, and Suppes (1967a, 1967b, 1969). See also the related work of Herbert Simon, which is most conveniently available in Simon (1977).

[4] This technical suggestion, in the set-theoretic approach, essentially goes back at least to the 1955 dissertation of Adams. A similar suggestion was made independently by Ryszard Wojcicki in 1969; cf. Wojcicki (1973, 1974a).

[5] Sneed's discussion of Ramsey-eliminability was available in a mimeographed form already in 1968. It was commented on by Tuomela in his Stanford dissertation in 1969 (cf. Tuomela, 1973) and by Simon and G. Groen in 1973 (cf. Simon, 1977, Ch. 6.6). See also Swijtink (1976).

[6] See Stegmüller (1976a), p. *X*.

[7] See Stegmüller (1974, 1975, 1976a, 1976b, 1977, 1978a, 1978b).

[8] See also Sneed (1978) and Balzer (1978).

[9] See Sneed (1976), Stegmüller (1976b), and Kuhn (1976).

[10] For other attempts, see Popper (1963), Suszko (1968), Giedymin (1970), Niiniluoto and Tuomela (1973), Nowak (1975), Scheibe (1976), Törnebohm (1976), Krajewski (1977), Niiniluoto (1978b), Tuomela (1978a), Rantala (1978b) and Przełecki (1978).

[11] For reviews of the Sneed–Stegmüller approach, see Przełecki (1974), Wojcicki (1974a), Moulines (1975a), Feyerabend (1977), Tuomela (1978b) and Rantala (1978a).

[12] See Balzer and Sneed (1977). Here Pot(X) is the power set of X.

[13] Stegmüller (1976a) p. 108, avoids this problem by making the more specific assumption that the elements of a matrix are of the form $\langle D, f_1, \ldots, f_n \rangle$, where $D \neq \emptyset$ and each $f_i (i = 1, \ldots, n)$ is a real-valued (partial) function on D.

[14] In a many-sorted structure, one may divide the domain into disjoint sets, such as the set of nonmathematical objects, the set of real numbers, etc. Many-sorted structures can be reduced, if one wishes, to the ordinary one-sorted structures.

[15] This treatment was used by Carnap in his later work on inductive logic (see Carnap, 1971); it is also a basic tool in modern 'abstract logic' which studies the properties of 'logics' with as few assumptions about syntax as possible. Note also that the characteristic function of $M \subseteq M_p$ is a proposition in the sense of Montague.

[16] Scheibe (1978) makes this point concerning the Sneedian notion of constraint C.

[17] If the law M is not an elementary class, it does not have a first-order formulation. If such 'laws' have to be considered in dealing with examples of scientific theories, more powerful (such as higher-order or infinitary) languages are needed.

[18] Harris allows in L^* for quantification over the elements z of M_p and over the elements of the domains of z in M_p, but in the above example this sort of duality in quantification is avoided.
[19] Sneed's criterion of T-theoreticity concerns functions (see Sneed, 1971, pp. 31–33); Stegmüller gives a criterion also for relations (Stegmüller, 1976a, p. 53). It is not clear whether one can speak of theoretical and non-theoretical domains of models – even if in the general definition of cores in Balzer and Sneed (1977) this seems to be presupposed. For discussion about Sneed's criterion, see also Kamlah (1976) and Tuomela (1978b).
[20] A classical argument that the use of such instruments as magnifying glasses, microscopes, etc. is theoretical relative to optics is given by Duhem (1954), pp. 153–158.
[21] One might, of course, define K-theoreticity by referring, instead of the existing theoretical applications $E_{SC,t}(K)$, to all the possible theoretical applications of K. In this case, one should refer also to the possible expositions of these possible applications. If this makes sense, then the claim that a certain term is K-theoretical would be a strong metascientific hypothesis which is true or false independently of the already known applications of K. However, even in this situation it is still the case that the opinion of the members of SC at time t as to the truth or falsity of this metascientific hypothesis is relative to the existing theoretical applications of K at time t. If the members of SC change their opinion about the status of this hypothesis, then core K will change to another core.
[22] Some arguments for the latter alternative have been presented also by Harris (1978).
[23] Cf. Tuomela (1973) and Rantala (1977).
[24] For example, 'double star' is an optics-theoretical but not astronomy-theoretical term in astronomy.
[25] Cf. the distinction between substantive and methodological correspondence rules in Sellars (1963). Cf. also Tuomela (1978b).
[26] For a different treatment of operationalization, see Wojcicki (1973, 1974a).
[27] This means that quantification within members of Γ will be restricted to set I.
[28] For example, the theory of Boolean algebras as applied to switching circuits seems to be, in this sense, a theory of mathematical physics.
[29] For a related way of distinguishing (extralinguistic) theories and their (linguistic) formulations, see Suppe (1974), pp. 204–205.
[30] These assumptions are usually abstracted from the characteristics of certain particular mathematical objects.
[31] For students of mathematics, it is very impressive to see e.g. how easily one can introduce a topological structure into various sets and how much information the basic axioms concerning topological spaces (and their specializations) contain.
[32] This view is a little naive, since it ignores the fact that natural laws have, due to their 'lawlike' character, counterfactual force and thereby refer also to other 'possible worlds' besides the actual one. It is not clear to me whether Sneed's requirement that the domains of the intended applications of physical theories should be connected (Sneed, 1971, p. 255) involves a restriction to the parts of the actual world only.
[33] Another difficulty is the following: it has been argued that the Ramsey-sentence technique which has attracted many instrumentalists is equally well a tool for a scientific realist since it involves the assertion of the existence of certain theoretical entities (cf. Hempel, 1965, p. 216; and Tuomela, 1973). But even if the Ramsey-sentence approach may commit one to accept the existence of certain theoretical *functions*, it is

not clear whether the existence of these functions commits one to accept the existence of certain 'real' physical *properties*. If the mass function exists, we can still think *either* that mass is a physical property *or* that masses are just coefficients to be introduced into calculations. (As a historical remark about the Ramsey-technique, let us note that Henri Poincaré used, in his discussion about thermodynamics, the formulation 'There is something which remains constant' for the principle of the conservation of energy; see Poincaré, 1952, p. 127.)

[35] See Sneed (1971), pp. 293–294; Sneed (1976), pp. 130–131; Stegmüller (1976a), pp. 191–195; Balzer and Sneed (1977).

[36] See Moulines (1979) and Stegmüller (1978b).

[37] In the first edition of his *Structure of Scientific Revolutions* (1962) Kuhn used the word 'paradigm' to cover several different sorts of things which a group of scientists may share. In 'Postscript 1969' to the 2nd edition (1970), he distinguishes between two such things which he now calls 'exemplars' and 'disciplinary matrices'. See also Kuhn (1974) and Suppe (1974).

[38] In his slightly different account, Moulines visualizes normal science as a "living net changing and growing in different directions over historical time". He requires essentially that, in a theory-evolution $\{N^t\}$, $t = 0, 1, \ldots,$ every theory-element in N^{t+1} must be a specialization of some element in N^t. (This type of requirement will be criticized below.) He does not assume that the paradigmatic theory-element $\langle K_0, I_p\rangle$ is the historically first theory-element in the development of the theory, and he allows that I_p is divided into several 'homogenous' subsets each of which contains its own paradigmatic examples.

[39] These questions are separate: there are those who think that Kuhnian normal science does not exist (see Feyerabend, 1977).

[40] Let K' be obtained from K by adding new non-K-theoretical components to the members of M_{pp} and M_p. Then there may exists between theory-elements $\langle K, I\rangle$ and $\langle K', I'\rangle$ interesting relations which do not seem to be definable by the relations of theoretization, specialization, and reduction. This contradicts the claim of Balzer and Sneed (1977), p. 200. Together with theoretization and reduction relations, this relation may be useful in giving an account of the method of 'idealization and concretization' (cf. Nowak, 1975, and Krajewski, 1977).

[41] For attempts to clarify Kuhn's notion of a 'puzzle' by means of the logic of questions, see Kleiner (1970) and Girill (1973). Much further work in this direction is needed. Cf. also Niiniluoto (1976).

[42] Cf. Suppe (1974), pp. 135–151.

[43] Cf., for example, Poincaré (1952), p. 163.

[44] See Lakatos (1970); cf. also Laudan (1977). This is connected with the idea that research traditions can be defined by means of 'cluster concepts', so that the successive elements within a tradition bear only a family resemblance to each other. This could be used in the definition of the 'theoretical claims' of Kuhn-theories.

[45] For details, see Sneed (1971), Stegmüller (1976a), Mayr (1976), Sneed (1976), Balzer and Sneed (1977).

[46] For interesting comments on this question, see Kuhn (1976).

[47] For attempts to measure theory-distance, see Tuomela (1978a) and Niiniluoto (1978b).

[48] Note that there is no absolute way of counting the number of solved problems, since

the solution of several different problems may be reducible to the solution of one and the same problem. This fact is taken into account in the Sneedian concept of reduction.

[49] See also Niiniluoto (1977, 1978b). The basic idea is the following: Let g be a first-order generalization, and let $m(g, C_i)$ be the distance of g from a constituent C_i, where constituents are the strongest consistent generalizations expressible in the language of g. If C_x is the true constituent, then the truthlikeness of g is defined by $m(g, C_x)$. As C_x is normally unknown to us, the most reasonable estimate for the degree of truthlikeness of g on some evidence e seems to be the expected value

$$\text{ver}(g/e) = \sum_i P(C_i/e)m(g, C_i)$$

where the sum is taken over all constituents. A step from g to g' is progressive if and only if $m(g, C_x) < m(g', C_x)$, and this step seems progressive on e if and only if $\text{ver}(g/e) < \text{ver}(g'/e)$.

[50] This is, in fact, not more surprising than the observation that one can travel from one place to another by several different routes. This shows that progress branching is compatible with the view that there is one specified goal of science (cf. Stegmüller, 1978a, p. 59); it is incompatible with this view only if we deny the possibility that the branches later merge together. However, it has been argued, already by William Whewell in 1840, that the most prominent feature of scientific progress is just the merging of different paths in scientific truth-seeking (cf. Niiniluoto, 1978a).

[51] As some attempted solutions may be so misleading that they are worse than the non-existence of a solution, this value might as well be defined by the difference between the truthlikeness of proposed laws and that of a tautology.

BIBLIOGRAPHY

Achinstein, P.: 1968, *Concepts of Science*, Johns Hopkins Press, Baltimore.

Balzer, W.: 1978, 'Incommensurability and Reduction', in Niiniluoto and Tuomela (1978), pp. 313–335.

Balzer, W. and Sneed, J.: 1977, 'Generalized Net Structures of Empirical Theories, I', *Studia Logica* **36**, 195–211.

Beth, E., 'Semantics of Physical Theories', in H. Freudenthal (ed.), *The Concept and the Role of the Model in Mathematics and Natural and Social Sciences*, Reidel, Dordrecht, 1961, pp. 48–51.

Bogdan, R.J. (ed.): 1976, *Local Induction*, Reidel, Dordrecht.

Butts, R.E. and Hintikka, J. (eds.): 1977, *Historical and Philosophical Dimensions of Logic, Methodology and Philosophy of Science*, Reidel, Dordrecht.

Carnap, R.: 1971, 'A Basic System of Inductive Logic', in R. Carnap and R.C. Jeffrey (eds.), *Studies in Inductive Logic and Probability*, Vol. 1, University of California Press, Berkeley, pp. 35–165.

Church, A.: 1956, *Introduction to Mathematical Logic*. Princeton University Press, Princeton.

Duhem, P.: 1954, *The Aim and Structure of Physical Theory*, Princeton University Press, Princeton.

Feyerabend, P.: 1977, Review of Stegmüller (1973), *The British Journal for the Philosophy of Science* **28**, 351–369.

Frege, G., *Begriffschrift* (1879). English transl. in J. van Heijenoort (ed.): 1967. *From Frege to Gödel: A Sourcebook in Mathematical Logic 1879–1931*, Harvard University Press, Harvard.

Giedymin, J.: 1971, 'The Paradox of Meaning Variance', *The British Journal for the Philosophy of Science* **22** 30–48.

Girill, T.R.: 1973, 'The Logic of Scientific Puzzles', *Zeitschrift für allgemeine Wissenschaftstheorie* **4**, 25–40.

Harris J.H.: 1978. 'A Semantical Alternative to the Sneed–Stegmüller–Kuhn Conception of Scientific Theories', in Niiniluoto and Tuomela (1978), pp. 184–204.

Hempel, C.G.: 1965, *Aspects of Scientific Explanation*, The Free Press, New York.

Hempel, C.G.: 1970, 'On the "Standard Conception" of Scientific Theories', in M. Radner and S. Winokur (eds.), *Minnesota Studies in the Philosophy of Science*, Vol. IV. University of Minnesota Press, Minneapolis, pp. 142–163.

Hesse, M.: 1974, *The Structure of Scientific Inference*, Macmillan, London.

Hintikka, J. and Tuomela, R.: 1970, 'Towards a General Theory of Auxiliary Concepts and Definability in First-Order Theories', in J. Hintikka and P. Suppes (eds.), *Information and Inference*, Reidel, Dordrecht, pp. 298–330.

Kamlah, A.: 1976, 'An Improved Definition of "Theoretical in a Given Theory" ', *Erkenntnis* **10**, 349–359.

Kleiner, S.: 1970, 'Erotetic Logic and the Structure of Scientific Revolution', *The British Journal for the Philosophy of Science* **21**, pp. 149–165.

Krajewski, W.: 1977, *Correspondence Principle and the Growth of Knowledge*, Reidel, Dordrecht and Boston.

Kuhn, T.S.: 1962, *The Structure of Scientific Revolutions*, The University of Chicago Press, Chicago, 2nd ed. 1970.

Kuhn, T.S.: 1974, 'Second Thoughts on Paradigms', in Suppe (1974), pp. 459–482.

Kuhn, T.S.: 1976, 'Theory-Change as Structure Change: Comments on the Sneed Formalism', *Erkenntnis* **10**, 179–199. (Also in Butts and Hintikka, 1977, pp. 289–309.)

Lakatos, I.: 1970, 'Falsification and the Methodology of Scientific Research Programmes', in I. Lakatos and A. Musgrave (eds.), *Criticism and the Growth of Knowledge*, Cambridge University Press, Cambridge, pp. 91–195.

Laudan, L.: 1977, *Progress and Its Problems*, Routledge and Kegan Paul, London.

Lehrer, K.: 1977, 'Social Information', *Monist* **60**.

Levi, I.: 1976, 'Acceptance Revisited', in Bogdan pp. 1–71.

Mayr, D.: 1976, 'Investigations of the Concept of Reduction, I', *Erkenntnis* **10**, 275–294.

Montague, R.: 1961, 'Deterministic Theories', in R. Washburne (ed.), *Decisions, Values, and Groups, II*, Pergamon Press, Oxford, pp. 325–370.

Moulines, C.-U.: 1975, Review of Sneed (1971), *Erkenntnis* **9**, 423–436. [1975a.]

Moulines, C.-U.: 1975, 'A Logical Reconstruction of Simple Equilibrium Thermodynamics', *Erkenntnis* **9**, 101–130. [1975b.]

Moulines, C.-U.: 1976, 'Approximate Application of Empirical Theories: A General Explication', *Erkenntnis* **10**, 201–227.

Moulines, C.-U.: 1979, 'Theory-Nets and the Dynamics of Theories: The Example of Newtonian Mechanics', *Synthese* **41**, 417–439.

Niiniluoto, I.: 1976, 'Inquiries, Problems, and Questions: Remarks on Local Induction', in Bogdan, pp. 263–296.

Niiniluoto, I.: 1977, 'On the Truthlikeness of Generalizations', in R.E. Butts and J.

Hintikka (eds.), *Basic Problems of Methodology and Linguistics*, Reidel, Dordrecht, pp. 121–147.
Niiniluoto, I.: 1978a, 'Notes on Popper as Follower of Whewell and Peirce', *Ajatus* **37**, 272–327.
Ninniluoto, I.: 1978b, 'Truthlikeness: Comments on Recent Discussion', *Synthese* **38**, pp. 281–329.
Niiniluoto, I.: 1978c, 'Verisimilitude, Theory-Change, and Scientific Progress', in Niiniluoto and Tuomela (1978), pp. 243–264.
Niiniluoto, I. and Tuomela, R.: 1973, *Theoretical Concepts and Hypothetico-Inductive Inference*, Reidel, Dordrecht.
Niiniluoto, I. and Tuomela, R. (eds.): 1978, *The Logic and Epistemology of Scientific Change* (Acta Philosophical Fennica **30**), North-Holland, Amsterdam.
Nowak, L.: 1975, 'Relative Truth, the Correspondence Principle and Absolute Truth', *Philosophy of Science* **42**, 187–202.
Poincaré, H.: 1952, *Science and Hypothesis*, Dover, New York.
Popper, K.R.: 1963, *Conjectures and Refutations*, Routledge and Kegan Paul, London.
Przełecki, M.: 1969, *The Logic of Empirical Theories*, Routledge and Kegan Paul, London.
Przełecki, M.: 1974, 'A Set-Theoretic versus a Model Theoretic Approach to the Logical Structure of Physical Theories', *Studia Logica* **33**, pp. 91–112.
Przełecki, M.: 1978, 'Commensurable Referents of Incommensurable Terms', in Niiniluoto and Tuomela (1978), pp. 347–365.
Przełecki, M. and Wojcicki, R. (eds.): 1977, *Twenty-Five Years of Logical Methodology in Poland*, Reidel, Dordrecht.
Rantala, V.: 1977, *Aspects of Definability* (Acta Philosophica Fennica **29**), North-Holland, Amsterdam.
Rantala, V.: 1978a, 'The Old and New Logic of Metascience', *Synthese* **39**, 233–247.
Rantala, V.: 1978b, 'Correspondence and Non-Standard Models: A Case Study', in Niiniluoto and Tuomela (1978), pp. 366–378.
Scheibe, E.: 1973, *The Logical Analysis of Quantum Mechanics*, Pergamon Press, Oxford.
Scheibe, E.: 1976, 'Conditions of Progress and the Comparability of Theories', in R.S. Cohen *et al.* (eds.), *Essays in Memory of Imre Lakatos*, Reidel, Dordrecht.
Scheibe, E.: 1978, 'On the Structure of Physical Theories', in Niiniluoto and Tuomela (1978), pp. 205–224.
Sellars, W.: 1963, *Science, Perception and Reality*, Routledge and Kegan Paul, London.
Shapere, D.: 1974, 'Scientific Theories and Their Domains', in Suppe (1974), pp. 518–570.
Simon, H.A.: 1977, *Models of Discovery*, Reidel, Dordrecht.
Sneed, J.: 1971, *The Logical Structure of Mathematical Physics*, Reidel, Dordrecht.
Sneed, J., 1976, 'Philosophical Problems in the Empirical Science of Science: A Formal Approach', *Erkenntnis* **10**, pp. 115–146. (Also in Butts and Hintikka, 1977, pp. 245–268.)
Sneed, J.: 1978, 'Invariance Principles and Theoretization', in Niiniluoto and Tuomela (1978), pp. 130–178.
Stegmüller, W.: 1970, *Theorie und Erfahrung*, Springer–Verlag, Berlin–Heidelberg–New York.

Stegmüller, W.: 1973, *Theorienstrukturen und Theoriendynamik*, Springer-Verlag, Berlin–Heidelberg–New York.

Stegmüller, W.: 1974, 'Theorienstruktur und logisches Verständnis', in W. Diederich (ed.), *Theorien der Wissenschaftsgeschichte*, Suhrkamp, Frankfurt am Main, pp. 167–209. English transl. 'Logical Understanding and the Dynamics of Theories', in W. Stegmüller, 1977, *Collected Papers*, II, Reidel, Dordrecht, pp. 150–176.

Stegmüller, W.: 1975, 'Structures and Dynamics of Theories: Some Reflections on J.D. Sneed and T.S. Kuhn', *Erkenntnis* **9**, 75–100. Also in *Collected Papers*, II, pp. 177–202.

Stegmüller, W.: 1976a, *The Structure and Dynamics of Theories*, Springer–Verlag, New York–Heidelberg–Berlin.

Stegmüller, W.: 1976b, 'Accidental ("Non-Substantial") Theory Change and Theory Dislodgment', *Erkenntnis* **10**, 147–178. Also in Butts and Hintikka (1977), pp. 269–288.

Stegmüller, W.: 1976, 'Normale Wissenschaft und wissenschaftliche Revolutionen – Kritische Betrachtungen zur Kontroverse zwischen Karl Popper und Thomas Kuhn', *Wissenschaft und Weltbild* **29**, 169–180. [1976c.]

Stegmüller, W.: 1978a, 'A Combined Approach to the Dynamics of Theories', *Theory and Decision* **9**, 39–75.

Stegmüller, W.: 1978, 'The Structuralist View: Survey, Recent Developments and Answers to Some Criticisms', in Niiniluoto and Tuomela (1978), pp. 113–129.

Suppe, F.: 1972, 'What's Wrong with the Received View on the Structure of Scientific Theories', *Philosophy of Science* **39**, 1–19.

Suppe, F. (ed.): 1974, *The Structure of Scientific Theories*, The University of Illinois Press, Urbana.

Suppes, P.: 1957, *Introduction to Logic*, Van Nostrand, New York.

Suppes, P.: 1967, 'What is a Scientific Theory', in S. Morgenbesser (ed.), *Philosophy of Science Today*, Basic Books, New York, pp. 55–67. [1967a.]

Suppes, P.: 1967, *Set-Theoretic Structures in Science*, mimeographed, Stanford University, Stanford, [1967b.]

Suppes, P.: 1969, *Studies in the Methodology and Foundations of Science*, Reidel, Dordrecht.

Suszko, R.: 1968, 'Formal Logic and the Development of Knowledge', in I. Lakatos and A. Musgrave (eds.), *Problems in the Philosophy of Science*, North-Holland, Amsterdam, pp. 210–222.

Swijtink, Z.: 1976, 'Eliminability in a Cardinal', *Studia Logica* **35**, 73–89.

Toulmin, S.: 1972, *Human Understanding*, Vol. I, Princeton Unviersity Press, Princeton.

Tuomela, R.: 1973, *Theoretical Concepts*, Springer, Vienna.

Tuomela, R.: 1978a, 'Scientific Change and Approximation', in Niiniluoto and Tuomela (1978), pp. 265–297.

Tuomela, R.: 1978b, 'On the Structuralist Approach to the Structure and Dynamics of Theories', *Synthese* **39**, 211–231.

Törnebohm, H.: 1976, 'On Piecemeal Knowledge-Formation', in Bogdan, pp. 297–318.

van Fraassen, B.: 1970, 'On the Extension of Beth's Semantics of Physical Theories', *Philosophy of Science* **37**, 325–339.

Wojcicki, R.: 1973, 'Basic Concepts of Formal Methodology of Empirical Sciences', *Ajatus* **35**, 168–196.
Wojcicki, R.: 1974a, 'Set-Theoretic Representations of Empirical Phenomena', *Journal of Philosophical Logic* **3**, 337–343.
Wojcicki, R., 1974b, 'Comments on Przełecki', *Studia Logica* **33**, 105–107.

ADDITIONAL BIBLIOGRAPHY

After the completion of this essay in 1978, a number of new studies and applications of the structuralist view have appeared. They include:

Balzer, W.: *Empirische Geometrie und Raum-Zeit-Theorie in mengen-theoretischer Darstellung*, Kronberg, 1978.
Balzer, W.: 'On the Status of Arithmetic', *Erkenntnis* **14** (1979), 57–85.
Diederich, W. and Fulda, H.F.: 'Sneed'sche Strukturen in Marx' "Kapital" ', *Neue Hefte für Philosophie* **13** (1978), 47–80.
Moulines, C.U.: 'Cuantificadores existenciales y principios-guia en las teorias fisicas', *Critica* **10** (1978), 59–88.
Moulines, C.U. and Sneed, J.: 'Suppes' Philosophy of Physics', in R.J. Bogdan (ed.), *Patrick Suppes*, D. Reidel, Dordrecht, 1979, pp. 59–91.
Sneed, J.: *The Logical Structure of Mathematical Physics*, 2nd revised edition, Reidel, Dordrecht, 1979.
Stegmüller, W.: *The Structuralist View of Theories: A Possible Analogue of the Bourbaki Programme in Physical Science*, Springer-Verlag, Berlin, Heidelberg, New York, 1979.

V. N. SADOVSKY

LOGIC AND THE THEORY OF SCIENTIFIC CHANGE

The discussion of J. Sneed's and W. Stegmüller's logical reconstruction of Kuhn's conception of sciences change leads inevitably to the formulation of more general questions such as: what form, from the modern point of view, should be characteristic of the theory of change of scientific theories? What role belongs to logic, logical formalization of the content (intuitive) sentences of this theory during construction of such a theory? Certainly, these questions are not new. They essentially touch upon the central problems of the philosophy of science, and each influential methodological conception provided its own answer. And the rational reconstruction of Kuhn's views developed by Sneed and Stegmüller and, in particular, the discussion of these issues between Kuhn, Sneed and Stegmüller that took place in 1975 at the Fifth International Congress of Logic, Methodology, and Philosophy of Science shed new light on these topics which are fundamental for the philosophy and history of science. Thus it is appropriate to consider this continuing discussion.

First of all it should be noted that the expression 'The Sneed–Stegmüller "rational reconstruction" of Kuhn's views' is not unambiguous at the present stage of discussion. In 1971 and in the following years the notion of Sneed's conception of logical structure of mathematical physics did not produce any uncertainties, since at that time there existed the only presentation of this conception in Sneed's well-known book *The Logical Structure of Mathematical Physics* (1971). Strictly speaking, this conception did not bear directly on Kuhn's theory of normal and revolutionary science. (It is well known that in this work Sneed only formulated an assumption about the closeness between his conception and Kuhn's theory.) The notion of of Sneed–Stegmüller's logical formalization of Kuhn's conception was also sufficiently definite for some time after 1973; it was based on the canonical presentation of this reconstruction in Stegmüller's book *Theorienstrukturen und Theoriendynamik* (1973, English translation 1976). Since 1976, however, this trend of research has undergone

J. Hintikka, D. Gruender, and E. Agazzi (eds.), Pisa Conference Proceedings, Vol. I, 49–61.

certain modifications, the most important of which are the following two. First, Sneed himself in his paper at the Congress of LMPS in 1975 (mentioned above) introduced a number of changes in his theory (published in 1977); they consisted in attributing the fully developed formal approach directly to Kuhn's theory of the structure of scientific revolutions, in the replacement of the conception of a theory's "expanded core" by the conception of theory-nets and in the development of the concepts of specialization and theoretization. This trend of research continued in the paper authored by Sneed together with Balzer and published in 1977 in *Studia Logica*. Second, various possible revisions and expansions of Sneed's approach have been discussed, most recently by M. Przełęcki, C.-U. Moulines, J. H. Harris, I. Niiniluoto and some other authors. (For a brief review of these modifications of Sneed's approach, see I. Niiniluoto's contribution to the present conference.) Thus, taking into account the development of the above set of problems during the last 7–8 years, as well as the fact that the concept of 'Sneed–Stegmüller "rational reconstruction" of Kuhn's views' may be now ascribed a meaning which is far from unambiguous, we deem it appropriate to distinguish between the original position of Sneed (which will be denoted as Sneed, 1971), the interpretation of this position by Stegmüller (Stegmüller–Sneed, 1973), the subsequent development of Sneed's formalism by Sneed himself and his group (Sneed–Balzer) and, finally, a great number of various modifications offered most recently (the structuralist conception of the scientific theories, 1975–1978).

All the above forms of logical reconstruction of the structure and dynamics of scientific theories represent certain *formal constructions* and as such they are either based on some or other content (intuitive) ideas concerning scientific theories or interpreted later in intuitive terms. These two levels of any formal construction (*the formal level* and *the intuitive, content level*) should be taken into account in the development and evaluation of any formal construction and, naturally, they should be fixed explicitly in the analysis of Sneed–Stegmüller formalism.

There are sufficiently complex relations between the above two main levels of any formal construction, especially for rich formalized constructions. The intuitive-content ideas the formalization of which is strived for are, as a rule, described in formal terms only partially – highly typical in this respect is the very structure of Sneed's paper of

1977 "Describing Revolutionary Scientific Change: A Formal Approach": each formal statement of this paper is accompanied by the exposure of the intuitive ideas that lie in the basis of such statements. Besides, it is emphasized both by Sneed and later by I. Niiniluoto (1980) that even the intuitive singling out of (1) the set of all possible models for the full conceptual apparatus of a theory, including theoretical components (M_p), (2) the set of the corresponding non-theoretical models (M_{pp}) and (3) the set of those potential models which are not excluded by the fundamental law of the theory (M), which is basic in Sneed's approach, is only partially expressed in his formal logical apparatus. On the other hand, the development of the formal apparatus is to a great extent autonomous with respect to the corresponding content level of the scientific knowledge and, as is well known, contains a possibility of obtaining paradoxical consequences. The failure of Popper's conception of verisimilitude (Popper, 1963) is a vivid recent illustration of this: Popper's criterion of the verisimilitude of scientific theories which is intuitively quite acceptable, – the theory B is more closely similar to the truth than the theory A if and only if either (1) the truth-content of B but not its falsity-content exceeds that of A, or (2) the falsity-content of A but not its truth-content exceeds that of B – if formalized, as is shown by P. Tichy (1974), D. Miller (1974) and J. Harris (1974), leads to an unacceptable consequence: if A and B are false theories, neither of them can be closer to the truth than the other.

The above complexities of the relations between the formal and the intuitive levels of scientific knowledge require (in the analysis of Sneed's formal construction) a distinction at least between the content-intuitive interpretation which is *really* expressed in this formalism and Kuhn's conception of the development of science the logical formalization of which appears as the *objective* of these studies. Obviously, these two components by no means coincide.

Finally, in order to conclude the construction of the conceptual scheme in the frame of which, as it seems to us, one should analyze the Sneed–Stegmüller approach, it is necessary to emphasize specially that, according to all researchers who work in this field, Sneed's understanding of the scientific theory represents a rejection and a definite alternative to the so-called standard conception of the scientific theory which was adopted everywhere in the forties–sixties. The standard conception of a scientific theory is based on its representation as a set of sentences closed under deduction. As this conception was being developed, especially at the end of the sixties, many of its essential difficulties and limitations were disclosed (see, in particular, F. Suppé, 1974, in whose works this conception appears under the name "received view"). According to Sneed, in order to understand a theory, it is necessary to construct in a set-theoretic way a certain mathematical structure and to indicate the set of its applications (both actually realized at present and potentially possible in future). Working out the relations between the standard view and Sneed's (in Stegmüller's (1973) terminology, "statement view" and "non-statement view") conceptions of the scientific theory in detail is an important aspect of evaluation of the limits and possibilities of Sneed–Stegmüller's approach. It should be noted, in addition, that the classical standard conception is rejected today, as a matter of fact, by all investigators, and other forms of its replacement or at least modification, alongside with the rejection of it on the part of Sneed, are suggested, for instance, the model-theoretic approach developed by M. Przełęcki, V. Rantala, R. Tuomela, the semantic approach of Bas van Fraassen, F. Suppe and others and some other conceptions. It is obvious that the adequate understanding of Sneed's conception needs the clarification of its relation to these approaches that are in any case not strictly standard.

The above makes it possible to suggest the following conceptual scheme of the rational evaluation of Sneed–Stegmüller's rational reconstruction of Kuhn's views (see the diagram on the next page).

This scheme depicts in a visually obvious way the main relations that are subject to revision and evaluation. We shall call them: (1) the relation of Sneed–Stegmüller's formalization of the structure and dynamics of scientific theories taken in its general form to the standard conception; (2) the evolution of development of Sneed's formalism; (3) the relation of Sneed's formalism in its various forms (Sneed, 1971; Stegmüller–Sneed, 1973; Sneed–Balzer; the structuralist conception, 1975–1978) to those intuitive conceptions about scientific knowledge which lie in the basis of this formalism and which are actually described by it; (4) the relation of Sneed–Stegmüller's conception of the scientific theory to other non-standard approaches in the methodology of science; (5) the relation of Sneed–Stegmüller's conception to Kuhn's theory of normal and revolutionary science;(6) the relation of Kuhn's conception to the subsequent development of the theory of growth of scientific knowledge (Lakatos, Toulmin, Agassi, Feyerabend, Laudan

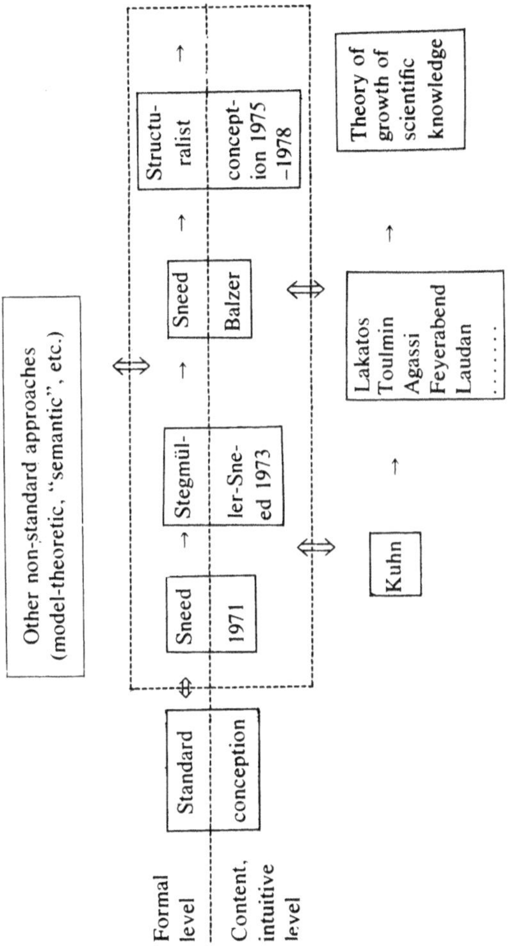
Formal level
Content, intuitive level
Standard conception
Sneed
1971
Stegmül-
ler-Sne-
ed 1973
Sneed
Balzer
Structu-
ralist
concept-
ion 1975
–1978
Other non-standard approaches
(model-theoretic, "semantic", etc.)
Kuhn
Lakatos
Toulmin
Agassi
Feyerabend
Laudan
......
Theory of
growth of
scientific
knowledge

and others). The analysis of the last relation implies the singling out of at least the contours of the most adequate (from the point of view of the researcher who conducts such an analysis) conception of the structure and growth of scientific knowledge as well as the determination of the role and place of logical formalization in such conception. In what follows I shall make some remarks concerning the above relations but the main emphasis will be on the two last of them.

What are the most important specific features that are typical for Sneed–Stegmüller's conception of the scientific theory as compared to the standard approach? Apparently, these features are: (1) the contrast, in Stegmüller's terminology, between a "statement view" and a "non-statement view"; (2) relativization of the relation "empirical-theoretical" not with respect to a certain language but with respect to the given theory; (3) the introduction into the definition of a theory of the set of its applications (both present at a given time and possible in future) which are determined by the basic mathematical structure of the theory; (4) the formal specification of the scientific community that accepts a given theory. It is only in the paragraph (4) that the effect of Kuhn's ideas is revealed explicitly; it can be implicitly (for the corresponding interpretation of a theory's set of applications) noticed also in (3) (indeed, one can single out in the set of applications of a certain theory I the subset of paradigm examples I_0, which is an analogue to Kuhn's paradigm). All other specific features of Sneed's conception are irrelevant with respect to Kuhn's theory, and hence Sneed's formalism is quite *general in nature.* This fact is mentioned by Sneed himself: "the formalism is independent of his [Kuhn's] theory and flexible enough to accommodate many other views", and further: "the formalism may be extended in the other direction as well" (Sneed, 1977, p. 259). This formalism, as it is developed by the present time, contains no means (which is emphasized correctly in Niiniluoto's paper, 1980) for distinguishing between mathematical and empirical theories. It is is natural that this circumstance limits essentially the heuristic and normative possibilities of the conception under study. In what follows we shall discuss again the generality of Sneed's formalism, and now let us consider carefully the relation of this conception to the standard approach.

The pathos of Sneed–Stegmüller's approach to the scientific theory consists in its *nonlinguistic property* (the rejection of the two-level theory language conception and of the understanding of the theory as a certain language). It is obvious, however, that no matter how much we could stress the fundamental importance of establishing the relations not between the linguistic but between set-theoretic entities that are expressed in Sneed's works through sets M_p, M_{pp}, M, C, I and I_o for the understanding of a scientific theory, we shall still need a certain language for the description of the theory, at least at the meta-level. And if one takes into account the results obtained in the study of relevant problems by J. Harris, M. Przełęcki and others, one would completely agree with Niiniluoto's statement: in the aspect of the relation between "statement view" and "non-statement view" "the structuralist conception of theories is essentially equivalent to a suitably amended statement view of theories" (Niiniluoto, 1980). This conclusion, however, applies only to one aspect (aspect 1) of the four main specific features of the Sneed–Stegmüller conception mentioned above. The point in question is the improvement of statement view, and therefore the contrasting expressed in paragraph (1) is preserved, but within the limits described. And the most significant features, from the point of view of establishing the relation between Sneed–Stegmüller's conception and the standard approach, are the features (2), (3), and (4). The specificity of multi-application approach is discussed in detail in Niiniluoto's paper; therefore, I shall concentrate my attention on the feature (4).

As is well known, the concept of *scientific community* introduced by Kuhn into the historical and scientific context led to a very heated argument, and Kuhn's description of the functions of scientific community in the evolution of science resulted in his being accused of subjectivism and irrationalism. All this now is to a considerable degree a matter of the past, and Kuhn's innovation is recognized everywhere as a substantial achievement in bringing the research problems of the history of science, the philosophy of science, and the sociology of science closer together. In Sneed–Stegmüller's formal construction the scientific community appears as the concept of holding by a person p at a time t of a certain theory T. The relation of the person to a given theory or to some of its essential components, first of all, to the set of paradigm examples I_o or to the set I_t^p of applications of a given theory which is recognized by a person p at a time t, turns out to be an essential component of the definition itself of the scientific theory (the theory as understood by Kuhn) – see definitions 15, 16, and 17 in Stegmüller's book (1973). The standard conception whose conscious objective is the total elimination from the theory of science of all non-methodological components excludes absolutely such an approach. And the acceptance of the latter introduces the *historicity* even in the definition of the scientific theory – indeed, this definition implies that the variation of t is accompanied by the variation of the set I_t^p, but I_o turns out to be a constant component of any I_t^p.

Let us now discuss in detail the relation of Sneed–Stegmüller's conception to Kuhn's theory of normal and revolutionary science. Both Sneed and Stegmüller have no claims other than those of partial reconstruction of Kuhn's theory (compare "I have used the theory-net formalism in an attempt to partially reconstruct Kuhn's theory of science", Sneed (1977), p. 259). This is certainly true, but the question is which of Kuhn's ideas fall out from this reconstruction. As it seems to me, there are quite a number of them. First of all, I would like to mention the fact that, according to Kuhn, the paradigm is essentially heterogeneous as far as its composition is concerned. In the second edition of the *Structure of Scientific Revolutions*, when he revises the meaning of the term "paradigm" and suggests, as a matter of fact, to replace it with a more appropriate term "disciplinary matrix", Kuhn singles out its four most important components: symbolic generalizations, metaphysical beliefs, values and exemplars. All of them play an important role in Kuhn's picture of the evolution of scientific knowledge, and the rejection of even a single component among them impoverishes this picture considerably. And the Sneed–Stegmüller formalization contains explicitly only the exemplars (in the form of the set I_o), at any rate, it is exactly in this way that the paradigm concept is introduced by Stegmüller and Sneed. Certainly, it is possible to say that the fundamental law of the theory defined through sets M and C formalizes the symbolic generalizations of the displinary matrix, although it is not clear to what extent this formalization is adequate, and thereby also performs a certain paradigm function, but it is obvious as well that at least two components of the disciplinary matrix – the metaphysical statements and values – are not expressed in the above formalism. The main difficulty here is associated with the fact that in Kuhn's paradigm the scientific proper and philosophical (metaphysical and methodological) elements coexist, and formalization is applied only to certain *homogeneous* (in this case, scientific proper) *entities.*

As we have already noted, in Kuhn's theory an important place is occupied by the concept of scientific community, the structure and mechanisms of functioning of communities of scientists that share a common paradigm. The Sneed–Stegmüller formalization contains this aspect of Kuhn's views only in its extremely abstract form: the scientific community is reduced here to the relation of a single scientist to one or another theory. At best, the temporal variation of the relation of this scientist to the given theory is taken into account. To put it differently, the explication of the structure and dynamics of the scientific community is replaced by subjecting to *formalization of the attitude of the individual scientist with respect to the scientific theory* exemplifying this community.

The third important aspect of Kuhn's ideas which is not reflected in the Sneed–Stegmüller formalism pertains to the mechanism of revolutionary changes in science. Various types of reduction of theories established by Sneed and Stegmüller provide, if one is permitted to use such an expression, the *static dynamics of theories*, i.e., the set of relations between theories that are logically possible, for which one theory is reduced to another or is replaced by another, and so on. And the nature of Kuhn's picture of the revolutionary changes in science, for all its being quite obviously incomplete, is essentially dynamic, while its Sneed's formalization deals only with the explication of some final results of scientific revolutions.

When discussing the boundaries (those that are reached and, in general, possible in principle) of the logical formalization of Kuhn's theory describing the growth of scientific knowledge, one can formulate a natural question: and why, as a matter of fact, is the present discussion of these issues limited to include nothing but Kuhn's conception? Does this concept really lead to the necessary and sufficient understanding (at the content level) of the phenomena of scientific change studied in it and is our problem really reduced only to its logical revision? The answer to this question is obvious: certainly, no. The problems of logical formalization appear not only with respect to the methodology of Lakatos' research programs (many aspects of which, incidentally, are very close to those of

Kuhn's conception of paradigms) but also with respect to the conceptions of the growth of science formulated by Toulmin, Agassi, and others, and with respect to the theory of scientific progress by Laudan suggested recently (1977). It appears to me that the fact that these theories have not up to now become an object of the corresponding logical studies is a disappointing misunderstanding.

Inasmuch as the invariable or relatively invariable for a certain period of time core of scientific knowledge ("hard-core" of Lakatos' research program, Laudan's core of the research tradition, and so forth) is introduced in the above conceptions, the latter, because of the generality of Sneed–Stegmüller's formalism mentioned earlier, can be logically reconstructed in the frame of the formal apparatus suggested by Sneed. The problem is to actually realize such formalization having in mind the fact that each such conception contains obvious specific elements (the rejection of the phenomenon of normal science, the study of evolving populations of the scientific concepts, the acceptance of the variations of the core of research traditions that are frequently quite significant, and the like) which would require the methods of formalization of their own.

At the same time, certain fundamental difficulties of realization of such projects are already clear today. Let us mention just two of them. The main tendency in the development of the conceptual apparatus which is capable of providing the adequate picture of the structure and evolution of scientific knowledge consists, after the failure of the anti-metaphysical orientation of logical positivists, in the accounting, at a deeper and deeper level, for the role of metaphysical and methodological components as relatively stable determining factors of the growth of knowledge. As we have already mentioned, this point is explicitly taken into account by Kuhn. It manifests itself to a still greater extent in the interpretation of Lakatos' "hard-core" and Laudan's core of research tradition (in this case the research tradition is reduced, as a matter of fact, to "certain metaphysical and methodological commitments which, as an ensemble, individuate the research tradition and distinguish it from others" (Laudan, 1977, pp. 78–79)). And if this is so, the logical formalization is facing now the problem of *rationalization of heterogeneous composition* (the scientific, philosophical, value-normative, and such) of the factors that determine the scientific progress, which has as yet no solution.

The second fundamental difficulty is concerned with the formalization of *the mechanisms of development*, in particular, of the revolutionary change of science. All attempts of such formalization that have been undertaken up until now pertain only, in Sneed's words, to establishing relations between "products of science" (Sneed, 1977, p. 249) and totally abstract themselves from the methods of obtaining such products.

The above should be by no means interpreted as the sceptical attitude to the possibilities of logical reconstruction of the structure and development of scientific knowledge. Everybody knows the advantages of the rigorous formal description, and it is absolutely not necessary to stand up for them again. The construction of logical theories is not an end in itself, however, it always pursues one or another meaningful objective. In this respect, the difficulties in the logical reconstruction of the structure and dynamics of scientific knowledge that are experienced now are mostly associated with the fact that, as was noted earlier, the study is applied only to Kuhn's conception the methodological and historical-scientific inequacy of which (at any rate, with respect to many of its essential points) seems to be quite obvious today. This leads naturally to the problem of development of a more acceptable intuitive theory of change of scientific theories which should serve as the real basis of the subsequent richer logical formalization.

To conclude my paper, I will give a thesis formulation of the main principle which seems to me to be necessary for the foundation of the theory of change of scientific theories (for more detail see Sadovsky, 1977; Grjasnov and Sadovsky, 1978). These principles proceed from the Marxist understanding of science and take into account the results of the development of the specific problems of the methodology of science that have been obtained recently.

(i) The theoretical understanding of science is possible only if the *dynamic structure* of the scientific knowledge is constructed.

(ii) The nature of scientific knowledge has a *holistic character*; the division of scientific knowledge into the level of observation and the level of the theory is relative. Any observational statement is conditioned by the corresponding theory; it is "theory-laden".

(iii) *Philosophical* (ontological, metaphysical) *conceptions are closely related to the scientific proper* (specifically scientific) *know-*

ledge: philosophy not only stimulates science in a positive or negative way but the philosophical statements constitute an integral part of the "body" of science.

(iv) The dynamics of scientific knowledge is *not a strictly cumulative process.*

(v) The purpose of science is to achieve *the objective truth.* In order to describe the progress of science, it is necessary to take into account not only the successive approach to the truth but also the realization of *the following aims that are important epistemologically*: the obtaining of a better understanding of the phenomena under study, the solution of a greater number of scientific problems, the construction of simpler and more compact theories, and so forth.

(vi) The construction of the theory of science change is possible only when the relation between the internal logic of the development of scientific knowledge and the social conditionality of science and its evolution is established. Because of this *the synthesis of various approaches to the analysis of science*: the historical, the methodological, the "science of science", the psychological, the sociological, and the logical, appears as the proper method of the construction of such a theory. The main function of the logical approach consists in formalization of the available intuitive (content) ideas concerning science as it develops and in obtaining consequences from the construction of such theory for its further improvement.

Institute for
Systems Studies,
Moscow

BIBLIOGRAPHY

Agassi, J.: 1975, *Science in Flux*, (Boston Studies in the Philosophy of Science, vol. 28) D. Reidel, Dordrecht–Boston.

Balzer, W. and Sneed, J.D.: 1977, 'Generalized Net Structures of Empirical Theories I.', *Studia Logica* **36**, 195–211.

Feyerabend, P.: 1975, *Against Method*, New Left Books, London.

Grjasnov, B.S. and Sadovsky, V.N.: 1978, 'Problems of the Structure and the Development of Science in "Boston Studies in the Philosophy of Science"', in B.S. Grjasnov and V.N. Sadovsky (eds.), *The Structure and Development of Science*, Selection from Boston Studies in the Philosophy of Science, "Progress" Publishing House, Moscow, pp. 5–39 (in Russian).

Harris, J.H.: 1974, 'Popper's Definitions of Verisimilitude', *British Journal for the Philosophy of Science* **25**, 160–166.

Kuhn, T.: 1970, *The Structure of Scientific Resolutions*, 2nd, ed., University of Chicago Press, Chicago, Ill.

Kuhn, T.: 1977, 'Theory-Change as Structure-Change: Comments on the Sneed Formalism', in R.E. Butts and J. Hintikka (eds.), *Historical and Philosophical Dimensions of Logic, Methodology and Philosophy of Science*, D. Reidel, Dordrecht-Boston, pp. 289–309.

Lakatos, I.: 1970, 'Falsification and the Methodology of Scientific Research Programmes', in I. Lakatos and A. Musgrave (eds.), *Criticism and the Growth of Knowledge*, Cambridge, pp. 91–184.

Laudan, L.: 1977, *Progress and its Problems: Towards a Theory of Scientific Growth*, University of California Press, Berkeley, Los Angeles, London.

Miller, D.: 1974, 'Popper's Qualitation Theory of Verisimilitude', *British Journal for the Philosophy of Science*, **25**, 166–177.

Niiniluoto, I.: 1980, 'The Growth of Theories: Comments on the Structuralist Approach', above in this journal.

Popper, K.R.: 1963, *Conjectures and Refutations*. The Growth of Scientific Knowledge, Routledge and Kegan Paul, London.

Sadovsky, V. N.: 1977, 'Methodology of Science and Systems Approach', in *Systems Research, Yearbook 1977*, "Nauka" Publishing House, Moscow, pp. 94–111 (in Russian) (English Translation in *Social Sciences*, Moscow, **10**, 1979, 93–110).

Sneed, J. D.: 1971, *The Logical Structure of Mathematical Physics*, D. Reidel, Dordrecht–Boston.

Sneed, J. D.: 1977, 'Describing Revolutionary Scientific Change: A Formal Approach', in R. E. Butts and J. Hintikka (eds.), *Historical and Philosophical Dimensions of Logic, Methodology and Philosophy of Science*, D. Reidel, Dordrecht–Boston, pp. 245–268.

Stegmüller, W.: 1973, *Probleme und Resultate der Wissenschaftstheorie und Analytische Philosophie*, Band II, *Theorie und Erfahrung*: Zweiter Halbband, *Theorienstrukturen und Theoriendynamik*, Springer–Verlag, Berlin–Heidelberg. (English Translation – Springer-Verlag, New York, 1976).

Stegmüller, W.: 1977, 'Accidental ("Non-Substantial") Theory Change and Theory Dislodgment', in R. E. Butts and J. Hintikka (eds.), *Historical and Philosophical Dimensions of Logic, Methodology and Philosophy of Science*, D. Reidel, Dordrecht–Boston, pp. 269–288.

Suppé, F.: 1974, 'The Search for Philosophic Understanding of Scientific Theories', in F. Suppé (ed.), *The Structure of Scientific Theories*, Urbana, pp. 3–241.

Tichy, P.: 1974, 'On Popper's Definitions of Verisimilitude', *British Journal for the Philosophy of Science*, **25**, 155–160.

ZEV BECHLER

WHAT HAVE THEY DONE TO KUHN?

An Ideological Introduction in Chiaroscuro (but No Footnotes)

1. THE *A PRIORISM* OF THE EMPIRICIST TRADITION

Between the publication of the *Principia Mathematica* in 1911 and the outbreak of the war in 1914, Russell published two books. In these books he was engaged in establishing empiricism on a new basis, after having refuted the "old empiricists". This refutation was contained in the demonstration that

> there is general knowledge not derived from sense, and that some of this knowledge is not obtained by inference but is primitive. Such general knowledge is to be found in logic.
> [*External, p.* 51]

This demonstration, that "there must be some knowledge of general truths which is independent of empirical evidence, i.e. does not depend upon the data of sense", constituted a refutation of the empiricists in their traditional controversy with the rationalists, hence its historical importance. He added:

> This conclusion ... is important, since it affords a refutation of the older empiricists. They believed that all our knowledge is derived from the senses and dependent upon them.
> [*ibid.*]

In this line of innovation Russell was following his studies of Leibniz which he had published in 1901. And just as Leibniz strove to correct Locke's empiricism, so did Russell, by introducing into it a certain amount of *a priorism.* However, Leibniz was much more cool about his contribution, for he realized that Locke's *tabula rasa* was more a *façon de parler* than a basic tenet. So, after Leibniz points out the fictive status of this *tabula rasa* and sums up by declaring that "*Nihil est in intellectu quod non fuerit in sensu*"; *excipe*: "*nisi ipse intellectus*", he adds that this "fits quite adequately the opinions of the author of the *Essay,* who finds no small part of the concepts in reflective thought" (N.E. II, 1, §2). Nor did he find it worthy of comment when Locke came to mention, in Book IV, examples of

J. Hintikka, D. Gruender, and E. Agazzi (eds.), Pisa Conference Proceedings, Vol. I, 63–86.

knowledge which is both factual and *a priori* (IV, 3, § 14, 15), and his only comment on such an example (in IV, 7, § 5) is that true empiricists would deny its *a priori* status! Locke's empiricism, Leibniz clearly saw, was nothing of the sort which Russell describes as "older empiricism".

Compare now Hume's insistence on the sentimental, as opposed to sensational, character of some central concepts of any scientific discourse. Again, the allegedly revolutionary step to restore empiricism is here the argument that part of our knowledge is not given in sense experience but is conditioned by our nature. Hume's naturalism is his version of the *a priori* element in our empirical knowledge. But was this not a result of Hume's scepticism? Was not this part of his attack on the very possibility of science as an empirical system of knowledge?

There is a close historical connection between the empirical tradition and some versions of scepticism. However, empiricism was never sceptical in its overall orientation, even though it was born from, and nourished at first on, a militant opposition to Greco-Christian dogmatism. But even the sceptic part of Bacon's philosophy is already concerned with other issues. In a sense, Bacon attempts to incorporate into science dogmatism's most coveted asset: namely, its certainty. Yet it was his systematics of scepticism – his theory of idols – which introduced the notion of an inevitable existence of some *a priori* element in all our knowledge. He pointed out that one of these idols (and two of them, on other occasions) cannot be eliminated at all from our understanding which is, as a result, necessarily a somewhat distorting mirror:

And as the first two kinds of idols are hard to eradicate, so idols of this last kind [of the Tribe] cannot be eradicated at all. All that can be done is to point them out, so that this insidious action of the mind may be marked and reproved.

Since these idols "have their foundation in human nature itself", he went on to say in another place,

the human understanding is like a false mirror which, receiving rays irregularly, distorts and discolours the nature of things by mingling its own nature with it.
[*N.O.* I, §41].

There can be hardly a better example for the close connectior between skepticism and *a priorism*, and also for the close historica connection between both of these and the empiricist tradition. Yet there is more to it, for neither Bacon nor Hume – with whom I began – were sceptics by a long shot. This holds for the vast majority of empiricist philosophers, and constitutes a major point in any historically fruitful treatment of the empiricist tradition. The point is that though the tradition is not skeptic, it is nevertheless insistent upon the inevitable – or ineradicable – existence of an *a priori* element in all our knowledge.

2. NO EMPIRICISTIC PHILOSOPHY OF SCIENCE IS SCEPTIC

Let me first present, as before, an illustrative salad. Russell, that allegedly great feverish sceptic of our century, had little doubt concerning the orientation of his own empiricism in 1914, when he wrote that

> philosophical scrutiny, . . . though sceptical in regard to every detail, is not sceptical as regards the whole . . . Universal scepticism, though logically irrefutable, is practically barren; it can only, therefore, give a certain flavour of hesitancy to our beliefs, and cannot be used to substitute other beliefs for them.
> [*External*, p. 57]

Though Berkeley intended to refute scepticism along with "atheism and irreligion", as the subtitle of his *Treatise* indicated, Hume felt that he failed in this, and his argument to show this and to reject Berkeley's philosophy on this account, is almost identical with Russell's consideration:

> That all his [Berkeley's] arguments, though otherwise intended, are in reality merely sceptical, appears from this, *that they admit of no answer and produce no conviction.* Their only effect is to cause that momentary amazement and irresolution and confusion, which is the result of scepticism.
> [*Enquiry*, I, 12, p. 155n]

The "immense persuasive force" which physical science derives from its "astonishing power of foretelling the future", Russell explained, makes it impossible for us "as practical men, to entertain for a moment the hypothesis that the whole edifice may be built on insecure foundations" [ibid. 56]. So the starting point for any philosophy of science must be, he went on to say, the existence of indubitable elements in the foundations of our knowledge, which he identified as knowledge of facts of sense and of facts of logic. These could not be doubted without entering the realm of the pathological:

Real doubt, in these two cases, would, I think, be pathological. At any rate, to me they are quite certain ... Without this assumption, we are in danger of falling into that universal scepticism which ... is as barren as it is irrefutable. If we are to continue philosophising, we must make our bow to the sceptical hypothesis, and while admitting the elegant terseness of its philosophy, proceed to the consideration of other hypotheses.
[*ibid.*, p. 60]

This was also exactly Hume's view of the choice offered: either indubitability of the foundations of science or the pathological. So, in parallel with Russell, he rejects scepticism because it "admits of no answer" [Russell's "irrefutable"] and results merely in "irresolution and confusion" (Russell's "barren"), and in parallel with Russell he lays it down that some of our beliefs are "inevitable" and "indispensable" not because of any logical considerations, but simply because they are "natural" to us. To doubt what is "natural" to believe is, of course, to flirt with the unnatural (Russell's "pathological"). For example,

the sceptic ... must assent to the principle concerning the existence of body, though he cannot pretend by any argument of philosophy to maintain its veracity. Nature has not left this to his choice ... It is vain to ask: whether there be body or not? That is a point which we must take for granted in all our reasonings.
[*Treatise*, I, 4, §2, p. 187]

3. THE DISCOVERY VIEW OF PHILOSOPHY OF SCIENCE

It will be my suggestion that the features illustrated in the first two sections, namely the *a prioristic* and the anti-sceptical tendencies of empiricist tradition, stem from the fact that philosophy of science, as practiced in this tradition, is a vast apologetic enterprise. My hypothesis is that once it is viewed as a justification effort rather than as a discovery effort, these two features will appear as a natural and reasonable consequence. The main difference between these two views of philosophy of science is that the discovery view, which regards philosophy of science as an effort to discover the real and true nature (structure, aim, logic, etc.) of science and scientific activity, regards the products of this philosophy as true or false statements, whereas the apology view sees these as neither true nor false. Let me now present my reasons for adopting the apology view and rejecting the discovery view.

The main reason which leads me to this is that I cannot get a satisfactory answer to the following question: what is it for a statement in philosophy of science to be false? Take, for example, statements about the aim of science. Suppose one philosopher asserts that the aim of science is to make us happy, whereas another philosopher denies this and asserts that the aim of science is to make us wise. I wish to know which of them is ***right***. How am I to go about discovering this? Would it help to ask scientists their opinions about this? Obviously not, for we are concerned now with science and not with scientists. Also, why should they *know* this more truthfully than anyone else (seeing that they are concerned with velocities, mesons, and tensor fields rather than science)? At ***what*** – rather than ***whom*** – should I then look next? And what could possibly serve as a refutation of either assertion?

Or take statements about the "structure of science" (such as Duhèm's or Campbell's or Popper's or Kuhn's). Most philosophers used to assert that science has a deductive structure, and some assert that this is false and that science does not have a deductive structure. What does it mean to say that these are true and these are false statements? Again, no one means actual or practiced science is really deductive, nor does any one mean that it should be deductive, nor that people (scientists) regard it as "esentially" deductive. At most, philosophers may mean that scientific theories *may* be cast or recast into such and such logical mold. But, as we all know, scientific theories can be cast in any form we choose to cast them in, and this would hardly entitle us either to reject the deductive view as refuted, or to assert that scientific theories are strings of sentences arranged alphabetically and according to length, or that scientific theories do not have any structure but are, in fact, amorphous. And lest anyone suggest that these puzzles arise out of the fact that 'aim' is a notoriously difficult concept to apply to non-animate things, or that 'science' is too loose and general a concept and hence the indeterminateness of problems dealing with either its aim or structure, it should be observed that this same puzzle applies equally in regard to detailed properties of any particular theory: that there exists an upper limit to the velocity of signals is presupposed rather than inferred in the special theory of relativity (Reichenbach and Grünbaum affirmed this but Einstein in effect denied it); the deductive derivation of the universal law of gravitation from Kepler's laws contains a contradiction (Duhem, Popper and Lakatos affirmed this, but Newton denied it); Newton's laws of motion are special cases in the theory of relativity (Einstein affirmed this, but Feyerabend denied it), and so on and so forth.

Experts are ready with a yes/no answer to the question whether these and like statements are true (or false), but in fact – as any quick scrutiny of the literature will show – all that is actually offered in support of such answers is a proof that there is a way of constructing these theories that turns the assertions true (or false). But surely this is not all that they mean to say, for everyone knows fully well that other ways of constructing these theories exist. But what, then, is the surplus intention which would turn such assertions from empty deductive exercises into pregnant and philosophically interesting statements?

Obviously this difficulty stems, partly, from the circumstance that science, just as scientific theories, is not "given" but is, like Russell's entities, a logical construction. The crucial point, however, is not simply that it is not "given", but rather that not even its properties are "given" or "known". As the disputes within philosophy of science amply prove, philosophers cannot be said to "know" what are the specific properties, aims and structures of the entities denoted by 'science' and by 'scientific theory *x*'. Hence, whereas Russell could "construct" the concept 'cardinal number' to fit the properties demanded by arithmetic, and whereas Reichenbach could "rationally reconstruct" space-time in accordance with the properties demanded by relativity theory, philosophers are stuck with entities which demand a construction in order to philosophise about them, but which also demand that this philosophising be concluded ***before*** they are constructed (so that their properties be given before they be constructed).

This consideration indicates that if philosophical statements about science are to have interesting import, they cannot be regarded simply as logical derivations of porperties from ***some*** construction of science (or of a certain scientific theory). Any needed properties can be easily derived from the suitable construction. But even though the demand that statements in philosophy of science be either true or false cannot be accepted, yet I assume that they reflect more than mere logical consistency with some possible construction.

4. THE APOLOGY VIEW OF EMPIRICISM

It is my suggestion now that the philosophical intent or import – as distinct from the logical status – of philosophical statements about science, and especially of empiricism, is to be found in the hidden reasons for the selection of a particular construction of science rather than another equally possible construction. What are the possible reasons for such preference? Usually they are either socio-historical ("this is what Einstein really had in mind in 1905"), or essentialistic ("this captures the essence of the theory of gravitation", or, sometimes, "this cannot be a ***legitimate*** interpretation of this theory" or "of science"). Both should be rejected, for the first is irrelevant (who cares what Einstein had in mind?) and the second is obscure and incoherent. What characterises such considerations is what I called in the previous section the discovery view, namely, the belief that philosophical statements about science are either true or false. As against such truth-considerations, I suggest that one regard the selection of a particular construction because the properties derived from T_1, but not from T_2, are "***important***". And these are "important" for the following sole reason: they exhibit science as a "good" enterprise. Thus empiricism is based on an ideology which articulates the image which science ***must*** have, and it then assumes that science does in fact have this image, even as it is practiced in real life. This image determines the properties science must have, and these determine the choice of the construction to be put on any particular scientific theory. Finally, once the constructed theory is at hand, it serves as the object of logical analysis.

This is the surplus intention which saves philosophical analysis of science from triviality, but it does so by turning this analysis from discovery into apology. Empiricism is not an attempt to discover the true nature of science, but an effort to defend science against disparaging accusations by pleading that, firstly, these accusations are wrong and, second, even if they are right it is nevertheless for the best.

An apology, or a justification, or a defence (and I shall use these terms as technical and synonymous) has the grammatical form of a descriptive statement, but it does not describe any state of affairs in the world. In this sense, it is rather close to theoretical statements such as "a triangle has the sum of its angles equal to 180°", which does not describe how actual triangles should behave. And just as geometrical statements are not normative, but nevertheless are neither true nor false since they do not ***intend*** to describe any actual situation, so statements of the philosophy of science are neither true nor false because they have another function to perform, namely, to defend.

Now, the emphasis I suggest as fundamental is on the difference between explanation as the discovery of causal elements and causal connections which turn puzzling facts into old acquaintances, and defending, or apologizing, or justifying, as an action aimed solely at one target, namely, pacifying. What I have in mind is the effort to make your opponent see things your way, and it is very seldom that this can be achieved by specifying the true causes or "real nature" of the matter at hand.

5. INDUCTION AND ANTI SCEPTICISM: THE FIRST CONSEQUENCE

The basic ideological tenet of empiricism is this: science is the best manifestation of human intellect in its search for truth and the improvement of life. Since induction was usually taken as the basic method of empirical science but could not be explicated logically, empiricist philosophers came to the rescue. Their defence of science on this item, started by Hume, remained a central workshop for the defence business. Hume's vindication of science against the accusation of irrationality was, essentially, to point out that rationality as conceived by the sceptic is a myth and so science cannot be accused of irrationality. No accusation can possibly build on mythical standards. What actually happens, Hume argued, is that Nature takes care of her children and supplies them with an "instinctive" capacity which enables them to discern causality in the world and to differentiate between, on the one hand, connections between events which are merely accidental, and those, on the other hand, which are necessary. Thus "instinct", expressed in "natural belief", is generally unerring and is the sole and best basis for our inductive inferences. To accuse science of not being able to prove its inferences by logic is to completely miss the whole layout of human understanding and its connection with Nature. Science, as the best product of human reason, is based on the best and surest foundation that is ***realistically*** conceivable (i.e. in conformity with Nature). Hume's case for the defence was, then: science is not guilty; on the contrary, science should be praised and viewed as a paradigm of human reason in action; only human reason is instinctive, and its success is guaranteed by its essential "naturalness". To accuse science on account of the non availability of a logical distinction between necessary and accidental connections is to invent and then apply a myth to measure reality.

Popper's case rested on the demand to view science as a deductive activity all the way. Thus, he, in fact, accepted the standard of the accusation as valid, but went on to show that the description of science as inductive is mythical, and that science in reality fully conforms to the standard of the accusation. Thus Popper's apology is the converse of Hume's: What Hume rejected as mythical (the standard of accusation) Popper accepted as real, and what Hume accepted as real (science's inductivism) Popper rejected as mythical. This was the difference in their apologies, but their plea was identical: science is wholesome and a paragon of reason; science is not guilty.

What Russell took on to refute was the accusation that science cannot possibly be empirical, and this on two counts: first, it rests on sense-perceptions and these are notoriously "subjective"; second, it does not even rest on sense-perceptions, but is separated from them by a deep gulf. So, science cannot be empirical, but in so far as it is empirical it cannot be objective. The neglect of these accusations on the part of scientists endangers the validity of science (*Ext.* 79, *An. Mat.* 137), and Russell's philosophy in the two decades following the *Principia* was an effort to defend science against them. His major weapon was the logical construction which, in the *Principia*, he had already employed in order to defend the certainty of mathematics. Now he went on to show that space, time, and place are logical constructions (and ***not*** given sense-data) and summed up his effort on this point thus:

Such difficulties have made people doubtful how far objective reality could be known by sense at all, and have made them suppose that there were positive arguments against the view that it can be so known. Our hypothetical construction meets these arguments, and shows that the account of the world given by common sense and physical science can be interpreted in a way which is logically unobjectionable, and finds a place for all the data, both hard and soft. It is this hypothetical construction, with its reconciliation of psychology and physics, which is the chief outcome of our discussion.
[*Ext.* 79].

His logical construction for matter and causality had the same function, namely, to justify physics against the accusation of subjectivity:

Men of science, for the most part, are willing to condemn immediate data as "merely subjective", while yet maintaining the truth of the physics inferred from these data. But such an attitude, though it may be capable of justification, obviously stands in the need of it; and the only justification possible must be one which exhibits matter as a logical construction from sense-data ... It is therefore necessary to find some way of bridging the gulf between the world of physics and the world of sense.
[*Ext.* 81]

Thirteen years later he reiterated this same perspective. Russell is a singular case of a philosopher who was fully conscious of what he was doing. In fact he used on this occasion the term best suited for my purpose except for its ironic overtones (which made me reject it). The term was 'absolution'. The situation he describes is identical with the one he had treated in 1914: if perception is not trustworthy then physics may be false, and hence a rescue operation for physics is in order:

I do not say that physics in fact has this defect, but I do say that a considerable labor of interpretation is necessary in order to show that it can be absolved in this respect.
[*Analysis*, p. 137]

So, philosophy of science is here viewed as an act of absolution: by means of a very tricky interpretation, physics can be defended and presented as valid after all, in spite of its apparent sins. Russell's absolution of science in this case was difficult and involved a "considerable labor of interpretation" – of "observation" rather than of science. "Observation" was now interpreted as the outcome of abstraction, and it is through their common abstract status that science and observation are reconciled and the validity of science saved.

There is no point in asking whether "in fact" observation is abstract, whether matter is "really" a logical construction out of sense-data (which are themselves a logical construction!), or whether cardinal number is "really" the class of all classes which are similar to a given class. Logical construction is a defense instrument by means of which science is vindicated, and so the question is not whether these constructions are true or false – this is a senseless question – but only whether they succeed in absolving science.

These examples were meant to illustrate my thesis that the ideological business of empiricism is to defend science against accusations of various types; that these accusations are met by showing that science can be presented (rather than "by presenting") or viewed in a certain purified manner which will absolve it from the alleged sins; and, finally, that it follows from all this that empiricism must be antisceptical in its essence: its function is to fortify and praise, and not to weaken and bury.

6. SECOND CONSEQUENCE: THE *A PRIORI* OF BACON AND NEWTON

Very succinctly, the problem-situation for the empiricist program in the 17th century was this: it declared that it could surmount what seemed an impossible challenge, i.e., to discover invisible causes with certainty. Its solution was expounded by the program of analysis and synthesis which, between Bacon and Newton, aimed to become a deductive inference of causes from effects, the former invisible and the latter visible. This, however, was never quite satisfactory. Bacon was fully aware that his eliminative method ("by rejection and exclusion") was based not really on phenomena but rather on intuitive imagination, and Newton's critics argued that his "deductions" were faulty. Bacon saved his "new induction" by declaring that it was impossible to get rid of all the "idols" which corrupted human reason, but that his method was designed to produce merely greater and greater degrees of certainty as it proceeded. Newton was tougher and, stoutly rejecting his opponents' arguments, suggested that they settle, instead, for a lesser difficulty which he invented especially for this purpose: in passing from particular discovered causes to a universal law we are forced to generalise over space and time, and here we could eventually go wrong (for different time and space regions might be governed by different laws).

The pattern of these two early cases of apology is typical. Their first important feature is that they admit there is something wrong with science. Second, they propose that the source of the trouble is an evil which is (a) minor, and (b) inevitable. Prejudice and induction are minor evils for they do not destroy science; they merely make it somewhat more difficult to build and they have – as evils – no more than nuisance value: we may "point out" the prejudices, "so that [their] insidious action . . . may be marked and reproved" (*supra* p. 64).

And trouble about induction is equally to be put aside: "The meaning of conclusions made by induction", Newton wrote, "is that they are to be looked upon as general till some real exception appear. And in this sense gravity is to be looked upon as universal." However, when some such exception appears, nothing terrible happens. Above all, these conclusions from induction are not ***refuted*** by such "exceptions", they merely lose their generality but not their truth:

If no exception occur from phenomena, the conclusion may be pronounced generally. But if at any time afterwards any exception shall occur from experiments, it may then begin to be pronounced with such exceptions as occur.
[*Opticks* p. 404]

Inevitability is the second important trait, beside unimportance, of prejudice and induction. Prejudice, wrote Bacon, is either "very hard to eradicate" or decidedly inevitable, when it is rooted in the nature of the human species. By generalization and by it alone, Newton declared, we build our conceptual apparatus, scientific as well as everyday: "Thus it was that the impenetrability, the mobility, and the impulsive force of bodies, and the laws of motion and of gravitation were discovered". [*Principia*, p. 547]. His intent is clear: these concepts are not the result of logic or some deductive reasoning explicitly used. Rather, we gather them ("impenetrability", "mobility") consciously and with no felt effort – they are instinctive and natural. And so, if anyone wishes to insist on the uncertainty of the generality of mutual attraction, say, he would also have to regard impenetrability and mobility as equally uncertain, which would hardly do.

I shall return to the connection between the triviality and the inevitability of these and like factors in the empiricist tradition, but I wish now to insist a little on the implication of this inevitability. This stems (explicitly in Bacon and implicitly in Newton) from its inherence in human nature. So, just as rationalists declared that we had innate *a priori* knowledge, so Bacon and Newton imply that the human mind is by no means a mere fact-recorder but is rather an active contributor of nonfactual elements to our knowledge. The main difference between them is that whereas rationalists wish to build science on this *a priori* capacity alone, Bacon and Newton include it in their account only as the inevitable source of error. But both parties agree in this: that the human mind is anything but a *tabula rasa*. Locke's case and Leibniz's reaction (see *supra* p. 63) should now be weighed anew. What distinguishes the empiricist tradition is not that it denies *a priori* elements in our knowledge – it does not – but merely that it puts them to a different use, which I call their apologetic function: These *a priori* elements are the source of error in our knowledge, but they are trivial errors and the whole sorry business is an inevitable evil. This apologetic function sets the empiricist *a priori* apart as a separate species. It would appear again and again as the justification of the way science deviates from some ideal image of it, pleading that since the impediments to a full realization of the ideal are inherent and inevitable, it is idle to accuse real science of deviation. Accusations hold only where realizable possibilities exist. It follows that science as it exists is the full actualization of the best realizable conception of science:

> And although the arguing from experiments and observations by induction be no demonstration of general conclusions, yet it is the best way of arguing which the nature of things admits of.
> [*Opticks*, p. 404]

And Cotes wrote, in his preface to the *Principia*, that "this is that incomparably best way of philosophizing", echoing thereby Newton's dictum:

> These principles are deduced from phenomena and made general by induction: which is the highest evidence that a proposition can have in this philosophy.
> [Eddelston, p. 154–5]

7. HUME'S ARGUMENT

What was merely seminal with Bacon and Newton became central to Hume. His whole philosophy revolves around the idea that it is the myth of the intellectualist image of science that must be blamed for the accusations. To rectify the situation, this myth must be exposed as such, which Hume set out to do in his analysis of the fundamental concepts of science. His point is that these analyses prove that the comcepts are formed and applied by our "feeling" and not by any "rational" faculty. However this is not simply a brute fact, nor does it stem from any limitation of ours. Rather it is so because no other possibility logically exists. Since the idea of any cause does not "contain" the idea of its effect, deductive linkage is logically impossible, induction is thus seen to be both necessary and non-logical.

This Hume named "feeling": we are able to discover causal connections and distinguish lawful from accidental regularities only by an inborn "instinct".

Moreover, this feeling-instinct capacity is the best instrument which Nature could possibly endow us with for such a purpose, since it must be a fast-reaction instrument – which "instinct" and "feeling" are, but which deduction is not. Our innate instinctive capacity becomes the basis, and our science the extension, of a vast survival instrument. Viewed thus, science must now be regarded as the best available solution to a survival problem, and the fact that reason is the servant of feeling is as it should be in a harmonious Nature. The proof that our knowledge, our survival instrument, is founded on "feeling" is now merely a proof that Nature is rational in its care for its creatures. Thus, even though science is not intellectual, it is a rational device, and it is rational exactly because it is not intellectual. Naturalism and its correlate, anti-intellectualism is, therefore, Hume's defense of science and his justification of its success and high status.

The point I wish to make is that Bacon's prejudice, Newton's induction, and Hume's instinct have a common feature, and for this purpose I ignore their differences. This common feature is innateness, which implies both inevitability and nonintellectual (or natural) status. They all introduce a nonlogical element into science which is ineradicable. They are the main source of error but are, nevertheless, crucial to any successful science. Thus, their naturalistic *a priori* status – that is, their innateness and inevitability – makes them the pillar of apologetics which urge us to regard science as a human enterprise, a sweat and tears enterprise rather than a bloodless, purely intellectual structure. Since this is the only possible way open to us to construct science, clearly it is pointless and trivial to criticize science, on this account.

8. THE APOLOGY OF CONVENTIONALISM

It took several revolutions in science and the dawn of a new historicist era, culminating in Darwin's theory of progress, to transform the grand problem of empiricism from its initial one – i.e. how to discover hidden causes with certainty from phenomena – into the following: How to exhibit science as a respectable paradigm of knowledge, in the face of its ephemeral existence? This is different from the previous problem of apology only in the emergence of a clear historical dimension. The generalized structure of the apology now incorporated History, along with Nature, as the instruments of salvation, but this did not make a real difference to the structure of the apology as a whole: the historical dimension merely becomes an item of crucial ***evidence*** for the open-endedness of science, which fluid state the idea of progress is recruited to stabilise. The implicit premonition implied in the theory of the inherent source of error which is also the necessary basis of science, became now confirmed by history, so this was not really a shattering new defect in science. Hence the main thrust of empiricist apology remained unchanged: i.e. that science incorporates, willy-nilly, a human element which is the inevitable source of the erratic, or historic, existence of science.

The *a priori* – as the sum total of human prejudice, cultural pressure, discovery gimmicks, and nondeductive thought – became, after Kant, the principal justification argument of empiricism in the last two thirds of the 19th century. The ephemeral state of science and its open-endedness were now justified by arguing that such is the inevitable trait of anything human, and science is human. Being human, and not purely logical, it necessarily contains some random, arbitrary, subjective element. This was the new emphasis on subjectivism and arbitrariness which came to its peak with the establishment of pragmatistic empiricism and adopted by such philosopher-statisticians as Mach, Duhem, Le Roy, Poincaré, Ostwald and others.

But pointing out the inherent arbitrariness as an apology for the ephemeral state of science was only part of the defence. Because an effort had to be made to regard History as a beneficial factor which acts constantly towards the elimination of this subjective *a priori* element from science. The model of science which was established for this purpose was the core-shell model: History acts on science by separating its core – which is objective, purely factual, and usually mathematically formulated – from its shell, which is subjective and merely formal. The core is situated along the historical continuous transmission line which, like a kind of germ-plasm, is inherited and conserved since it is true, while the shell-somatic part is continuously eliminated or left over, being merely formal and devoid of truth-value. This notorious model exhibits the accumulative process of history as the correlative of the eliminative process: History cares for science just as Nature cares for man, and just as Nature installed instincts (instead of logic) as man's principal survival instrument, so History constantly acts to purify science of the inevitable side-effects of this rationality of Nature, and it does this by "eliminating and rejecting" the formal shell, thus objectifying science to the utmost limit logically possible. Progress, i.e., the continuous accumulation of the eternal core in ephemeral theories, was never put in doubt in this tradition. Science counts on History for that.

There was, however, a great deal of doubt as to our ability to predict the verdict of History: doubt as to our own ability to discern and separate the fact from the form, core from shell. This doubt was, again, an apology for human failure in its effort to help History, and the apology was formulated in what came to be named "conventionalistic" philosophy. Its main defence line was the argument that since recording facts implied a conceptual system and this in turn implied some subjective *a priori* element, the human effort to purify science of this element was doomed. And that is why we must wait for History and have confidence in it. Thus, that some conventionalism became a common symptom of empiricism at the turn of the century must be viewed as the natural outcome of the apologetic orientation of this philosophy and of its effort to present science as a rational and respectable enterprise despite its historic ephemeral existence. This is the reason why there is no single real, full-scale conventionalist among "conventionalist philosophers". This goes a long way, I suggest, to explain what looks at first like a plain contradiction within the empiricist tradition: namely, the cohabitation of faith in progress through history, and the proof that such a progress is impossible in principle. This convention-progress syndrome, seen by Bacon and flourishing in Kuhn, may be cured by the apology view of empiricism.

9. KUHN–A PROBLEM

Kuhn's theory says that science ***progresses*** by leaps and bounds across uncrossable chasms which separate one theory from the next. This looks like a flat contradiction: if t_1 is better than t_2 – which is what progress means – then t_1 and t_2 are comparable *vis-à-vis* this progress-criterion and hence cannot be "incommensurable", in this respect at least. Again, if t_1 and t_2 are truly incommensurable, then t_1 is as good as t_2 and hence no progress is made by the replacement. If

Kuhn's theory is consistent, then he must be re-interpreted so that either "progress" or "incommensurable" do not mean what they seem to.

Now, the choice here seems to be easy, for whereas his "progress" is qualified merely so as to exclude "truth" from its parameters, his "incommensurable" leaves untouched and unmentioned just those parameters which do determine his "progress". Hence, even though t_1 and t_2 may differ in regard to scope, predictive power, or exactness, say, they are nevertheless straightforwardly commensurable in just these parameters. And since "progress" means improvement in just these parameters, the inconsistency becomes merely apparent. What, however, becomes merely apparent along with it is the novelty of Kuhn's theory.

It is accepted all around that Kuhn's theory belongs in an ***antipositivistic*** movement that emerged after World War II. Its main theme was the inadequacy, *vis-à-vis* the history and *actualia* of science, of the notorious dichotomies of positivism, from among which the primary one selected for this onslaught was the context-dichotomy. The argument of the Young Turks, so the customary interpretation goes, was that in the *actualia*, or historic, time-dependent, science, there is no functional difference between the contexts of discovery and of justification. The implied novelty of this view was, allegedly, the apparently new status of the *a priori* in science: from being merely formal it became factual. The synthetic *a priori* was back, and science was no longer empirical as empiricists traditionally took it to be. This was, so it is usually claimed, the message and meaning of Kuhn's choice of the term "paradigm" (as equivalent to "dogma") for designating scientific theory-*cum*-its non-scientific body of presuppositions. A paradigm is held dogmatically, irrespective of experience and prior to it (*a priori*), yet it partly determines its content (synthetic). This means that the scientific enterprise is fundamentally irrational in its transformation mechanisms, for no rational dialogue is possible between parties which see the world differently and refuse experience when it refutes their party line. So the starting point, i.e. the rejection of the content dichotomy, led to the rejection of the rationality of scientific intercourse *via* the introduction into it of a kind of synthetic *a priori*. No more extreme plan of an anti-empiricist revolution could be drawn.

But if my previous point about the consistency of Kuhn's theory holds, then this revolutionary picture of Kuhn is in trouble. For if, for the sake of consistency, incommensurability is restricted to such aspects of a paradigm as are not involved in progress considerations, then this incommensurability is of no consequence in Kuhn's theory of scientific change. The fact that scientists talk ***through*** each other is of no importance for scientific change if ***what*** they talk about does not concern and cannot influence such a change anyway. Since the irrelevancy of cross-paradigmatic discourse does not arise from the paradigms' incommensurability, surely this irrelevancy would remain even were they commensurate. But then the whole incommensurability affair would turn out to be an idle wheel in Kuhn's theory. Which, so we somehow feel, it is not.

10. SOLUTION BY APOLOGY

My solution is to suggest that Kuhn's once revolutionary theory has at long last (after being, first, absurd, and then sensible but false) reached the old-hat stage. I suggest regarding it as a proper empiricist philosophy, namely, an apology for science. As such it cannot possibly be an attack on science and an exposition of its irrationality. Situated squarely in the midst of the empiricist tradition which I sketched before, it is rather a defence of the rationality of the scientific enterprise – of its final product as well as its history. Its problem is to argue that even though science incorporates a great deal of *a priori*, it regularly succeeds in shaking it off and in expanding and growing into states which are constantly progressing according to strict empiricist criteria. Moreover, its *a priori* part is vital in regulating the pace of this progressive movement, guarding it against too fast a growth (which was regarded as a menace to science ever since Bacon asserted that his method's aim was, partly, to slow down reason, or "hang weights" on it "to keep it from leaping and flying". *N.O.* I, 104). Thus governing the proper growth-rate, the *a priori* value gives way only when the pressure of ***empirical*** findings is great enough to break through its check, demanding thereby the institution of a new *a priori* pressure-value in the form of a scientific revolution.

Thus Kuhn's incommensurability of paradigms is, after all, meant to designate a merely regulative rather than a truly constitutive element in scientific theory. For had he taken it as designating truly a content-element, he would have had to state that ***all*** paradigms are equally good in *all* regards – which he cannot and does not do, since progress is real with him. Hence, even though the battle cry was actually "destroy the external-internal curse", the real outcome of the battle is something quite different. For instead of the expected outcome, namely, that even external factors are content-determining – and thus internal in fact, which is what eliminating the dichotomy meant to achieve – instead of this, the actual result is the external-internal dichotomy as real as before. With one difference: it was shown to be a blessing rather than a curse.

This Kuhn showed by exhibiting the rationality of the external factor: only because it has such a decisive grip on science – i.e. only because it is an *a priori* element – can it and does it succeed in regulating the proper pace of science. The Humean structure of this gambit is unmistakable: exactly as the rule of "instinct" in human life proves the rationality of Nature in its care for man (for logic and facts, if they were given exclusive reign, would have destroyed him through sceptical paralysis); even so the rule of the paradigm in scientific life proves the rationality of scientific community, as well as the derived rationality of History (for logic and facts, were they given exclusive reign, would have destroyed both through sceptical paralysis or falsificationist rush).

So just as the empiricist tradition argued that science is rational because external factors were merely external (and inevitable), so Kuhn argues that science – or rather the Scientific Community – is rational because it creates a flow and ebb (with periodic motion) in the grip which external factors exert over science. For some time and some purpose they get internalised, and then they lose their hold and get externalised, thereby determining the periodic eruption of revolutions to be just right. And, most important, Kuhn's theory of the causes of such eruptions is a singularly enlightening item concerning his empiricistic ideology, his apologetic technique, and his conservative philosophy of science.

11. THE NEMESIS OF APOLOGY: KUHN'S COPERNICUS

That a scientific revolution is preceded by a crisis is a vital principle with Kuhn. For were it otherwise, i.e. were revolutions *sudden* eruptions, they would also be irrational. To exhibit the rationality of the scientific community it is necessary to have a crisis for each revolution.

Next, a Kuhnian crisis is created only by a ***factual*** pressure, i.e. a factual misfit which just won't conform. Now Kuhn is most explicit in his insistence that whatever else may be the circumstances, a factual misfit is the unexceptional law, and the essence of the crisis.

The empiricism of this tenet is crucial for my empiricist interpretation of Kuhn. For were he intending to argue for the dogmatic character of theories and for their complete independence from experience – what "paradigm" apparently came to signify – he would have had no need at all for factual misfits, observational troubles and predictive failures. But he is quite unequivocal about the necessity of these to precipitate a crisis. Hence he says, in effect, that people ***see*** that their paradigm, their manner of seeing, is in trouble: they are not blind to its vulnerability, and it ***is*** vulnerable. Moreover, since the trouble is ***always*** an observational misfit, the paradigm is ***experimentally refutable*** and ***refuted.***

In fact, this deep belief in the falsificationist thesis of empiricism is the most plausible explanation for a very curious episode in Kuhn's writings. And I stop now to recount it not because I think he was wrong – I really don't know – but only because it serves to illustrate the depth of his empiricist commitment. Kuhn illustrates his thesis about the central role of anomaly in a crisis by the Copernican revolution. Anomaly in the physical sciences, is always an observational misfit, and awareness of anomaly is

> the recognition that nature has somehow violated the paradigm expectations that govern normal science.
> [*Structure*, p. 53]

Now, after he makes his point that "awareness of anomaly plays a role in the emergence of new . . . phenomena", he goes on to say that

> a similar but more profound awareness is prerequisite to all acceptable changes of theory. On this point historical evidence is, I think, entirely unequivocal.
> [*ibid.* p. 67]

So his thesis is simply that a necessary condition for any "acceptable" theory-change must be: first, an anomaly, i.e. an observational failure which resists explanation, and, second, an awareness of this failure as a failure. Bacon, Carnap, and Popper would be very pleased. Now, his next sentence is:

The state of Ptolemaic astronomy was a scandal before Copernicus' announcement . . . furthermore . . . the awareness of anomaly has lasted so long and penetrated so deep that one can appropriately describe the fields affected by it as in a state of growing crisis.
[*ibid.*]

In the next pages he treats this anomaly as a "discrepancy" between theory and observation, and the normal period between Ptolemy and Copernicus is described as a prolonged effort, which constantly failed, to eliminate these discrepancies. And it was this constant failure to explain observational discrepancies which was responsible for the crisis – primarily and essentially. Kuhn is ready to admit that there were several non scientific factors which also served to enhance this crisis, but he insists that they were merely secondary causes. He mentions, in this category, "social pressure for calendar reform . . ., medieval criticism of Aristotle, the rise of Renaissance Neoplatonism", but he puts all these aside as mere "external factors" which determine such factors as the "timing of the breakdown, the ease with which it can be recognized, and the area in which . . . the breakdown occurs. Though immensely important, issues of that sort are out of bounds of this essay." [*Ibid.* p. 69]. And all this comes to support his main tenet:

Technical breakdown would still remain the core of the crisis, [*ibid.*]

by which he means failure to explain observational discrepancies. This thesis is of the utmost importance to Kuhn's theory. The notion that what are usually called "external factors" are truly external – not at the core but rather at the shell of crisis – and that the core is ***always*** an observation failure, this good old falsification-empiricism, is crucial to his portrait of the scientific community as rational and "scientific". Hence his need for a "scandal" prior to Copernicus. The absence of such a scandal would be a dark blemish on the rationality of science.

And here comes trouble. For standard histories of astronomy do not record any crisis, or any scandal, in professional astronomy preceding 1543. Kuhn's only mentioned case is a pronouncement by Domenico de Novara (with no reference), but he refers the reader to Dreyer's *History* chaps. XI–XII, and then to his own *Copernican Revolution* pp. 135–43. Now, in Dreyer I could not find any trace of crisis, try as I did. In fact, Dreyer regarded the situation as a quiet normal science scene, with an effort at an improvement here and there but no more. And for a good reason, too. He regarded the Ptolemaic astronomy as an admirable achievement which was almost completely successful in its predictive function:

> That the system as a whole deserves our admiration as a ready means of constructing tables of the movements of the sun, moon and planets cannot be denied. Nearly in every detail (except the variation of the distance of the moon) it represented geometrically these movements almost as closely as the simple instruments then in use enabled observers to follow them ...
> [Dreyer, p. 200]

As for those discrepancies like the moon's distance-variation (and Dreyer forgets Venus' case, to which Osiander refers in his preface to the *De Revolutionibus*) there is a ready explanation in the empiricist historiography why they did not create any "scandal", which Dreyer adopts too: Astronomers could not be bothered with the system's truth-value, and falsification never interested them for they all – starting with Ptolemy – regarded the system as false, but efficient. Osiander's preface is a most illuminating expression of this situation.

Kuhn's reference to his own previous work is even more curious. For though he had stated there that the Ptolemaic system failed to give "results that quite coincided with naked eye observation" (*Copernican*, p. 140), and that "a perceptive astronomer might well wonder" about the chances of a successful repair job, yet Kuhn also wrote that no one before Copernicus saw the matter as a failure:

> For the *first time* a technically competent astronomer had rejected a time-honored scientific tradition for *reasons internal to his science*, and *this* professional awareness of technical fallacy inaugurated the Copernican Revolution. A felt necessity was the mother of Copernicus' invention. But the feeling of necessity was *a new one. The astronomical tradition had not previously seemed monstrous.*
> [*Copernican* p. 139] (my italics, Z.B.)

Now, "monster" is what Copernicus called contemporary Ptolemaic traditional astronomy. But he did not mean by it anything beyond "ugly" and "non-harmonious". He decidedly did not mean "inaccurate", and in fact he never mentioned any observational discrepancy as the reason for his own crisis. This is quite a pregnant point and is very conspicuous in his carefully worded letter to the Pope. This is, in fact Copernicus' apology, justifying his outlandish act before the Pope and his fellow astronomers. He refers to the *uncertainty* in astronomy concerning the year's length, the *un-*

certainty of astronomers about the right theoretical devices (epicycles, excentrics, concentrics), the use of the equant, the inability to deduce the harmonic structure of the cosmos, and finally the ugly form of present day astronomy which he likens to a monster.

The absence of observational failure among the charges of this apology is conspicuous. And for Kuhn it is detrimental. But Kuhn never hesitates. He goes ahead anyway and one page after citing Copernicus' apology he sums it up thus:

Diffuseness and continued inaccuracy – these are the two principal characteristics of the monster described by Copernicus.
[*Copernican* p. 141]

Whatever "diffuseness" might designate, Copernicus' own use of "monster" does not mean "inaccurate", by his own testimony.

In sum: Even though Kuhn refers us to his own work for testimony about crisis, in fact he had stated in that work the following: (a) There might have been a crisis if anyone wished to go into one; (b) but in fact there was none, for no one saw the situation this way; (c) Copernicus was the only one who records a crisis. Surely this is a far cry from "scandal" and "deep awareness of a technical breakdown".

My purpose, let me remind you, was merely to illustrate the depth of Kuhn's commitment to the empiricist core-shell model with its falsificationist rationality. It is possible that the singular way in which he treats his sources – and I picked this case simply because among these sources there is a book he wrote five years earlier – suggests that an in-depth historical research about the pre-1543 situation would be here of no relevance. The empiricist philosopher has an enterprise to absolve, not any truth to discover.

And so, what Kuhn develops is the empiricist core-shell model, with a vengeance. For the pulsating motion which he argues for, which periodically loosens the grip of the *a priori* paradigm and lets experience assert it is time for a change, this pulse between core and shell follows the laws of a pre-established harmony in the history of science: there is time for the *a priori* and there is time for the *a posteriori* and the scientific community knows when, and all is for the best. This is the kind of empiricism's scientidiocy which, demanding the abolition of the descriptive-normative dichotomy, has traditionally legitimised the sleek naturalistic passage from science's *actualia* to its methodology.

University of Haifa.

REFERENCES

Russell, B.: 1954, *Analysis of Matter*, Dover, N.Y.
Kuhn, T.: 1959, *Copernican Revolution*, Random House, New York.
Dreyer, J.L.F.: 1953, *History of Astronomy*, Dover, N.Y.
Eddelston, J. (ed.): 1850, *Correspondence of Newton and Cotes*, London.
Hume, D.: 1902, *Enquiry Concerning the Human Understanding*, Oxford.
Russell, B.: 1960, *Our Knowledge of the External World*, Mentor.
Leibniz, *Nouveaux Essais*.
Bacon, F.: *Novum Organum*.
Newton, I.: 1952, *Opticks*, Dover, N.Y.
Newton, I.: 1966, *Principia*, University of California.
Kuhn, T.: 1962, *Structure of Scientific Revolutions*, University of Chicago Press, Chicago.
Hume, D.: *Treatise of Human Understanding*.

ROBERT E. BUTTS

COMMENT ON ZEV BECHLER'S PAPER 'WHAT HAVE THEY DONE TO KUHN?'

Zev Bechler takes exception to my thesis of the functional identity of philosophy and science.[1] I think that his views on what follows from the rejection of this identity are both philosophically errant and politically artless, and therefore dangerous. Especially bothersome is his vigorous support of the contention that the intellectual residue available to philosophers of science is *apology*: philosophy of science is not an intellectual discipline at all; philosophers of science cannot investigate, they can only apologize for science. But apology is not substantive, it is the result of apologizing – apologetics is a *social role.* On this (corrupt) reading of the role of philosophers of science they cannot even criticize. The kind of role classically apportioned to critics of literature and the other arts cannot be their role, *because what they say can never be determined to be either true or false, even in principle.*

I suspect that no one would even entertain this unaffectedly simple view of philosophy of science who had not also accepted an equally oversimple understanding of apologetics, especially when embedded in an institutionalized structure. The epistemological features of apology are not difficult to reveal. An apology for a view (doctrine, article of faith, belief) is an epistemological cop out, an explicit abandonment of any attempt to deliver the evidence for that view in a rationally defensible manner. Contrasted with the presentation of an evidenced case, an apology is always an epistemological second best, a substitute known to be defective on relevant points of argument or presentation of evidence. Epistemologically an apology is an expression of regret, an ingredient in an act of contrition in the face of failure: "I have done what I can; I am sorry that I cannot do better."

When the epistemology of apology becomes institutionalized a methodology is introduced: it is the methodology of rhetoric. The formal art of expression replaces the formal rule structure of presenting evidence and reasoning. Artificial eloquence replaces natural conviction. The methodology of rhetoric generates the politics of persuasion; in historically tragic and unhappy cases the politics of

J. Hintikka, D. Gruender, and E. Agazzi (eds.), Pisa Conference Proceedings, Vol. I, 87–91.

persuasion gives way to the ideology of power, which in turn yields the anguish of the exercise of force. Zev Bechler argues that philosophers of science should eagerly accept this epistemology, this methodology, this politics, this ideology, and this exercise. Philosophy of science is the apologia of science.

Fortunately for the philosopher of science there is a special feature of institutionalized apology totally ignored by Zev Bechler that completely saves his field of study from such drastic perversion. *Scientists for the most part think that they have absolutely no need of philosophers of science; they offer the philosopher of science no protection whatsoever.* And protection is of the essence of the quasi-contractual exchange involving the apologist and the figure (or institutionalized dogma) for whom (or for which) he apologizes. The Pope protects his priests and theologians, the Prime Minister of Canada protects his cabinet members, the union leader protects his workers. The model is *very* familiar: the wise father protects his *loyal* sons. And they must be loyal; the protection is guaranteed only up to a point. It is quickly withdrawn when any fundamental part of the whole apologetic structure is threatened. Priests can be defrocked, theologians made to recant, cabinet ministers forced to resign, and workers quietly dismissed.

Ask yourself: does science (as working scientists) protect its philosophers? Please suppress the unbridled laughter just long enough to consider that in this case the matter of the contractual protection is exactly the opposite. For scientists do not accept that there can be anything offered on their behalf that is epistemologically second best. At the point where their dogmas are threatened they become businessmen, they will show that their product *sells better* than available technological alternatives. Furthermore, they are successful businessmen. A fair sum of money for scientific work and ancillary activities is included in the budgets of every industrialized nation in the world. And so it is that scientists resist and often reject the overtures of philosophers of science, almost all of whom, by comparison, are engaged in very small business enterprises indeed.

Think in this connection of the new appointment of a young philosopher of science to a university faculty hitherto without such philosophical representation. The young philosopher is full of enthusiasm for this field and the good it can do the future progress of enlightened science. With the praise and encouragement of his teachers still ringing in his ears, the young philosopher approaches the scientists with an expectation of unquestioned support – in his heart-of-hearts he envisages interdepartmental courses, interfaculty research programmes, a new journal, a series of high quality books, and the like. The scientists reject him. Far from welcoming his apologetic skills, they regard him as a *threat*. If left uncontrolled, he will reveal to science students and others the fact that the scientists fail to discharge one of their own obligations, the obligation to criticize their own un-selfcritical domain of inquiry, an obligation that carries with it concerted scholarly attention to exactly those questions of logic, methodology, epistemology, and ethics so dear to the young philosopher. The scientists argue: the important thing is to do science, not to talk *about* it; one small experimental result or one theorem proven is worth a whole library shelf full of treatises telling us about experimentation and about the logic of proof. It matters not at all that everyone sees through this selfseeking argument. Threats must be put down, those who threaten must be punished. The young philosopher retreats, longing for the day when he will be joined by another of his own kind. He will henceforth seek to establish his own fraternity.

A welcome apologist cannot threaten; that would be gross disloyalty. But the fact that scientists themselves react as if to a threat carries an important message. The scientific establishment has all that it requires to put aside unfavourable propaganda or to banish would be apologists. What it cannot put aside – and hence seeks to ignore – are the results of a kind of inquiry in which it should itself be engaged. Scientists, on pain of losing their entitlement to be enthroned at the top of the hierarchy of inquirers, must respect all inquiry. Anything that passes as second best they have the ready means of buying out of business. (We begin to see how deeply the artificial separation of science and philosophy – fields of inquiry identifiable as one – cuts.) These institutional parameters of the scientific enterprise are awesome, touching as they do all the most basic questions of public and professional morality. Nevertheless, they reveal that philosophy of science is a form of investigation: inquiry that has a respectable subject matter, and *that subject matter is science itself*. I will not emphasize the positive features of this revelation. Surely any reader of this comment can supply a list of the contributions *to science* made by philosophers of science. Instead, I invite you to notice one other feature of Zev Bechler's position, the lack of epistemological integrity in philosophy of science resulting from the fact that there is no general criterion for determining when a philosophical claim is either true or false.

From the fact that each form of inquiry has a subject matter it does not follow that each form of inquiry possesses a decision criterion for true/false statements. If we generalize this requirement of an epistemic criterion, then it can be shown, or many think it can be shown, that science itself lacks such a criterion.[2] Scientists, just like other inquirers, try their ideas to see how far they will go, and the success of their inquiries never crucially depends upon considerations of truth or falsity.[3] If truth telling were the leading desideratum in science, then all earlier scientific theories now discarded must be taken to have been completely without success of any kind, and that conclusion would totally abolish any sense in which the work of scientists is cumulative, any sense in which successful science *leads to* better results.

Finally, if what I have been sketching is a correct view of the science/philosophy functional identity, then it follows that Zev Bechler cannot be correct when he suggests that there can be no sceptical philosophies of science. If the philosopher were an apologist, then of course he could not temper his propaganda with scepticism; banishment from academe would be the result. It is because the philosopher is a scientist and must put his case in the best evidenced way possible that he *must be* sceptical. It is no part of inquiry to turn inquiry itself into dogma. To question the epistemological credentials of science is not to cease to be a philosopher of science: it is precisely to engage in the highest form of science/philosophy. Bergson's pessimism about science is just as much philosophy of science as is Whewell's optimism about science. And if all negative, critical, and sceptical approaches to science must be ruled out of court as not being philosophies of science, then we yield the institution to a dogmatic science that exacts a high price: *it will be its own apologist.* An unphilosophical science is as bad as an unscientific philosophy. Cut the cord that unites science and philosophy, and the scientists will totally dominate the intellectual domain. Banished from that domain, unemployed philosophers will become, not apologists, but unwholesome double agents working both for science and for the society that it dominates. The philosophical goal will not be apology, it will be something infinitely more sinister: *it will be intellectual espionage.*

The University of Western Ontario

NOTES

[1] See my contribution to the second volume of these proceedings, 'Methodology and the Functional Identity of Science and Philosophy'.
[2] This point was made independently at the conference by David Gruender in his comments on Zev Bechler's paper.
[3] I am here adopting, without argument, a position that would turn out to be much like that of Larry Laudan in *Progress and Its Problems, Toward a Theory of Scientific Growth*, (Berkeley & Los Angeles, 1977), Chapter 1 and *passim.*

JOSEPH D. SNEED

COMMENTS ON BECHLER, NIINILUOTO AND SADOVSKY

Bechler's view of philosophical accounts of empirical science as 'apologetic' rather than descriptive I find interesting and challenging. That some philosophers who have offered accounts of empirical science thought the subject of these accounts was a 'good thing' and that the accounts are more polemical than descriptive or analytical seems clear. Galileo's writings on scientific methodology appear to me to have this character. That *every* account of empirical science from Aristotle to Hempel (and beyond) may accurately be viewed in this way seems considerably less evident. But I could be convinced by a careful study of the texts and their historical context. That philosophical accounts of empirical science must *be* (in some sense) 'apologetic' seems to me to be false. I take this strong version of the 'apology thesis' to be the one Bechler is defending.

What seems to me basically implausible here is this: Empirical science is just one among many kinds of things people do collaboratively – produce plays, buildings, manufacture automobiles. Sociologists apparently produce descriptions of at least some of these activities without 'evaluating' them. All but the most extreme proponents of the 'value ladenness' of the social sciences would admit this. Even the extremists in this respect find themselves hard pressed to argue that sociology, at this mundane level, *could not* be value free. At best they show that sociologists do in fact either explicitly or implicitly (through the aspects of the activity they emphasize) take an evaluative position. Now what is so peculiar about empirical science in this respect? Why should it, more than other human activities, elude dispassionate description?

More directly to the point, why should the products of empirical science – roughly, empirical theories – elude dispassionate description when the products of others apparently do not? It does not seem to be impossible to say generally and descriptively what 'a production of a play' is. To be sure, any reasonably precise general characterization is going to be a bit arbitrary 'at the edges'. What we are willing to countenance as 'a production of a play' will be determined in good

J. Hintikka, D. Gruender, and E. Agazzi (eds.), Pisa Conference Proceedings, Vol. I, 93–104.

part by what kind of a (sociological) 'theory of theatre' we intended to embed it in. This is certainly a *kind* of value judgment. But, clearly it *need not* entail anything at all about the social value of the institution of theater. Why should the concept of 'scientific theory' be different from the concept of 'production of a play' in this respect?

It may be true that most people who would choose to spend considerable professional energy on constructing a 'theory of theatre' would also be inclined to have strong attitudes (either pro or con) about the social value of this institution. One may even see how the 'pro-bias' might dominate at least among researchers likely to produce detailed, sensitive studies. People who have had some firsthand involvement with theater are more likely to be in a position to describe it accurately and also more likely to think it's generally a 'good thing'. But were we seriously questioning the social value of theater (perhaps with a view toward making adjustments in the level of public financial support), it would seem that we could surely find researchers to describe it for us in an acceptably neutral way.

Bechler seems to think that there is something peculiar about scientific theories that makes them less amenable to 'objective description' than other products of human activity: '... scientific theories are not "given" but are ... logical constructions. The crucial point is not simply that they are not "given", but rather that not even their properties are "given" or "known" ....' Now this seems clearly false to me. Pre-theoretically we know just as much about scientific theories as about productions of plays. Specific scientific theories are taught in universities, discussed at conferences and expounded in literature; just as specific productions of plays occur in theaters and are reviewed in newspapers. Of course it might be that the concept of a 'scientific theory' cannot be made precise enough to play a role in a theory of empirical scientific activity. Just as the naive concept of a 'production of a play' might not lend itself to incorporation into a theory of theater. But surely it is not *a priori* evident that this is so. The proof of the utility of the concept lies in the theory in which it appears. We do not yet have a very good 'theory of science' (nor of theatre, so far as I know) but I see no reason to think we could not have one. Nor do I see reason to think that the concept of 'scientific theory' could not play a role in it.

Niiniluoto's sketch of a formal-linguistic version of a theory-element core (pp. 8–10) I find illuminating. He has made very clear what is required for a linguistic formulation of the 'structuralist view' of empirical theories. I am not entirely convinced that his suggestion for dealing with constraints is the *only* possibility for a linguistic formulation of this feature. Kanger and others have suggested to me that a modal logic is the most natural formal-linguistic tool to be used. The fact that informal discussions of truth conditions for statements involving theoretical concepts (e.g., Mach's 'definition' of mass) are often formulated as counterfactual conditionals lends some plausibility to this suggestion.

It seems clear that some questions about the logical structure of empirical theories can be most effectively (perhaps only) treated in a formal-linguistic formulation. Among these are perhaps 'foundational questions' for the structuralist view – being clear about the ontology that underlies the view. That naive set-theory is not adequate has now become clear to me. The collections of models of set-theoretic predicates (e.g. M_p and M_{pp}) are not 'strictly speaking' sets and the structuralist's practice of cavalierly treating them as if they were is at least a bit unfastidious even if no contradictions are apparent. Rigorous adherence to the formulation Niiniluoto suggests would avoid this. Regarding these models as 'species of "structures"' in the sense of Bourbaki or as 'categories' are alternative approaches. A formal-linguistic approach along these lines might also lead to a more satisfactory formulation of the concept of Ramsey, eliminability of theoretical concepts from theories in which non-trivial constraints operate. Similarly, a linguistic formulation might make it easier to be clearer about 'what it is that's theoretical' when we talk about theoretical concepts, though other approaches to ontological foundations provide clarity here as well.

However, it is not clear to me how a linguistic formulation of a specific theory-element core for a 'real-life' theory might be useful. If one could be produced at all (say for classical equilibrium thermodynamics), I suspect it would be forbiddingly complex and add little to our understanding of the theory and its links to other theories. It could of course put us in a position to answer the Ramsey eliminability question for the theoretical concepts of this theory. But questions of this sort seem to be most interesting to philosophers with some kind of metaphysical axe to grind about the status of 'theoretical entities'. I doubt that those concerned with describing the development of thermodynamics would be too interested in learning whether entropy was Ramsey-eliminable.

My doubts about the usefulness of linguistic formulations of cores of specific theories should not be taken to mean that I regard research in this direction as unproductive. Quite to the contrary, I think it is one very promising line to take in putting the structuralist view of empirical theories on a sound foundation. Foundational questions are important; so is a sharper understanding of how theoretical concepts work. But, more important, in my view, is the treatment of further specific examples of empirical theories with the apparatus we have at hand. That the foundations of this apparatus are a bit insecure should make us cautious and attentive to foundational results. But it should not deter us from going ahead exploiting an apparutus that seems to work. In this respect, those of us interested in descriptive reconstruction of empirical science should follow the example of our colleagues among the mathematicians and physicists.

Niiniluoto is correct in his criticism of my exposition of K-theoreticity (p. 13). I now think it would have been better for me to discuss this criterion after laying out my account of the logical structure of a theory's empirical claims. Not doing so was occasioned by my desire to motivate the discussion of logical structure by tieing it to the traditional 'problem of theoretical terms'. Niiniluoto's definition of K-theoreticity (p. 11) appears to be essentially satisfactory though a bit infelicitously formulated (see below). I am not sure what the subsequent discussion of 'actions' adds. Why we need 'actions' as well as (or as replacements for) semantical entities like statements or propositions to understand 'presupposition' as it's used in his definition is not clear to me.

Niiniluoto suggests (p. 16 ff.) that we add to a theory element a set of 'intended theoretical applications' J and consider a 'theoretical claim' $J \in AT(K)$. Despite his detailed arguments in support of this, I remain unconvinced that this can be made intelligible – let alone that it is useful in describing any empirical theory I know about.

Before examining Niiniluoto's discussion in more detail, let me make it clear that I have no philosophical prejudices against 'taking theoretical concepts seriously'. I am not a 'closet nominalist' who secretly cherishes the hope that we may yet demonstrate conclusively the 'in principle eliminability' of all theoretical concepts and thereby expose the vulgar naïveté of empirical scientists.

In fact, I suspect that there are 'hard cases' of non-Ramsey-eliminability where we could not do this. On the other hand I do think we must *try* to construe the claims of empirical theories in a way that their truth conditions are clear. To do otherwise would simply be uncharitable toward empirical scientists. Like the rest of us, they may sometimes utter nonsense – even professionally. But to assume at the outset that they do seems inappropriate. The Ramsey-sentence form and modifications of it involving constraints provide a construal of claims involving theoretical concepts with clear truth-conditions. Inter-theoretic relations like reduction and equivalence provide other possibilities, but they too ultimately come to rest on a (modified) Ramsey sentence. We *may* discover other ways to satisfactorily construe claims with theoretical concepts that have nothing to do with Ramsey sentences. And too we may find theoretical concepts being used in ways we can make no good sense out of at all.

Niiniluoto offers 'six arguments to defend the theoretical claims and their testability'. The first (p. 17) argues that in making the distinction between K-theoretical and K-non-theoretical concepts, we have assumed at least that K may have successful theoretical applications. Surely, we have assumed that $\bar{r}^{-1}(I) \cap AT(K) \neq \emptyset$. But we have not assumed that there is some specific $J \subseteq AT(K)$ or $E_t(K) \subseteq J \subseteq AT(K)$. That Niiniluoto refers to $E_t(K)$ in his definition of K-theoreticity (p. 11) as '*the* existing theoretical applications' is a bit misleading. Better would be '*a* set of theoretical applications accepted at some t'. This preserves the essential features of the definition and avoids giving the impression that users of K have some fixed set of theoretical applications in mind as they go about applying K. Rather, they discover what the appropriate theoretical applications are in the process of using the theory. Users of (classical or relativistic) particle collision mechanics don't know at the outset what the masses of particles are. They discover this as they use the theory to explain particle collisions.

Niiniluoto's second argument is more interesting. He points out that if theory-element core K is a theoretization of K' and K' does not provide the possibility of unique determination of its theoretical concepts on its range of intended applications then the truth conditions of $I \in A(K)$ are unclear. This is so. And this is (I think) why such examples of theoretization are not 'found in nature'. If there is this kind of theoretical ambiguity in K' what we do is 'aggregate K' before theoretizing to K. That is, we lump together the K'-theoretical structures that are 'equivalent' in the sense of being equally good theoretical emendations of members of $A(K')$. We then let these equivalence classes be the non-theoretical structures of K. Intuitively, this is the same as requiring the laws and constraints of K to be invariant under the equivalence relation determined by K'-theoretical non-uniqueness. For details see Sneed (1978).

I find Niiniluoto's third argument ambiguous at the phrase '... apply some $K' \neq K$ to Z'...'. To apply K' to Z' would seem to require that either $Z' \in M'_{pp}$ (my understanding of 'apply') or $Z' \in M'_p$ (Niiniluoto's suggested 'theoretical' application). Were $Z' \in M'_{pp}$, then K' would be related to K by something like theoretization. It's hard to see how K' could then help us determine Z'. Thus I assume Niiniluoto means $Z' \in M'_p$. Thus $M'_p \cap M_p \neq \emptyset$, so M_p and M'_p must be structures of the same type (species, category). In fact this suggests that both K and K' are infelicitously formulated. What we really have is a third K_o of which both K and K' (slightly reformulated) are specializations with intersecting (but not nested) laws and constraints. In this case theoretical function values in K', were they uniquely determined on $M' \cap M$ (more precisely on $AT(K) \cap AT(K')$, would have to be carried over to K. But this is just a special case of the well-understood way in which specializations (of K_o in this case) afford additional methods of determining theoretical concepts and 'overarching' constraints carry them to other specializations.

Niiniluoto's fourth argument (p. 18) is the most interesting to me. It calls attention to the fact that inductive and probabilistic concepts have not yet been integrated into the structuralist framework. I find it difficult to judge the plausibility of Niiniluoto's treatment of his example because the concept of 'inductive reasons' (p. 20) is not intuitively very clear to me. (This is definitely not to say I think it could not be made clear – to somebody at least. It's just an admission of ignorance.) Probabilistic concepts are more familiar to me and (I think) do roughly the same job.

From the point of view of probability theory a natural way to treat the example is this. Roughly, the non-theoretical structures are all possible distributions of relative frequencies of observable attributes. The intended applications are those distributions we observe in people in societies we are interested in. The theoretical structures are all possible probability distributions over observable attributes-*cum*-P-phobia. The law picks out distributions in which P-phobia and the observable attributes are correlated positively with unspecified values of the correlation coefficients. The constraints say that the values of the correlation coefficients remain constant across applications. The Ramsey-functor intuitively takes us from a 'doubly theoretical' *ideal* probability distribution over entities involving the theoretical concept P-phobia to observable relative frequencies. Conversely, observable relative frequencies allow us to 'estimate' the value of the theoretical concept – the values of the correlation coefficients. How to precisely formulate this intuitive idea is not obvious. That's why I say we don't yet have a satisfactory integration of probability into the structuralist view.

A 'real-life' example which requires this is Mendelian genetics. Here the non-theoretical structures are (generally non-Markov) stochastic processes over manifest (observable) traits in a species. The discrete 'time parameter' of the process is 'generations'. The theoretical structures are *Markov* processes over state descriptions involving theoretical, genetic properties 'connected with' observable traits. Statistical prediction of relative frequencies of observable traits in successive generations from genetic properties and statistical estimation of genetic properties from observed relative frequencies of traits should be modeled formally by something like a Ramsey-functor.

Niiniluoto's fifth argument (pp. 20-1) is difficult for me to understand. It appears that there can be no $O: M_p \rightarrow M'_p$ in which elements appearing in $O(x)$ are non-K'-theoretical, for all members of M'_p *per definition* contain K'-theoretical elements. Suppose we drop this requirement. Then O is just a candidate for a 'strong reduction' relation. But Niiniluoto requires nothing that would assure that it is. Niiniluoto's claim (8) '... gives support to (4) ....' presumably in some inductive manner. Because the many-one-ness of O means that (8) does not entail (4). But, in the absence of Niiniluoto's (on my understanding inconsistent) requirement (ii), the truth conditions for (8) are no more apparent than those of (4).

Perhaps what Niiniluoto really wants here is $O; M_p \rightarrow M'_{pp}$. That is, O is a mapping of K-theoretical structures into K'-non-theoretical structures. In the case that O is factorable through r (i.e., there is an $O': M_{pp} \rightarrow M'_{pp}$ such that $O = O'$ or) O would simply be a candidate for a (weak) reduction relation. But if O is *not* factorable through r we would have an intertheoretic relation of a type not yet investigated.

In the non-factorable case what we have is roughly this. Both M_p and M'_p are theoretical structures 'over' M_{pp}. M'_p projects onto M'_{pp} via the familiar Ramsey-functor that simply 'wipes out' theoretical concepts. M_p projects onto M'_{pp} via O which does not tell us how it decides which member of M'_{pp} it will associate with members of M_p. Nevertheless we should apparently regard $\langle K, O\rangle$ and K' as two theories about M_{pp} and require them to be compatible in the sense that $\bar{\bar{O}}(AT(K)) = \bar{\bar{r}}(AT(K')) = A(K')$. Further, if we are to take $\bar{O}(J) \in \bar{\bar{O}}(AT(K))$ as evidence for $J \in AT(K)$, Niiniluoto seems we should also take $\bar{r}(J') \in \bar{\bar{r}}(AT(K'))$ as evidence for $J' \in AT(K')$. But the latter seems not a very plausible thing to do because here we have come to expect that even though $\bar{r}(J') \in A(K')$, J' there is no reason to expect that $J' \in AT(K')$. That is just to say that there will always be plenty of theoretical emendations of $\bar{r}(J')$ which fail to satisfy the laws and constraints of K'. Seen in this way, it appears unlikely that an O relation of this kind would support the sort of inductive inference that Niiniluoto suggests.

Another peculiar feature of the inter-theoretic relation provided by such an O is this. We appear to have two theories about M'_{pp} and one theory about M_{pp}: i.e.

$$K = \langle M_{pp}, r, M_p, M, C\rangle$$
$$K' = \langle M'_{pp}, r', M'_p, M', C\rangle$$
$$K_O = \langle M'_{pp}, o, M_p, M, C\rangle$$

Isn't it remarkable that the same theoretical structures – M_p – are useful in talking about both M_{pp} and M'_{pp}? Yet we see no relation between M_{pp} and M'_{pp}. This suggests that such intertheoretic relations are not very likely to be 'found in nature'.

Niiniluoto's sixth argument is based on the claim that '. . . that a term is K-theoretical is historically relative and can change when new applications of the core K are found'. On my account of the *normal* historical development of a theory based on K, this is not true. What is true is that our methods of determining K-theoretic concepts change as new applications are found in which the K-theoretical concepts appear in new laws. In this way the values of K-theoretical concepts associated with I may become more uniquely determined. This may have the effect of enhancing the testability of claims in a theory K' which is a theoretization of K. But it does not fundamentally alter the testability status of *K's* own claims. Of course the claims of *K* become 'stronger' (intuitively, at least, in a way that Popperians might laud) as the 'net' of specializations of *K* grows. But they are still claims about $I \subseteq M_{pp}$ rather than about some $J \subseteq M_p$. There can also be non-normal (not necessarily 'revolutionary') developments in that inter-theoretical relations between *K* and other theory cores are postulated. These too can produce new ways of determining *K*-theoretical concepts for members of *I*. But those discovered thus far still require us to assume we are dealing with members of $A(K)$ before we can use them.

I find Niiniluoto's analysis of the motivation behind the 'multi-application approach' illuminating. To it I would only add this. Whether or not the multi-application approach is fruitful may be dependent on the theory one is about reconstructing. Some theories may have just one (big) intended application. Others may have lots of little applications, but operate under the assumption (rigorously formulatable as a constraint) that these can all be embedded in one big application with the same structure. Theories of measurement and physical geometry appear to be like this. Other theories may have lots of little applications and no commitment to viewing them as substructures of a larger structure. Dalton's 'atomic theory' of combining weights in chemical reactions appears to be a clear-cut example of this kind of theory. There is just no plausible way to aggregate two chemical reactions producing a 'bigger' chemical reaction.

Niiniluoto's suggestions for modifying what Balzer and I have said about 'Kuhn-theories' appear to be sound. I certainly think there is much room for improvement here. The only reservations I have are about permitting modifications of K_O within the same normal scientific tradition. K_O is intended to model the consensus about fundamental concepts that holds the tradition together. If you see it apparently changing within the same tradition this suggests that you're wrong in identifying the fundamental concepts. I would not however insist on this. There are clearly examples of multiple K_O's within the same tradition – e.g. different formulations of classical particle mechanics. Perhaps we need something more than just K_O's to provide an adequate account of what's conceptually central to a tradition. Moulines' idea of a 'theory-frame' might be a more effective tool here (Section V, these proceedings).

Niiniluoto is correct in pointing out (p. 35) that the structuralist account of normal science is incomplete in that '... it does not contain a theory of belief formation in scientific communities ...' At best the structuralist view characterizes the objects of belief (i.e. propositions, statements) that are shared by members of scientific communities. *How* they come to be shared is surely a very interesting problem. An adequate account of this might put the establishing of scientific traditions within the scope of deliberate social policy.

I also find Niiniluoto's suggestions for a 'theory of progress' interesting. In particular, his suggestion for estimating 'The value of a theory ...' attacks a problem of fundamental importance whose solution would be of obvious practical value. It would be very useful to see work in this direction carried to the point that we were able to see (at least in principle) how to calculate the expected value of social investments in supporting a scientific tradition. What one apparently needs is some kind of probability measure over alternative possible development path that describes the 'internal dynamics of the tradition' as well as some measure of the 'social value' of solutions to problems. Showing how to come up with such a probability distribution would be very deep insight into the nature of empirical science.

Sadovsky discusses at some length the relation between formal and 'content intuitive' accounts of the structure and development of scientific theories – a point also raised in discussion by Professor North. In response I should like to briefly sketch my own views on this.

Very roughly, I view the relation between informal (or 'content intuitive') accounts of scientific theories and their formal counterparts as something like the relation between theoretical and experimental activities in an empirical science. Of course informal accounts like those of Kuhn, Lakatos and Laudan are not analogous to experimental activities like determinations of the speed of light. Rather, they are attempts to give a coherent overview of the results of empirical investigations – case studies – in the history and sociology of science. In doing so they (inevitably?) formulate general claims that have implications beyond their 'data base' of case studies. It is the *concepts* employed in these general claims that are the raw material for formal accounts of the nature of empirical science.

What do such formal accounts add to our understanding of scientific activity? First, at the level of general claims about the nature of empirical science they *can* provide the means for introducing more precision. Formal accounts can provide means of making subtle distinctions that remain masked in the vagueness of everyday language. For example, the net-formalism for Kuhn-theories that Balzer and I have developed suggests several (equally plausible?) reconstructions of Kuhn's concept of 'paradigm'. Whether these distinctions prove to be useful ones remains to be seen. In general, the subtlety of expression provided by formal accounts can suggest unnoticed *possibilities* to empirical investigators and thereby make informal accounts more effective guides to empirical research. This is rather like the theoretical physicist's noting the existence of unexpected solution possibilities to a system of equations believed to describe a certain phenomenon which suggest new lines of experimental investigation. Until the theoretical account is expressed in the highly 'formal' apparatus of differential equations these additional possibilities remain unnoticed.

Second, at the level of case studies, I think a *good* formal account of the structure of empirical theories can be a useful heuristic guide. It can provide a kind of taxonomy of the intellectual moves empirical scientists typically make and thus *help* the historical investigator find his way through the maze of mere description. I emphasize 'help'. At best the formal accounts available now describe the intellectual products of empirical scientific activity. This is surely only a part of what historians and sociologists of science are (and ought to be, in my view) concerned with. Even so restricted, I think the *explicit* use of formal methods for describing particular scientific theories will be of limited use to the historical investigator. I surely do not expect that historical investigators will report their findings in something like a net-formalism.

Explicit formalizations of specific theories like thermodynamics and particle mechanics have different uses. First, they provide a test of the adequacy of the formal account. Second they may, as an added dividend, clear up some 'philosophical puzzles' surrounding the theory's fundamental concepts. That the concepts in the formal account are really adequate can only be demonstrated by carrying through applications to specific cases in some detail. Once the adequacy of these accounts is 'confirmed' in this way they can be applied in a much more free and loose way as heuristic and informal-descriptive tools. The added dividend that may accompany these detailed applications could also be of use to the historical investigator in situations where discussion of these philosophical puzzles played some role.

Sadovsky also notes that '... the development of the formal apparatus is to a great extent autonomous with respect to the corresponding content level of the scientific knowledge and ... contains a possibility of obtaining paradoxical consequences.' (p. 51) I agree; and, while I too am concerned about ruling out paradox, I am somewhat more concerned about avoiding 'sterility'. Once theoreticians (in any discipline, so it seems) have a reasonably rich, tightly formulated formal apparatus there is a strong tendency for them to focus their energies on working out its 'internal problems' without regard for its applications. In a way that we understand only partially, mature scinetific traditions seem to operate to avoid the sterility of theoretical development. Theory and experiment work in a somewhat independent, yet symbiotic, manner both contributing to the 'progress' of the enterprize. I venture to hope that the 'science of science' is on the verge of this kind of maturity.

REFERENCES

Moulines, C.V.: 1980, 'An Example of a Theory Frame: Equilibrium Thermodynamics', in the second volume of these proceedings.

Sneed, J.D.: 1978, 'Theoritization and Invariance Principles', in I. Niiniluoto, and R. Tuomela, (eds.). *The Logic and Epistemology of Scientific Change*, (Acta Philosophical Fennica 30), North-Holland, Amsterdam.

B. G. YUDIN

THE SOCIOLOGICAL AND THE METHODOLOGICAL IN THE STUDY OF CHANGES IN SCIENCE

One of the main arguments put forward by numerous opponents of T. Kuhn claims that he mixes unjustifiably the methodological and the sociological aspects of the problem in his analysis of scientific revolutions. Even his adherent M. Masterman reproaches him for his use of the key concept 'paradigm' in various meanings that include both sociological and methodological ones. And, in general, in the analysis of science the postulate of the inadmissibility of mixing the above aspects or approaches (understood in this context in the broadest sense of the word) is accepted very frequently as something self-obvious which does not require any justification.

In this case it is implied that scientific knowledge is considered by methodology in the internal way: that it is interested in the meaningful aspects of knowledge, in its content, in that element of scientific knowledge which can be reconstructed rationally. Sociology, on the other hand, approaches knowledge in the external way, abstracting itself exactly from these aspects. It is well known, though, that different researchers distinguish between the internal and the external in application to science in essentially different manner, which is explained, apparently, by the differences in their understanding of science itself, which are by no means always explicated.

Such opposition of the sociological and methodological approaches to the study of science, when they actually appear as mutually exclusive, is fraught with a number of undesirable consequences. It turns out, first of all, that the knowledge obtained on the basis of one or another approach cannot actually be related or compared in any way and can neither be confirmed nor disproved by the other. A situation of dichotomy is formed: either one has to base one's argument on methodological considerations or on sociological ones, while the possibility of constructive synthesis and mutual complementation of both approaches understood in this way turns out to be unattainable from the very beginning.

At the same time, the more and more frequent appeals (implicit or explicit) from each approach to the other give evidence of the fact that problems appear in the frame of each of them for the solution of

J. Hintikka, D. Gruender, and E. Agazzi (eds.), Pisa Conference Proceedings, Vol. I, 105–109.

which the proper means of their approach are insufficient. And since none of the alternative approaches by itself makes it possible to solve satisfactorily such problems as the genesis and development of science, the regularities of the change of theories, etc., it is natural that in the actual study aimed at the solution of real problems in the field of the history of science representatives appear in one or another way that are typical for both approaches. One of the methods used in this case is associated with the 'internal' (understood now as methodological) being regarded as the initial method in the study of science, and the 'external' (respectively, sociological) being brought in at the discontinuity points when the 'internal' criteria and factors do not operate. According to I. Lakatos (1971), '... rational reconstruction or internal history is primary, external history only secondary, since the most important problems of external history are defined by internal history'. He adds that the external history explains the pace, localization and identification of historical events, or the causes of deviation of the real course of events from that which should be expected according to rational reconstruction.

In such explanations, however, the external turns out to be unorganized, introduced not in a systematic manner but *ad hoc*, because from the very beginning it appears as something additional, complementary. It is for this reason that the whole construction is deprived of harmony and natural quality. As a result of this, the appeals to 'synthesis' sometimes lead to giant sociological crutches being brought in as obviously disproportionate supports for weak products of 'rational reconstruction'; the literature devoted to the analysis of the Copernican revolution or Darwin's evolution theory contains many such examples. And although Lakatos speaks against such 'universal' externalist explanations extremely sharply, his position contains no criteria and foundations that would allow one to dissociate oneself from such constructions and subject them to criticism.

One more argument in favor of the possibility of formulating such a problem as the interconnection between the methodological and sociological approaches in the study of the functioning and development of science can be reduced to the following. If the main directedness of the methodological analysis of science is spoken of in most general terms, it is easy to state that in the center of attention of such an analysis is the exposure of those methodological norms, initial assumptions, cognitive attitudes, etc., which guide one or another researcher in one or another scientific direction. But the foundations themselves that constitute the methodologist's position as the specific research position, the peculiarity of his approach are associated exactly with the fact that the researcher who is engaged in a certain concrete scientific problem by no means always nor totally and completely realizes these norms, prerequisites, assumptions on which his argument is based. And most frequently the adoption by a researcher of some or other methodological norms and being guided by them does not result from rational choice on his part. It is obvious that even the methodologist's studying the primary sources is conditioned at least partly by the fact that the methodological aspects of the scientific work should, as a rule, be *exposed, disclosed* in the course of special study, that it is both possible and necessary for the methodologist to 'read out' from these texts not only what their authors had in mind but something else as well.

At the same time, so far as the methodological norms are not only unconsciously used by the researchers but circulate within the limits of the scientific community and are transmitted from one generation of researchers to the other, it is clear that in this case we are dealing with mechanisms of the functioning and development of scientific knowledge that are *social* by their nature. In this connection we would like to say the following. In his postlude to the second edition of *The Structure of Scientific Revolutions* Kuhn distinguishes between the sociological and, as it were, operative, i.e., methodological meaning of the term 'paradigm'. In the second case he considers paradigm not only from the purely methodological but also from the sociological point of view: as a *generally recognized model*; the point in question is thus the functioning of the paradigm inside a scientific community. And on the whole, when one abstracts oneself from the social mechanisms, which is frequently both possible and necessary, it is, therefore, a result of the special procedure of *idealization*, the construction of the specifically methodological point of view.

Thus, one can see that inside the methodological approach proper there is no sufficient ground for the separation of what should be the object of rational reconstruction, from the external history and for the exposure of points at which the rational reconstruction should be complemented, because of logical, and not pragmatic necessity, by the data and explanations from the external history. In exactly the same way it contains no means for justification of exactly what layers and spheres of the external history can, generally speaking, be brought in for explanation. As one can see, neither the methodological component of the concrete scientific research activity nor the work of the methodologist itself possess the necessary degree of logical completeness and purity which is implied by the prerequisites of the narrow rationalism.

It seems to us that, on the basis of the above, it would make sense to doubt the fact that the postulate about the original opposition between the methodological and the sociological methods of studying science is both natural and necessary. The seemingly natural character of this postulate is apparently frequently the result of just following the already formed tradition. And when we reject the necessity to accept this postulate precisely as such, it becomes possible to formulate the question about the internal, closer interactions between methodological and sociological elements in scientific study itself. Finally, by not accepting this postulate we are facing the problem of exposing the sociological contents which, as it were, are hidden behind the methodological categories and, on the other hand, in what way the methodological means and norms are realized in sociological mechanisms of the functioning and development of science, filling them with their contents.

Certainly, the point in question is not their disordered and eclectic mixing but the representation of both of them as different projections of science which are, however, mutually connected and related. It would be of interest from this point of view to attempt to expose the dual (both methodological and sociological) nature of certain basic concepts of the methodological approach to the study of science. To put it differently, we do not, for instance, perceive the fact that there are both methodological and sociological elements in Kuhn's concept of 'paradigm' as evidence of the inadequacy of this concept (certainly, under the condition that this circumstance is controlled by reflection). The point is that there is a whole group of concepts among those used by the methodologist such that their contents pertain, entirely or partially, to the fixation, description and explanation of some or other aspects of the *activity* of the scientist, and not just the already objectivized results, the products of this activity. It is possible to include in their number the same concept of paradigm, Lakatos' concept of research program, the concepts of approach, orientation, the picture of the world, explanation, justification, proof, demonstration, verification, etc. Such concepts constitute the skeleton of the methodologist's apparatus or, in other words, they determine his specific position. But the same concepts fix the connections of what is done by researcher, his determinacies as the subject of the activity and of what is intersubjective, either in its being given to the researcher or in the imperative nature of those requirements which make it possible to speak of the common significance of each new result of cognitive activity. It is the latter circumstance which enables one to speak of the sociological loadedness of these methodological concepts.

We would like to note in conclusion that such concepts as 'paradigm', 'research program' (to a somewhat lesser degree), etc., describe the structure of the maintenance of theories, instead of their change, and are oriented towards the exposure of norms that ensure the stability of knowledge. Meanwhile, one can hardly present the activity in science in *its essence* only as the preservation of the already available knowledge; it always contains as its impulse the tendency to go beyond the known. Obviously, there should be, also, methodological norms (which again have a sociological meaning) that control and condition exactly the *increase* of knowledge. The necessary condition for the possibility of at least correct formulation of the problem of theory *change* and the *structure* of this process is the exposure of these norms and the establishment of their hierarchy.

REFERENCE

Lakatos, I.: 1971, 'History of Science and its Rational Reconstructions', PSA 1970, Boston Studies in the Philosophy of Science, Vol. 8, ed. by R.C. Buck and R.S. Cohen, D. Reidel, Dordrecht, pp. 91–136 (see especially p. 105).

SECTION II

THE EARLY HISTORY OF THE AXIOMATIC METHOD

M. V. POPOVICH

CONCERNING THE ANCIENT GREEK IDEAL OF THEORETICAL THOUGHT

1. The logical reconstruction of the theoretical systems of ancient science permits an appraisal of the character of the theoretical achievements of antiquity in a new way. The reconstruction of the structure of ancient logical systems, begun by J. Łukasiewicz and his pupils, can be compared with the reconstruction of the Euclidean geometry during the intensive development of geometry in the 19th and 20th centuries, first of all in the works of D. Hilbert. Extremely interesting results, in particular, were obtained concerning the method of analysis, as it was realized in ancient texts [1], [2].

It seems that historical retrospection can be the aim of logical advancement only to a small extent, as the reconstruction of Euclidean geometry has been only a subordinate result of the development of contemporary geometry. To make the picture of the past more complete, it would be useful to supplement the reconstruction of the structure of ancient theoretical thought with an analysis of general notions about the ideal of scientific thought prevailing in the epoch of creation of the axiomatic method. This is the main task pursued in the present paper.

The premises I am starting from can be characterised in this way: I doubt whether the intimate mechanism of reasoning differs essentially in different societies and in different epochs. That's not we are talking about, to find differences between "ancient Greek syllogisms" and "ancient Indian syllogisms". The logical mechanism of solution in theoretical, in particular mathematical, tasks is generally the same. At least more obvious are the differences of a higher order manifesting themselves when a solution of problems satisfying a certain society in a certain epoch ceases to satisfy another society in another epoch. For instance, the result known as the "Pythagorean theorem" was obtained by the ancient Greeks as well as by the ancient Chinese. From the point of view here formulated it is of no importance how this result was being obtained by the Greeks or by the Chinese. It is important to know why solutions which had satisfied the Chinese didn't satisfy the Greeks.

J. Hintikka, D. Gruender, and E. Agazzi (eds.), Pisa Conference Proceedings, Vol. I, 113–124.

Needless to say, such an analysis is in any case important not so much for the history of science as for the self-evaluation of modern scientific knowledge. Our criteria of selection of acceptable theoretical constructions, including our criteria of exactness and demonstrativeness, are historically dependent and historically confined, and self-knowledge is possible only in comparison with other scientific communities.

2. Achievements of contemporary comparativistics enable one to maintain that extremely archaic structures of outlook and notions, frequently descending from common Indo-European roots, were, firstly, unexpectedly complicated and advanced, and, secondly, had a lively cultural background in ancient society to a more considerable extent than they have been thought to have. Well, the idea of the world fire as the universal substance and the eschatology of world fires, inherent in the ancient Indo-European thought, still lived until recent times in the thought of many people. The stability and habitualness of this idea explains well-known passages in Heraclitus. (Here other ancient ideas, particularly Stoic ones, could also be mentioned.) Democritos' idea that the image of thing is created between a thing and the eye as the result of emanations from the eye to the thing, will stay unintelligible until we take into account the fact that for the archaic outlook "to touch by eye" was not sharply differentiated from touching by hand (many ethnographic materials witness to this). That's one reason why the ideal of theoretical knowledge formed in ancient Greece ought to be considered against the background of archaic notions.

It is universally recognized that the social system of democracy in the Greek politics was a factor of great importance, which determined a new understanding of cognition. (See, in particular, [3], p. 18 and ff.) From following custom as established by God (θέμις – "supposed") the society goes over to following law as being humanly established (νόμος – "law"). This change in outlook manifested itself, particularly, in the desacralisation of the cosmological symbolism in Greek architecture (see [4], [5], [6]).

It is necessary to mention that, as was ascertained by S. Vikander, in the ancient Indo-European community there already existed a strict opposition of the practical sphere to the sacral sphere (for example, continuous sacral exchange between members of the community had been opposed to their practical dealings; mythological personages embodied both the sacral idea and its physical action – cf. Indra and Vayu in the Indian mythology). From another side, desacralised sphere in the ancient Greek conscious incompletely corresponds to the modern notion of the "secular" world.

In archaic consciousness the world was thought to be an action filled by some inner sense, an action which as modelled in ritual performances and prophecies. World was thought of as the action, *θέᾱ*, performance, spectacle, in which man takes part and at the same time apprehends, sees, observes it. (Common Indo-European $*\sqrt{dh\bar{a}u}$; compare Sanscrit *dhīḥ* – "thought", "wisdom", *dīdhyē* – "I observe", "perceive", "think"). The most important act in this model of the world-spectacle was to put objects in their own places, i.e., to put the world in order (*θέσις* – action of putting something in its own place, without which the world, *θέᾱ*, is unthinkable). Indicating or drawing lines is the most important function in the regulation of the world. The Indo-European word $*\sqrt{di\acute{k}}$ means "imperative legal order", "instruction"; from here we have in Sanscrit *diştih* – "designation", *diç – dik* – "direction", "designation", and in ancient Greek *δικέ* – "world legal order". On this ground the notion of demonstration – *δείκνυμι* (I), – "to make somebody to see" (see [7], [8], [3]) was first developed.

Desacralisation of cultural life showed itself, particularly, in the liberation of ideas about "indication" as a means of understanding the sense of cosmic *θέᾱ* from ritual load. A. Szabó showed that the terminology of the geometric doctrine of proportion has its origin in the theory of music [9]. Probably in this case more general reasons are to be considered, than the influence of Pythagoreans. Desacralised, secular art, both architecture, theatre, and music, together with philosophy and mathematics, was confronted with the ritual, sacral sphere of culture. The general notion for the whole secular sphere was *θεωρία* as "action of seeing", "looking", and *θεώρημα* as "an object of sight", "something that one sees", "spectacle" (the same terminology is used in sacral sphere: compare *θεωρός* – "ambassador on prophecies and games", *θεωρίς* – "Sacral boat, brought *theoros* to Delos").

In connection with this the verb *δείκνυμι* – "show", "indicate" – too attains the significance relatively free from the sacral sense, but at the same time common for the whole secular culture. A. Szabó, who discovered the evolution of this notion, notes three meanings of the term δείκνυμι: [1] "to show, to make somebody to see, to indicate; [2] to explain; [3] to demonstrate. In the beginning period, I believe, there were not three but one and the same meaning. It is interesting that Socrates, according to Xenophon, raised the traditions of his δεἰκνυμι to Odysseus because of his ability to convince ("to show", "to demonstrate by words"), based on what is generally recognized and unquestionable. In this case one can see the most ancient scheme of the future axiomatic constructions. I believe that the criteria of the correctness of the δείκνυμι for the first time didn't permit proper geometrical demonstrations from verbal explanations of the essence of the world.

To the end of antique epoch Lucian ridiculed the groundless explainings of natural philosophy:

> And doesn't it demonstrate the stupidity and full ignorance of philosophers speaking about things not so clear but insisting on their correctness and denying the contrary view to have any significance, they hardly do not swear, that Sun is scorching globe, that Moon is populated, that stars drink water scooped by Sun from the sea as in the well rope, and distributes it between them equally? ((10), p. 277–278).

Thales from Miletus belonged to such natural philosophers, to whom probably belonged the first geometrical demonstrations ((10), p. 67). I believe that their criteria of the correctness of the δείκνυμι for the first time didn't permit to separate proper geometrical demonstrations from verbal explanations of the essence of the world.

But sacral "explaining" and desacralised "indication by words" are strictly opposed. Very interesting from the point of view of the development of semiotic ideas is the substantiation of legitimacy of the two types of understanding of the world given by Plutarch, a near contemporary to Lucian.

"But, as I suppose, nothing prevents both scientists and speculators from being right, because the one explains the causal connection and the other the purposive one.... Those who think that finding the causal connection in some events they prove these events couldn't be signs, overlook that in reasoning in such a way they deny existence not only of divine signs but of any artificial signs as, for instance, signs made by iron disks or by fire, or the definition of time by the length of the shadow of sun dials." ([11], p. 200.)

3. The second progress of great importance in ancient consciousness was the separation of the indication of empirical data from the

indication of abstract essences and mentally performed actions. It is this progress that the formation of proper geometrical demonstration is connected with. The activity of Eleatic philosophers had outstanding significance for Greek mathematics, very well shown by A. Szabó [9]. *θεώρημα* becomes an abstract construction, *θεωρια* a mental construction generating abstract objects.

This process, connected with geometrisation of the Greek mathematics, is well investigated. Strong and weak sides of the mathematical methods of ancient Greeks are connected with their ideals of acceptability of mathematical constructions that found their expression in the geometrisation of mathematics. Geometrical methods were confined by the fact of incommensurability, the discovery of which, according to Szabó, led to a "new technology of demonstration and to antiempirical transformations alien to visual methods" ([9], p. 287). Along with the feat of paradoxes of infinity and continuity generated by these methods, they paralyzed the possibilities of Greek theoretical thought, as was excellently shown by V.F. Kagan when he analyzed the legacy of Archimedes [11].

Judging from these generally recognized statements, the assertion can be formulated that the ancient Greek ideal of theoretical knowledge is derived from the notion of thought as a process of construction, a notion that ought to be distinguished clearly from the habitual notion in European science of thought as a calculation. In addition to this thesis I want to put forward some considerations about another step which is in my opinion of first rate significance for the formation of the ancient ideal of demonstrativeness. This matter is the opposition of analysis and synthesis, which for the Greeks was much more significant than the opposition of deduction and induction.

4. It is clear why induction is not opposed to deduction by Aristotle but considered (rather in passing) in the second part of his *Analytics* as a special case of syllogism: as ancient scientific thought didn't want to deal with infinities, induction can be identified with syllogism. But why did Aristotle avoid the term "logic", preferring instead of it "analytics"? Are there any other grounds except the ambivalence of the term λόγος which meant both "thought" and "word"? After all λόγος has had a long tradition up to this time! If the ancient image of scientific knowledge is oriented to mental construction, than, it seems, it is more natural to connect it with the notion of synthesis than with analysis.

The terms "analysis" and "synthesis" have in ancient Greek language the same sense as in the modern intuition: analysis is the separation of the whole into parts; synthesis is the joining of parts in a whole. But it is the word "synthesis" that is connected etymologically with the idea of theory: *σῦνθεσις* is derived from the notion of *θέσις*. The joint bringing up of some of "theses" to be considered as a whole (*ςὺν* – "with") was an action which from the point of view of the ancient Greek outlook had to be of much greater importance than a simple "separation of connections", "untieing", "solving" (latin "resolutio": "*ἄν* – *ἄλυσις*" – from "*ἄλυσις*", "chain", "connection").

A strong impact on the ancient outlook was made by Greek writing, which was the first one to realize the phonetical principle. In particular, this paradigm of significant whole, formed of elements, each of which has no significance alone, lies at the bottom of Greek atomism. This paradigm relies upon the image of the connection between words and letters. Of course, in order to make words of letters, it is necessary to divide the word in sounds; but the idea that the world consists of some first elements, *ἀρχαι*, is very ancient alone, and mystery, as it seemed, consists in how the whole is made of *ἀρχαί*.

Original ideal *δείκνυμι* just coincided with mental construction as connecting the parts in a whole by means of verbal explanation, i.e. with *σύνθεσις*. True change comes in the time of Plato and is connected with his opposition to arbitrary explanation – substantiation, characteristic of natural philosophers; the substantiation that excludes certain possibilities and hence explains why the rest is necessary. Analysis precludes splitting a situation in something similar to "possible worlds" with a view of discovering among them the impossible ones and leaving only one possible, i.e. necessary, one. To judge from some ancient testimonies, the analytical method suggested by Plato consisted in accepting the unknown as a known. (See [12].) This means, in a certain sense, the possible, i.e. the unknown, "equality of rights" granting with the actual. From here takes its beginning Aristotle's understanding of necessity as impossibility to be otherwise. First a certain "to be otherwise" is postulated, and then its impossibility is stated. As was shown by J. Hintikka and U. Remes, the Greek analysis, as it was understood as a "resolution", had indeed been a combination of analysis and synthesis. It is interesting, too, that in accepting the unknown as known, according to O. Becker [15], Greeks were acquainted in a yet early period of development of mathematics with a form of semi-arithmetic, semi-geometrical method of solving arithmetic problems with the help of stones, perhaps of different colours and sizes. The high method of analysis has, consequently, its sources in the contemptible technology of counting. Nevertheless, essentially it means becoming aware of the splitting of the world in possible, impossible, and necessary situations, historically connected with the equality of rights of the known and the unknown. Thereby the rule of contraries is realized as a tool of theoretical analysis. This separates the ideal of apodictic necessity of theoretical demonstration from abstract natural-philosophical speculations. In this case the notion of thought as construction goes to background, but remains a ground of the ideal of theoretical knowledge.

But where is the source of Aristotle's dislike to terminology derivative from the word "λόγος"? In his *Analytics* Aristotle explains how logical and dialectical conclusions differ from proper analytical ones in containing unreliable, probable, "verbal" knowledge because they are based only on unanalyzed but commonly recognized theses. So, it seems that analytics is opposed to logic: dialectics is a reflection of an opposition of science to sophistics as verbal art; and as the point of departure in the term "λόγος" the meaning "word" is taken instead of the meaning "thought".

From Socrates undoubtedly begins that turn in the ancient Greek outlook which led to the logico-philosophical realisation of the ideal of analytically necessary knowledge. Alongside with this, Socrates's ideal of demonstrable knowledge looks still quite archaic. Maieutics of Socrates generated in term "induction". (But this is by no means the modern sense of the term: ἐπάγωγί of Socrates is, in essence, a prototype of deduction, because in search of counterexamples Socrates addresses data obviously known to the speakers beforehand.) To divert from it counterexamples taken from experience, we have a number of abstract possibilities – a scheme of the analysis. But it is characteristic that Socrates, according to Xenophon, treated abstract demonstrations in geometry sharply negatively as having only pragmatic, applied functions, the functions of counting. For this thinker first brought the problem of man to the centre of philosophy where its place is logically; and along with it we clearly see here the combination of courageous ideas of Socrates with his general conservatism, in this case in mathematics. This may appear symptomatical, because the Stoic tradition grew from Socratic schools as well as from the traditions of Platonists and peripatetics, but Plato and Aristotle didn't accept the anti-deductivism of Socrates in the sphere of mathematics. Along with this, it is the Stoic tradition that we are obliged to for the term "logic".

In Stoic philosophy we find an advanced semantics and, in particular, an opposition between a sign or name ("significative") and sense ("denoted"). But "λόγος", for the Stoics, is to the same extent "notions" as "words". Basing our views on later European traditions, we frequently modernize the meaning of the term "λόγος" and the ancient ideas of the name and the sense. Meanwhile, the archaic thought opposes "the name" and "the sense" in a way different from the modern one. In ancient Chinese philosophy the opposition of "name" and "business" ("min" and "sin") is associated with the pair "question" – "answer": name is the symbolic expression of certain circumstances, demanding an interpretation (a "business" or "sin") in the same way as a question. Indicating the sense of a given situation is the interpretation, in the same way as an answer (a "name" or "min"). That's why it is deeply false to identify the relation of "min" and "sin" with the relation of subject and predicate of the European tradition, although a distant connection is possible here. The essence of a subject is clarified by the practice of canonic questions and answers in puzzles of mythological content of different people. (For instance, Siberian Evenc-hunters ask: "What's the hole made in the wild deer's skin?". It is necessary to answer: "the sky". First there is a "business", second there is a "name".

Echoes of similar very archaic opposition we found in Heraclitus: "So, the name of the bow – the life, and the business – the death" ([14], fragment 48). The term "λόγος" does not come from an opposition of the inner sense and language expression; in Stoic philosophy, though acquainted with such opposition, λόγος still remains the inseparable "word-thought". Peculiarities of the sense of the term λόγος in the ancient texts were analyzed for the first time by A. Szabó, who came to the conclusion that originally it was a designation of each of two numbers, which were the limits of a διασθημα-interval in musical theory. Hence λόγος is a proportion, or a relation of two extreme numbers (proportion – ἀναλογία – "equality according to logos"). To the original senses of λόγος – "speech", "language", "thought" Szabó adds "number", "a number or community of things". The etymology of the word is related to that of λέγω, which means "to gather", "to choose", "to speak", "to enumerate" deriving from the Indo-European word $*\sqrt{\bar{l}\bar{e}g}$ – "to gather"; from here λόγος – "speech", "story", "intelligence", "calculation" [8]. (Compare Latin – legion = "people, gathered by words of command"). We might say that λόγος means "word-notion" in the sense of "set" (collection). It is the tradition that dominates in Pythagorean and Heraclitean philosophy, that Stoics return to. The term λόγος and the idea of the word-set, developing in the technology of counting and theory of proportions, generates sooner the practical and the theoretical *logistics* ([17], p. 132).

Unlike the traditions of Plato and Aristotle, which were classical for the ancient mind and in a definite way influenced the formation of logico-philosophical paradigms of axiomatic method in Greek science, the Stoic tradition is more connected with practice of rhetoric and logistical practice of calculation. (To nobody among Greek authors, except the Stoics, did it occur to calculate duration of the "great year", which exceeded in duration the number 10,000, which was for the Greek calculators the limit of great numbers!) One can argue with the statement of Łukasiewicz that Aristotle for the first time started using variables in the history of science (in logic); to an extreme extent, in the mind of ancient scientists verbal symbols used in syllogistics didn't differ from those accepted in geometry. As for the Stoics, they created the propositional calculus, where the same words "first", "second", etc. are used that played the role of variable quantities in the mathematics of Diophantos, too. (History of science knows arguments in favour of the first sign of variable ʃ α having its origin not from ἀριθμός "number", but from πρῶτος "the first"; see ([17], p. 146).

So, one can think that the opposition of analytics and logic goes together with the opposition of analysis as a choice between possibilities and verbal art, also with the opposition "theoretical analysis and the art of counting". Traditions of Babylonian algebra didn't fade in ancient Greek mathematics, and in its late period they were advanced by Diophantos. However, they didn't determine the general complexion of scientific thought with axiomatic method characteristic to it. Analogous tendencies in the field of logico-philosophical realizations of the ideal of theoretical thought are represented by the Stoics, whose legacy to the same time it is not characteristic of its main lines of development.

Because ancient influences seem to have reached Rome through the

Stoics more than through other schools and because of the incompetence of Roman translators, in Latin and, particularly in the Western European tradition of the Middle Ages the understanding of differences between logic, dialectics, and analytics inherent in ancient Greek tradition seems to have been lost. In later European university tradition the position is intensified by the fact that in courses of dialectics formal logic was usually taught, while in courses of logic the subject matter was something like epistemological commentaries on it. All this lead to wider understanding of the term 'logic' (for example, in Hegel). But the gist of the matter is the fact that the reception of logico-analytics in European Middle Ages was connected with entirely different general conceptions of knowledge and paradigms of demonstrativeness.

5. To compare ancient ideals of theoretical knowledge with modern ideas of demonstration, we can formulate the following statements.

(a) Even though the analytical method, as the ancients understood it, accepted the supposition of the completed character of certain initial constructions, and stood in opposition to synthesis as 'pure' construction, the main paradigm of thought as construction was preserved. Truly speaking, the idea of analysis alone – which meant to accept the unknown, i.e. the possible, as existing – is in conflict with idea of construction. Aristotle becomes clearly aware of this difficulty: on the one side, analysis must start from existing objects; on the other side, the initial definitions are what opens to us qualities of objects. Hence Aristotle's demand of definitions as constructions which at the same time characterize an object and prove its existence ("definition" as "statement, explaining why the thing is" [18], 93b).

(b) Modern methods for examining the demonstrativeness of scientific constructions are based on the idea of *logical calculations*. Under the incomparably higher capacity of these methods the loss of the "naive" idea of demonstration as construction is rather sad in some respects. It can be mentioned that modern logic with its mathematical ideas differs at very essential points from the original confidence of Hobbes, that thought is in essence the same as addition and substraction: Boole's algebra differs from the "usual" addition and multiplication in that it does not satisfy group axioms. Meanwhile, the notion of construction in a sense similar to the ancient one satisfies the group axioms in a way similar to the notion of the most fundamental operations of theoretical thought. Thus the modern ideals of demonstrativeness in knowledge are compatible with the notions of mental constructions.

(c) Modern notions of analytical truth, which are derived from the idea of logical calculation and which go far beyond the idea of "naive", intuitive character of analysis and synthesis, all still rest on the basis of ancient paradigms. It seems desirable to separate the ideas of the analytical and the synthetical from the ideas of deductive and factual (inductive) knowledge.

So, we can start from an intuition of analysis as a mental action, consisting in the construction of the new properties of given objects, and of synthesis as a construction of new objects with given properties. If the aim of the mental constructions is to build sets, then analytical mental action consists in indicating the way the elements of a given set are formed, or distinguishing sub-set from some basic set by means of forming general properties of the elements of the subset. Synthetical, in this sense, will be mental action which consists in indicating the condition or the parameter (i.e. the property) a given set satisfies. From here, under additional limits we may proceed to the commonly accepted definitions in logic and mathematics.

(d) In ancient science with its axiomatic method, and in ancient logic, which can be named an ideology of the axiomatic method, the rupture between demonstration and explanation did not exist, and a good definition was even considered as an explanation of bases and causes. In modern thought a rupture emerged between demonstration and explanation (description of the observed necessary consequences of certain statements and explanation of their sense). It would be interesting to discover to what extent the ideal of demonstration as logical calculation, which has dominated the mathematized part of modern natural science, is responsible for this.

Ukrainian Academy of Sciences

REFERENCES

[1] Remes, Unto: 1974, 'Ancient geometrical analysis and its use as a methodological model in antiquity,' in *Reports from the Institute of Philosophy*, University of Helsinki, No. 4.

[2] Hintikka, K. J. J. and U. Remes: 1974, *The Method of Analysis: Its Geometrical Origin and Its General Significance*, Boston Studies in the Philosophy of Science, Vol. 25, D. Reidel, Dordrecht–Boston.

[3] Кесссиди: Ф. Х.: 1972, От мифа к погосу, М.,
[4] Lethaby, W. R.: 1956, *Architecture, Nature and Magic*, London.
[5] Lethaby, W. R.: 1957, *Form and in Analysation*, London.
[6] Лелеков, Л. А.: 1976, Отражение некоторнх мифологических воззрений в архитектуре восточноиранских народов в первой половине I тысячелетия до Н.э. 'История и культура народов Средней Азии'. М.
[7] Герценберг, Л. Г.: 1972, Морфологичеекая структура слова в древних инлоирансних языках. Л.
[8] Boisacq, E.: 1938, *Dictionnaire Etymologique de la Langue Grecque*, 3 ed., Paris.
[9] Szabó, A.: 1969, *Anfänge der Griechischen Mathematik*, Budapest.
[10] Каган, В. Ф.: 1963, Очерки по геометрии. М.
[11] Плутарх.: 1961, Сравнительные жизнеописания. Т.I.М.
[12] Auerbach, M.: 'Platon a matematyka grecka', Lwów, *Odbitki z kwartalnika klasycznego* **6**.
[13] Becker, O.: 1957, *Das mathematische Denken des Antike*, Göttingen.
[14] Heraclitos von Ephesos: 1909, Berlin, Diels, 2, Aufg.
[15] Егāнян, А. М.: 1972, Греческая логистика. Ереван.
[16] Аристотель,: 1978, Аналитики I и П. Сочинения в 4-х томах. Т. 2, М.

KAREL BERKA

WAS THERE AN ELEATIC BACKGROUND TO PRE-EUCLIDEAN MATHEMATICS?

The basic ideas of a recent reconstruction concerned with the origin of deductive mathematics in the 6th and 5th centuries (cp. [9]–[12]) can be summarized briefly as follows: In its very beginning, there existed in Greece only a purely practical, empirical mathematics. The transformation of this empirical mathematics into theoretical mathematics, developed already in the pre-Euclidean period, could be achieved only by means of extra-mathematical reasons. This change, which resulted in the elaboration of Euclid's *Elements*, was, in its principal features, determined by the philosophy of two Eleatics: Parmenides and Zeno. Mathematics was, therefore, at least initially, a branch or an inherent part of dialectics.

These ideas are supported by various arguments. Some of them, because of the lack of sufficient historical sources, are a matter of discussion, whereas others, as I shall attempt to show, are implausible or wrong. From what will follow, one might, of course, conclude that my approach is too destructive. However, taking into account that the known historical sources do not give us enough information about the development of mathematics in the discussed period, it seems to me to be more appropriate to abstain from any polished explanation than to suggest an unwarranted theory.

Initially, I could simply refute the core of the view that without the philosophy of Parmenides and Zeno it would never be possible to construct such an elaborated system as that of Euclid's *Elements* ([12], 291) by asserting that Parmenides was younger than Pythagoras, and that he was originally a Pythagorean ([1], 311). Assuming that Pythagoras and the older Pythagoreans lived before Parmenides, it is obvious that their mathematical investigations could not have been influenced by the philosophy of the Eleatics. This view is widely accepted. Scholars such as Reidemeister claim that the Pythagoreans are the founders of scientific mathematics ([7], 52), D. J. Struik connects the origin of theoretical mathematics with the Pythagoreans and the Sophists ([8], ch. 3), K. v. Fritz ascribes to the Pythagoreans

J. Hintikka, D. Gruender, and E. Agazzi (eds.), Pisa Conference Proceedings, Vol. I, 125–131.

the recognition of incommensurability, which was evidently a very strong impulse for foundational studies in early Greek mathematics ([3], 91 etc.). This conception is supported by fragment 6 from Proclus which states that Pythagoras transformed geometry into a science by studying its foundations (archas) and by formulating its theorems in an abstract way. Even Szabó does not deny the abstract nature of the mathematics of the older Pythagoreans and admits that their teaching contained in Eucl. IX 21–36, X App. 27 is the oldest known deductive theory of Greek mathematics ([11], 121). In spite of this statement, he claims that the "highly developed logic" well-known to the Pythagoreans of the 5th century was taken over from the Eleatics ([11], 148) and that the Pythagoreans themselves could not develop mathematical investigations without their help.

Similarly, if Heracleitos preceded Parmenides ([11], 171; [4], 50), we have to adopt another explanation of the role of the Eleatics in the history of dialectics.

This counter-argument based principally on another chronology can be naturally questioned, since the known evidence is not so convincing as to decide with certainty that Parmenides preceded Pythagoras. This holds, of course, for the opposite chronology as well. Szabó, however, takes, his own view for granted without any doubts, because otherwise his claim that Pythagoras and the older Pythagoreans could not initiate the change from empirical mathematics to theoretical, would lose its strength.

Can we assert that Greek mathematics before Parmenides was just an empirical mathematics: only a sum of empirical knowledge? Can we make a sharp distinction especially in the case of mathematics between what can be understood as "empirical" and what as "theoretical"? I think that from the very moment when at least two different empirical problems had been handled by the same mathematical means, abstraction and, hence, theoretical deliberations were already present. That Greek mathematics in the pre-Eleatic period was not purely empirical is tacitly contained in the statement ([11], 115) that some rules for solving certain mathematical problems had been formulated in this period already. This does not imply conscious or systematic work in the field of the foundations of mathematics. It cannot be doubted that deductive systematization and conscious metatheoretical consideration presuppose a great amount of theoretical knowledge. That theory precedes metatheory is a fact well-known from history of science: an exact axiomatization of a new mathematical theory, as K. v. Fritz affirms ([3], 21), was never achieved before the theory was rather widely developed. The theory of mathematics in the pre-Eleatic period, as I suppose, was rather the work of mathematicians themselves. There are hardly any examples in the history of mathematics that philosophers who were not at the same time mathematicians have positively influenced the development of mathematics.

Szabó denies this view and argues that the Pythagorean mathematicians, who probably became interested even in foundational studies, could not develop the means necessary for this task. I think that it is necessary to admit, at least, one important circumstance which could have forced them to work in this direction as well: namely their discovery of incommensurability. Let me again draw a lesson from the history of science: scientists use methods and theories which are successful in their work, and only in the event something goes wrong do they feel the need of metatheoretical investigations. These investigations are naturally connected with general philosophical reflections.

Is Greek mathematics basically dependent on Parmenides' philosophy and methodology? Because a positive answer is connected with the problem of the relationship between mathematics and logic in antiquity, I shall make some remarks on this topic.

Szabó assumes that logic is prior to mathematics in general, and concludes from this opinion the priority of logic in the historical development as well. I need not point out that the question of priority, from the systematical point of view, is answered in the philosophy of mathematics differently. Neither from the historical point of view can an unambiguous answer be given. Taking into account that Aristotelian logic does not contain a logic of relations, although that is in fact the underlying logic of Euclid's *Elements*, and that no categorical syllogisms are utilized in Euclid's proofs; it seems very plausible that Aristotelian logic and the codification of ancient geometry are basically independent of each other ([5], 66). For our purposes it is, of course, more important to focus attention on what Szabó says about logic itself. Assuming that the use of indirect proof presupposes the principle of *exclusii tertii* ([11], 143), which is one of his basic arguments, not only in favor of the priority of logic, but also for the dialectical origin of deductive mathematics, he affirms that "the historical development of logic starts with the appearance of the indirect proof" ([11], 148). From the assumption that the indirect proof was developed by the Eleatics, he concludes that they were not only the inventors of dialectics, as Aristotle says about Zeno (having in mind the "art of argumentation" in contradistinction to dialectics as a philosophical method which he attributes to Plato), but of logic as well ([9], 57, [10], 283).

This conception confuses the difference between what might be called the "logic of thinking" and logic as a science. For the history of the logic of thinking there is no known beginning. The property of thinking implies some kind of logical pattern which, under certain historical conditions became consciously reflected upon and gradually developed into the science of logic. E.g. the principle of contradiction is objectively determined by the everyday experience that something cannot be in one place and at the same time in another, that one cannot attribute and, at the same time, deny a property to a certain thing, etc. The intuitive use of the principle of contradiction and, analogously, the utilization of the principle of *tertium non datur* do not imply that there was already elaborated some theory of logic. The origin of logic as a science begins with the systematic analysis of logical forms.

There were before Aristotle various anticipations. The historical sources confirm that Gorgias (cf. fr. B2 from Sextus) formulated very precisely – in contradistinction to what we find in the poem of Parmenides – the principle of contradiction. Nevertheless, I shall avoid speaking about the logic of the Sophists, just as I object to the phrase "logic of the Eleatics".

According to my view, it is not necessary to assume some developed logic as a background of the practical use of indirect proof. Further, the procedure used in dialectics, eristics, and rhetoric was rather a *reductio ad absurdum* which "shows the falsity of an assumption by deriving from it a manifest absurdity", then an indirect proof which "establishes the truth of an assertion by showing the falsity of the opposite assumption" ([6], 162). Because of the difference between the practice of the dialecticians when refuting the theses of their opponents, and the indirect proof as applied in mathematics, there is strong indirect evidence for the assumption that the indirect proof was rather developed by the mathematicians themselves. Szabó denies this possibility ([12], 292). But how is one to explain that it was Parmenides who discovered and developed this proof-method? For this crucial question he finds only a very unsatisfactory explanation: the method of indirect proof arose from the criticism of the cosmogony of Anaximenes with respect to the cosmogony of the Milesian philosophers ([12], 292, 336).

If Greek mathematics had followed Zeno, Euclid and his predecessors – maybe Hippocrates of Chios or Theudios of Magnesia – would never have developed mathematics as a deductive system. This insight was, for Aristotle, a fact beyond any doubt. The logical background of Zeno's paradoxes could not serve as an underlying logic for any deductive system. His paradoxes and analogously those of the Chinese dialectician Kung-sun Lung (4th century B.C.), who formulated, among others – independently of Zeno – the Paradox of the Arrow, represent the prehistory of logic as a science. The arguments of the Sophists and the origin of paradoxes in this period are one of the relevant impulses that led to the analysis of logical forms and – under further conditions found only in ancient Greece – to the elaboration of the science of logic by Aristotle.

Let me quote in this connection G. Vlastos ([13], 377) who stresses that "there is nothing in our sources that states or implies that any development in Greek mathematics (as distinct from philosophical opinions about mathematics) was due to Zeno's influence". Vlastos is even more sceptical about Parmenides' place in the history of logic. According to his opinion, the poem of Parmenides could not offer a logical theory.

This becomes obvious when reading the extant text of the poem without prejudice. The test of the poem related to the two basic "ways of truth" ([2] fr. 4, 5) – "*hopós estin te kai hós uk esti mé einai*" (the first one) and "*hós uk estin te kai hós chreón esti mé einai*" (the second one) – is translated variously. Indeed, the interpretation of the "estin" in this text is a question of dispute, exemplified by the following translations: "*dass das Seiende ist und dass es unmöglich nicht sein kann*" – "*dass es nicht ist und dass dies Nichtsein notwendig sei*" ([2], ed. 1912); "*dass IST ist und dass Nichtsein nicht ist*" – "*dass NICHT IST ist und dass Nichtsein erforderlich ist*" ([2], ed. 1971); that *It is*, and that it is impossible for it not to be" – "That *It is not*, and that it must needs not be" ([1], 1973); "that it *is* and cannot not be" – "that it *is not* and must necessarily not be" ([4], 48). The third way, consisting in a conjunction of the first and the second one "is ruled out by pointing to an alleged contradiction in it" ([4], 48).

Szabó quotes the three ways in an already interpreted manner, namely [1] "*to on esti*" respectively "*to mé on uk esti*", [2] "*to on uk esti*" respectively "*to mé on esti*" and [3] "*to on esti kai uk esti*" ([9], 54), and uses the following translation: [1] "*das Seiende ist*", [2] "*das Seiende ist nicht*" and [3] "*das Seiende ist und ist auch nicht*" ([11], 149). He takes it for granted that the three ways are the three basic principles of logic in its oldest pre-aristotelian form: namely the *principium identitatis*, the *principium contradictionis*, and the *principium exclusii tertii* ([9], 57, similarly [11], 109, 149). This interpretation is wrong. There is no principle of identity as "*arché*" in Aristotle's metaphysics or logic, the formulation of the third way, in fact, violates the principle of contradiction, and to maintain that it is the principle of excluded middle, is evidently an elementary logical error.

There is a lot of confusion even in his conception of dialectics. On one side, he says "that the dialectics of the Eleatics is nothing else than the utilization of indirect proof" ([12], 291) and that dialectics – as understood in the context of his monograph – is the art (*Kunst*) of argumentation ([12], 313, note 74, similarly 365); but on the other one he speaks about the "dialectics of thinking" ([9], 57; [10], 283) and considers dialectics – especially in comparison with mathematics – as a science ([12], 329). Now, there is a well-known difference between "art" and "science" in Greek philosophy, and dialectics was not understood as a science – as "*epistémé*". It is, therefore, unjustified to speak about technical terms in dialectics. It is wrong to consider mathematics as a branch, as a later developed special domain of the older dialectics ([12], 320, 329, 345). Were this true, mathematics could be equally well considered as a special kind of rhetoric or eristics simply because they deal with argumentation as well.

All these failures cannot be compensated by very interesting etymological analyses, which have to prove the dependency of the basic metamathematical terms on terms used in dialectical argumentation. Neither is this argument convincing. It is a truism that all technical terms in Greek philosophy and science were coined by analogy, by a shift in meaning, etc. from the everyday Greek language. Similarly, we have to admit the transfer of terms from one domain to another one. What is important is the adopted meaning of the given term, its definitory codification, or its explication. The discrepancy in the terminology of Euclid, Aristotle and Proclus clearly exemplifies that the metamathematical terminology in ancient Greece was not fixed. Because of the terminological evolution in the work of one and the same author, it is sometimes very difficult to elucidate the intended meaning of a given term. We cannot, therefore, draw relevant conclusions from the current usage of words. Neither is it appropriate to derive from the similarity of words used in different contexts the conclusion that they have a similar meaning or that this similarity implies some conceptual relationship.

To conclude, I want once more to emphasize that any modern reconstruction of pre-Euclidean mathematics cannot present in respect to our fragmentary sources a unified picture of what happened or might have happened in the given period. In any case, the ambitious prospect of a new explanation must not sacrifice consistency with historical facts and confirmed theoretical interpretations.

Czechoslovak Academy of Sciences

REFERENCES

[1] Burnet, J.: 1930, *Early Greek Philosophy*, London.

[2] Diels, H. and Kranz, W.: 1912, *Fragmente der Vorsokratiker*, Berlin (Zürich–Dublin, 1971).

[3] Fritz, K. v.: 1955, *Die ARXAI in der griechischen Mathematik*, Archiv für Begriffsgeschichte, Bonn, Bd. I, pp. 13–103.

[4] Furley, D.J.: 1967 'Paramenides of Elea', *The Encyclopedia of Philosophy*, Vol. 6, The Macmillan Comp. & The Free Press, New York and London, pp. 47–51.

[5] Mueller, I.: 1974, 'Greek mathematics and Greek logic', in J. Corcoran (ed.), *Ancient Logic and Its Modern Interpretations*, D. Reidel, Dordrecht/Boston, pp. 35–70.

[6] Polya, C.: 1957, *How to Solve It*, Doubleday, Garden City.

[7] Reidemeister, K.: 1949, *Das exakte Denken der Griechen*, Hamburg.

[8] Struik, D.J., 1956, *A Concise History of Mathematics*, London.

[9] Szabó, A.: 1954, Zur Geschichte der Dialektik des Denkens', *Acta Antiqua* **2**, pp. 17–62.

[10] Szabó, A.: 1954, 'Zum Verständniss der Eleaten?' *Acta Antiqua* **2**, pp. 243–289.

[11] Szabó, A.: 1956, 'Wie ist die Mathematik zu einer deduktiven Wissenschaft geworden?' *Acta Antiqua* **4**, pp. 109–152.

[12] Szabó, A.: 1969, *Anfänge der griechischen Mathematik*, Akadimiai Kiadó, Budapest.

[13] Vlastos, G.: 1967, 'Zeno of Elea,' *The Encyclopedia of Philosophy*, Vol. 8, New York and London, pp. 369–379.

JAAKKO HINTIKKA

ARISTOTELIAN AXIOMATICS AND GEOMETRICAL AXIOMATICS

Professor Szabó deserves credit for calling our attention to the interplay of philosophical and mathematical influences in the development of Greek axiomatics. It is this interplay that lends a special flavor to much of the early as well as some of the later history of the axiomatic method. I believe, however, that in the last analysis the total picture of the early development of axiomatics will turn out to be quite different from the one Szabó paints. My reasons for this belief are nevertheless subtler than one might first expect. Professor Szabó finds the true ancestors of the central mathematical methodology of the Greeks, including the axiomatic method, in the Eleatic dialectic. In so doing, Szabó *prima facie* misses a large part of the interdisciplinary interplay with which he is dealing. Most other historians of the axiomatic method would give the pride of place on the philosophical side of the fence to Aristotle, who is sometimes called the first great theoretician of the axiomatic method and whose ideal of a science was by any account explicitly and self-consciously axiomatic. Szabó admittedly discusses Aristotle, but gives the Stagirite short shrift, dismissing him as having played no real part in the development of the axiomatic methods actually used in mathematics.

Even if one believes that Aristotle's actual influence on mathematics was negligible, his views seem to merit close attention by all historians of the axiomatic method. The importance of these views lies of course in the fact that Aristotelian axiomatics is by far the most fully developed object of comparison on the philosophical side with the uses of axiomatic method in Greek mathematics. Szabó's procedure in downgrading Aristotle's role might thus seem unfortunate. Isn't he throwing by the board one of the most important sources of the very development he is dealing with?

It appears that some of Szabó's arguments for assigning Aristotle to the historical limbo he occupies in Szabó's story are in fact mistaken. Other pieces of evidence he presents are ambivalent, but not nearly as strong grounds for indicting Aristotle as Szabó seems to think.

J. Hintikka, D. Gruender, and E. Agazzi (eds.), Pisa Conference Proceedings, Vol. I, 133–144.

Without trying to present an alternative total picture of the development of the axiomatic method, I will begin by presenting a number of corrections to Professor Szabó's account, mostly to what he says about Aristotle.

All page references not otherwise specified will be to Arpad Szabó, *The Beginnings of Greek Mathematics*, D. Reidel, Dordrecht, 1978 (translation of *Anfänge der griechischen Mathematik*, R. Oldenbourg Verlag, München and Wien, 1969), or to Jaakko Hintikka, 'On the ingredients of an Aristotelian science,' *Nous* **6** (1972), 55–69.

(i) Szabó claims (p. 188) that the nontechnical meaning of *deiknymi* as "sichtbarmachen" is relevant to mathematical and philosophical usage. This claim is made dubious by the fact that Aristotle's use of *deixis*, *deiknymi*, and related terms is already quite sophisticated. As usual, Aristotle is not consistent, and does not use strict technical terms. However, there is unmistakable contrast in Aristotle between *deixis* and *apodeixis*. (See Hintikka, pp. 56–57.) Typically, the former could be used by Aristotle of all and sundry persuasive "showings," while the latter was used of logical (syllogistic) inferences from appropriate permises. If further evidence is needed, one of the clearest examples of the Aristotelian distinction is found in *Post. An.* *II*, 7, 92a34–b4, where *deixis* is the more general notion which comprises as special cases at least *apodeixis*, *epagoge*, and *aisthesis*. What some of the requirements of *apodeixis* are according to Aristotle is seen from *Post. An.* *I*, 2, 71b20–25. There is a strong, albeit tacit, assumption in the latter passage that a syllogistic proof is a necessary condition of an argument's being an *apodeixis*, even though Aristotle denies it is a *sufficient* condition. It is eminently clear that a syllogistic proof cannot be a necessary condition for *deixis* for Aristotle. (It is interesting to see that the demonstrative part of a Euclidean proposition was typically called in the late ancient mathematical and philosophical usage *apodeixis*, not *deixis*.) The Aristotelian evidence thus strongly suggests that the two terms *deixis* and *apodeixis* were fairly clearly separated from each other at an early stage of the development of axiomatics, and the presystematic connotations of *deiknymi* are therefore scarcely valid evidence concerning mathematical and philosophical ideas of demonstration at the time of Aristotle, let alone Euclid.

Even though the precise meanings of *deixis* and *apodeixis* are not among the reasons why Szabó downgrades Aristotle, my observation shows that there is more in Aristotle than first meets Szabó's eye. In general, Szabó pays little attention to the interesting evidence for connections between Aristotle's terminology and Greek mathematics marshalled in B. Einarson, 'On certain mathematical terms in Aristotle's logic,' *American Journal of Philology* **57** (1936), 33–54 and 151–172.

(ii) An important point is missed by Professor Szabó when he dismisses a suggested comparison between Euclid's postulates and Aristotle's hypotheses by saying (p. 230, cf. p. 307 of the German ed.) that Euclid's postulates are not existential propositions, as Aristotle's hypotheses are supposed to be. This is doubly misleading. For one thing, it was claimed already in antiquity that the so-called postulates were indeed existential assumptions. (For evidence and discussion, see E. Niebel, *Über die Bedeutung der geometrischen Konstruktion in der Antike* (*Kantstudien, Ergänzungshefte,* vol. 76), Kölner Universitäts–Verlag, Cologne, 1959.) What is even more important is the striking fact that hypotheses were *not* claimed by Aristotle to be existential statements. When Aristotle speaks (in the passage referred to by Szabo and elsewhere) of "being or not-being" in characterizing hypotheses, he does not mean "existing or not existing." His phrase comprises equally *both* "existing or not existing" *and* "being or not being thus." (For evidence, see Hintikka, pp. 67–68; Richard Robinson, *Plato's Earlier Dialectic*, p. 101; and Charles Kahn, *The Verb "Be" in Ancient Greek*, D. Reidel, Dordrecht, 1973.) In general there is no trace in Aristotle of any real distinction between the "is" of predication and the "is" of existence. Both ideas are included in the Aristotelian *esti.*

That *einai* did not have an exclusively predicative sense in Aristotle is perhaps seen easiest from *Post. An.* II, 2–3. The same chapters show that Aristotle did not distinguish between the existential and predicative meanings of *einai* from each other at all. (For a general perspective on this issue, cf. also my paper ' "Is", semantical games, and semantical relativity,' *Journal of Philosophical Logic*, **8** (1979), 433–468.

(iii) One of the focal points of Szabó's book is his claim that the method of indirect proof (*reductio ad absurdum*) originated from the Eleatic dialectic and that it played an important role in the development of axiomatics. These are interesting claims which deserve a longer treatment than I can devote to them here. In this paper, I merely want to call attention to the fact that there is a lot of material in Aristotle that is relevant to Szabó's claims one way or other. In some ways, this Aristotelian material is likely to strengthen Szabó's hand in that it shows some connecting links between the Eleatic dialecticians and the later mathematical and philosophical ideas. At the same time, the Aristotelian evidence strikingly belies Szabó's cavalier dismissal of the Stagirite.

The evidence is not unambiguous, either. If we take Szabó's first claim first, it is relevant to note that the connection between reductive proofs and the dialectical method is very unclear in Aristotle. In any case, we (including Szabó) should not underestimate the tremendous importance of procedures that must be classified as dialectical or direct descendants of dialectical procedures in Aristotle. By these I mean of course the procedures Aristotle envisages for establishing the first premises of any one axiomatically built science. Are these dialectical methods connected by Aristotle with the idea of indirect proof? Aristotle several times connects the idea of a reductive proof with his definition of possibility. (See *An. Pr. I*, 13, 32a18ff.) According to the definition, what is possible must be capable of being assumed without any absurdities resulting therefrom. Hence, if the diagonal of a square is not necessarily incommensurable with the side, we must be able to assume that it has been measured commensurately with the side. But this quickly leads to a contradiction. (Cf. *Met. IX*, 4.) Hence the diagonal of a square is necessarily incommensurate with the side.

Can we find in these Aristotelian ideas traces of dialectic, Eleatic or otherwise? I cannot answer the question here to my own satisfaction. What I consider clear is that a connection cannot easily be excluded. For instance, indirect proofs are discussed by Aristotle in *Met. IX* in connection with Megarian views of determinism and denial of change. This already is something of a link between Eleatic dialectic and Aristotle, even though Aristotle clearly thinks of himself as opposing the Megarians. Elsewhere, I have shown that in spite of his avowed opposition, Aristotle came very close to sharing some of the key assumptions of the Megarians; see *Aristotle on Modality and Determinism* (Acta Philosophica Fennica, vol. 29, no. 1), North-Holland, Amsterdam, 1977. It is clearly here that the best evidence can undoubtedly be found for a continuous development from Eleatic dialectic to a full-fledged method of indirect proof.

(iv) Likewise it is not easy to see if there is any evidence to connect Aristotle's ideas of indirect proof with his axiomatic method. On most occasions, there does not seem to be any connection. However, *Phys. III*, 8, 200a14–29 suggests that Aristotle was aware of the connection between indirect proofs and geometrical axiomatics. Further inquiry is needed to see whether we can gather here any hard evidence for Szabó.

(v) In general, Szabó postulates far too readily influences by one thinker on another. Frequently similarities which apparently indicate influences are in reality symptoms of shared presuppositions. For instance, the Platonic assumptions concerning knowledge, ignorance, and opinion which Szabó mentions on p. 309 and following (pp. 422 ff, of the German ed.) can be viewed as consequences of a shared model of knowledge as a goal-directed process rather than on any influence in the usual sense of the word. (See on this shared model Hintikka, 'Knowledge and its objects in Plato,' in J.M.E. Moravcsik (ed.), *Patterns in Plato's Thought*, or in Hintikka, *Knowledge and the Known*, D. Reidel, Dordrecht, 1974.)

In the same way, several of the observations Szabó adduces as showing Eleatic influence on the development of mathematical argumentation or as showing Aristotle's isolation from this alleged mainstream development allow for different explanations. Indeed, Szabó is in many respects neglecting Aristotle quite unjustifiably. There is nevertheless one respect in which Szabó is in my judgement arguably right in his treatment of Aristotle. Aristotle is a good witness to the developments we are here interested in, and he had an interesting and in many ways sophisticated conception of the axiomatic method. Moreover, his ideas were undoubtedly influenced by mathematical paradigms. However, there is a very real sense in which Aristotle's conception of an axiomatically built science is foreign to real mathematical methods of argumentation. Aristotle believed that the tool by means of which all the deductions needed in an axiomatic science are effected (as we would say in our twentieth-century terminology, by means of which theorems are derived from axioms) is his syllogistic logic. This assumption colors Aristotle's whole philosophical theory of the structure of an axiomatically constructed theory (a science), as it is presented in the *Posterior Analytics*. It alienated him from mathematical practice, however, and led him to ideas quite foreign to what we are likely to find in mathematical axiomatics. Not only is it the case that syllogistic logic is insufficient to capture typically mathematical inferences. Since Aristotle's whole set of ideas concerning an axiomatically organized science is based on the syllogistic model, this model affects his theory of the structure of a science and not only his views of the individual deductive steps. In particular, Aristotle's syllogistic paradigm prejudges strongly his doctrine as to what the different starting-points (axioms and other primitive assumptions) of a science are. In my paper, 'On the ingredients of an Aristotelian science,' I have shown what the theory is to which Aristotle was led by his syllogistic paradigm. I argued in that paper that there are four kinds of primitive assumptions in any one axiomatically built science *apud* Aristotle:

(1) 'Axioms' (in Aristotle's narrow sense) common to all sciences.
(2) Atomic premises connecting two adjacent terms figuring in a minimal syllogistic premise.
(3) The widest premise of the science in question.
(4) Nominal definitions.

The need of postulating (2) follows from the fact that the most primitive premises of syllogistic inferences are the ones which postulate connections between terms closest to each other. The special status of (3) is due to the fact that only the existence of the widest term figuring in a science has to be postulated. The existence of everything else in the field of that science can be proved syllogistically. This widest term thus serves to define the field of the science in question. For details and for evidence, I must refer the reader to my *Noûs* Paper.

The syllogistically rather than mathematically motivated character of an axiomatic theory according to Aristotle probably explains Szabó's low opinion of Aristotle's significance. He does not seem to be aware of the letter of Aristotle's theory, as I argued earlier in this paper, but he may have caught some of its spirit. At least he is in this particular respect more perceptive than the scholars who have tried to understand the details of Aristotle's theory of science by starting from its alleged mathematical models. Even though Aristotle obviously also tried to push the mathematical sciences to the Procrustean or perhaps rather syllogistic bed of his theory of science (for instance, a generous portion of his examples are borrowed from mathematical sciences), in most cases they measure up very badly to his exacting model. Perhaps the most striking example is the important fact that in mathematics a surprising number of different conclusions can be proved from a small number of assumptions. Yet Aristotle acknowledges in so many words that in scientific syllogisms "the starting points (*arkhai*) are not much fewer in number than the conclusions" (*Post. An. I*, 32, 88b 4–5.)

There are nevertheless some features of Aristotle's theory of axiomatic science that seem to be better explainable by reference to mathematical practice than to the syllogistic model. Such features can also be used to throw light on the development of mathematical methods and modes of theorizing. Later, I shall give a couple of examples of such features.

Above, I said that the syllogistic character of Aristotle's ideal of an axiomatic theory probably explains Szabó's dismissal of Aristotle. This does not yet prejudice the question whether the character of Aristotle's theory justifies Szabó. It seems to me that we can in fact use Aristotelian evidence somewhat more extensively than Szabó does for charting and evaluating the history of Greek mathematical methodology. But in order to do so, we have to understand Aristotle better than has been accomplished so far in the literature. We have to understand Aristotle so well as to know which aspects of his doctrines are due to the syllogistic model and which ones are based on mathematical models. We must be able as it were to subtract the former from the totality of Aristotelian views in order to be able to reach the latter.

I believe that this can be done. In this paper, I can only outline a way of doing so, but I cannot trace all the consequences of the ideas I am about to outline. The fourfold distinction between different *arkhai* mentioned above and defended in my *Noûs* paper is a necessary first step. It must nevertheless be supplemented by insights I did not yet have when I wrote the earlier paper. There are in fact two important further features of Aristotle's theory which were not dealt with in my earlier paper and which seem to me to carry essentially further the line of interpretation begun in my 1972 paper.

(a) Aristotle has two senses of definition (*horismos*), the narrower and the wider one. The former restricts definitions to the "nominal" definitions (4) in the list (1)–(4) above. In the latter sense, not only (4) but also (2)–(3) are called by him definitions. Aristotle relies on the narrower sense in most of *An. Post. I* but on the wider one in most of *An. Post. II.* Hence Aristotle's discussion of the different kinds of definitions (and of the way these different kinds of definitions are reached) in *An Post. II*, 4–10 is in effect a discussion of the different ingredients (2)–(4) of axiomatic theories. The following comparison illustrates this parellelism.

An. Post. II, 7–10

. . . in some cases what a thing is is immediate (*amesa*) and a principle (*arkhai*); and here one must suppose, or make otherwise apparent, both that they are and what they are (which is what the arithmetician does, for he supposes both what the unit is and that it is). . .

but in those cases which have a middle term . . . one can . . . make them clear through a demonstration (*apodeixis*), but not by demonstrating (*apodeiknymi*) what they are (93b21–28).

The definition of immediate (*ameson*) terms is an indemonstrable (*anapodeiktos*) assumption of what they are (94a9–10).

One definition therefore is indemonstrable account of what a thing is; one deduction of what it is, differing in aspect from demonstration (94a10–13).

This is what sciences actually do for the geometrician assumes what "triangle" signifies (*semainei*) but proves that a triangle is (92b15–16).

An. Post. I, 10

Proper [to each science] too are the things which are assumed to be, about which the science considers what belongs to them in themselves, as e.g. arithmetic is about units and geometry about points and lines. For they assume these to be and to be *this*.

As to what are attributes of these in themselves, they assume what each signifies—e.g., arithmetic assumes what odd or even signifies, and geometry what irrational or inflection or verging signifies—and they show that they are (*hoti d'esti deiknyousi*) through the common items (*ta koina*) and from what has been demonstrated (76b3–11).

I call principles in each genus what cannot be proved (*me endekhetai deixai*) to be. Now what both primitives (*ta prota*) and the things dependent on them signify (*semainei*) is assumed; but that they are must be assumed for the principles (*arkhai*), and proved (*deiknymi*) for the rest. For instance, we must assume what unit or straight, and triangle signify, but that the unit and geometrical magnitude are; but we must show (*deiknymi*) that the others are (76a31–36).

In both columns Aristotle is clearly discussing the same thing, viz. the distinction between (2) and (3). (The translations are adapted from Jonathan Barnes' translation in the Clarendon Aristotle Series.)

If further evidence of the wide sense of *horismos* in *An. Post. II* is required, we can note that in *An. Post. II*, 3 90b24–28 Aristotle twice asserts that the starting-points (*arkhai, ta prota*) of a demonstration (*apodeixis*) are definitions (*horismoi*). This makes sense only if Aristotle is here considering (2)–(3) and not only (4) as definitions.[1]

(b) The other main additional insight needed here is connected with our earlier observation in (ii) to the effect that Aristotle never really distinguished the existential and the predicative force of 'is'. Hence the status of (3) as the only unprovable existential assumption of a scientific theory *ipso facto* meant that they were also in some sense the only unprovable predicative assumptions of a science. However, this unprovability cannot any longer mean unprovability in the sense that refers to *apodeixis* (cf. (i) above). It will have to mean impossibility to show the truth of (3) in the sense of some other variety of *deixis* than *apodeixis*.

Hence the question arises: Is there in Aristotle some nonsyllogistic sense of proving or 'showing' in which the widest (and possibly also the narrowest) premise of each science, and only they, cannot be proved by the normal means that science operates with? Now there is indeed in Aristotle a variant of non-syllogistic showing that fills the bill neatly. That is the procedure he calls *epagoge* in *An. Pr. II*, 23. Since it involves inverting one of the two premises of a syllogism, it involves terms that are *prima facie* wider or narrower than the extremes of the three syllogistic terms in question. Accordingly, it cannot be used to justify the widest premises (3) of an axiomatic science.[2] They are therefore unprovable in a double sense for Aristotle. They are the only premises of an axiomatic science that carry irreducible existential assumptions, and they must (possibly together with the narrowest premises) be justified by direct intuition (*noûs*) rather than by the systematic procedure of *epagoge* or 'induction' through which atomic premises (2) are established.

Moreover, the inversion of a syllogistic premise which is involved in *epagoge* according to Aristotle seems to be just what is involved in reaching definitory atomic premises according to Aristotle's account in *An. Post. II*, 8–10. The two can be shown to be structurally similar processes.

Thus we can summarize the ways in which the different kinds of basic ingredients of an Aristotelian science come to play:

(1) Common assumptions (axioms) have to be known before any knowledge can be acquired.

(2) Atomic premises are known through the particular kind of *deixis* which Aristotle sometimes calls *epagoge*. The same procedure is described in *An. Post. II*, 4–10.

(3) The generic premise: A generic premise is the widest premise of a science. It is known through *noûs*. The process of coming to know it is described by Aristotle in *An. Post. II*, 19.

(4) Nominal definitions are arrived at by stipulation.

What does all this entail for the early history of the axiomatic method in general and for Professor Szabó's views in particular? It is tempting to try to draw a parallel between Aristotle and Euclid as follows:

	Euclid	*Aristotle*
(1)	*koinai ennoiai* (common notions)	general axioms
(2)	definitions	atomic premises
(3)	postulates	generic premises

The main difference between this way of comparing Aristotle and Euclid to each other and most earlier comparisons is the Aristotelian analogue I am assigning to Euclid's definitions. Usually they are compared with what Aristotle calls definitions in *An. Post. I* (especially *I*, 10). It seems to me much more illuminating to try to assimilate them to what I have called atomic premises. If I am right, atomic premises are in fact called by Aristotle definitions, viz. in *An. Post. II*. Hence the sample evidence I gave earlier for Aristotle's dual usage in speaking of definitions also serves as indirect evidence for my Aristotle–Euclid comparison. Further evidence is likewise forthcoming; see for instance the passage quoted above from *An. Post. I*, 10, 76a31–36; *An. Post. II*, 3, 90b24–25; *Eth. Nic. VI*, 8, 1142a26 ff.

It also appears that the role of definitions was also on the mathematical side more important in antiquity than it is in modern mathematics. Some evidence concerning the way ancient mathematicians looked upon their own basic assumptions and upon the role of the definitions of the basic mathematical concepts is obtained from the criticisms levelled at Euclid's fifth postulate in antiquity. The gist of many of these criticisms is that Euclid's postulate does not give the essence of the straight line – or of any other geometrical concept, for that matter. (For instance, otherwise the objection that the converse of the fifth postulate is provable would not make much sense.) In slightly different words, the complaint was that the fifth postulate could not be conceived of as a definition.

My analogy between Aristotle and Euclid is inevitably only an approximate one, for reasons explained earlier. Even though I cannot examine the putative analogy as carefully as it deserves, the following failures can be registered here:

(i) Definitions do not typically operate as premises from which conclusions are drawn in Euclid, as definitions in the sense of class (2) have to do in Aristotle. Instead, the *apodeixis* part of a Euclidean proposition utilizes typically the common notions.

(ii) Euclidean definitions are not numerous enough to serve as syllogistic premises according to the Aristotelian scheme. (Cf. *An. Post. I*, 88b4–5.)

(iii) Euclidean postulates are not wider in scope than the definitions, unlike Aristotelian generic premises.

(iv) There does not seem to be any reasonable sense in which Euclidean definitions were thought of as being provable. They were not provable even in the loose sense of a *deixis* by *epagoge*.

It is not an accident, it seems to me, that most of these discrepancies turn on the Aristotelian idea that all primary premises of an axiomatic science can be thought of as definitions. As I tried to suggest in my *Noûs* paper, it was precisely the syllogistic model of scientific demonstration that encouraged Aristotle to assign such a sweeping role to definitions in an axiomatically constructed science. This illustrates vividly what I said earlier about the significance of the syllogistic model in alienating Aristotle's theory of the axiomatic method from the mathematical practice.

The basic reason for all these failures is Aristotle's syllogistic paradigm for his logic of science. When the effects of this particular paradigm are eliminated, however, several interesting comparisons are still possible between Aristotle and Euclid. Here I can only give selected examples, which exemplify the general points made above concerning the relation of Aristotle to the mathematical tradition.

(i) Aristotle's "common axioms" seem to be essentially tantamount to Euclid's "common notions". For instance, Aristotle's example in *An. Post. I*, 10, 76a40 is identical with Euclid's third "common notion".

(ii) Szabó is on the right track in emphasizing (e.g., pp. 302,

232–233; pp. 413, 310–311 of the German) the role of definitions as basic assumptions in Greek mathematical theories. He could have found much more evidence for this view in Aristotle.

The general historical significance of these comparisons cannot be discussed here.

NOTES

[1] This is all the more striking because in some of the quotes given earlier Aristotle is apparently restricting the term *arkhai* to (3) rather than both (2) and (3).

[2] I shall here overlook the interesting question as to what Aristotle is prepared to say of the narrowest syllogistic premises. Passages like *Eth. Nic.* VI, 8, 1142a26 ff are relevant to this question.

WILBUR RICHARD KNORR

ON THE EARLY HISTORY OF AXIOMATICS: THE INTERACTION OF MATHEMATICS AND PHILOSOPHY IN GREEK ANTIQUITY

The manner of the origins of deductive method both in philosophy and in mathematics has exercised the thoughts of many notable scholars. Unique to the ancient Greek tradition were conscious efforts to comprehend the process of thinking itself, and these inquiries have since developed into the philosophical subfields of logic, epistemology and axiomatics. At the same time, the pre-Euclidean Greek mathematicians turned to the problem of organizing arithmetic and geometry into axiomatic systems, in effect setting a precedent for the field of mathematical foundations. Did these two movements in the history of thought arise independent of each other, through some extraordinary coincidence? It would appear far more likely that this common adoption of deductive method resulted through interaction between the fields of philosophy and mathematics. But if this is so, to which shall we attribute priority?

The present contribution will examine arguments of the Hungarian philologist, A. Szabó, who has adopted a most provocative position on this issue. In brief, he maintains the pre-Euclidean mathematicians accepted a formal deductive methodology as a direct response to the example of the Eleatic dialecticians, Parmenides and Zeno. Indeed, by describing the fifth-century Pythagorean number theory as "perhaps the greatest and most lasting creation of Eleatic philosophy", Szabó subordinates early mathematics to Eleaticism.[1] He maintains further that the subsequent formal studies in geometry resulted from the effort, only partially successful, to establish there a deductive system comparable to that effected for number theory. In particular, Euclid's *Elements* are interpreted as a direct answer to difficulties raised by the Eleatics.

Szabó thus clearly advocates assigning to philosophy priority in the emergence of deductive method. Yet one need not search long to find equally firm advocates of the opposite view: that mathematics owes no such debt to philosophy, either then, or indeed ever.[2] I shall seek here to establish a position between these extremes. But my intuition is that the latter view is closer to the accurate portrait of early Greek

J. Hintikka, D. Gruender, and E. Agazzi (eds.), Pisa Conference Proceedings, Vol. I, 145–186.

mathematics. In the first part of this paper I shall take up Szabó's arguments and indicate certain major areas where his position is unsatisfactory: notably, in its reading of the early mathematical literature and in its account of the alleged dialectical motives of Euclid.[3] In the second part I shall undertake to describe the interaction of geometric and philosophical studies in antiquity through three examples: the relation of Aristotle and Euclid on the first-principles of mathematics; the contributions of Eudoxus to proportion theory; and the formal style of Archimedes. We shall thus see that the interaction between the two disciplines hardly fits the simplified picture of a unilateral influence; nor did this interaction always contribute to the advancement of research in either field.

I. THE ORIGINS OF DEDUCTIVE METHOD

Szabó's thesis of the Eleatic influence on the rise of deductive method in early mathematics is reminiscent of certain prominent, but now largely discredited, views. I refer to Tannery's thesis that Zeno's paradoxes had been directed against a naive number-atomism among the early Pythagoreans and to the thesis of the pre-Euclidean "foundations crisis", argued by Hasse and Scholz and others.[4] But Szabó is attempting something more subtle, for the question he addresses is more profound: what were the origins of the deductive method itself and what was the cause of its being adopted into mathematics through the notion of "proof"?

Szabó opposes the view that deductive method came to characterize Greek mathematics through some spontaneous evolution internal to it.[5] Older traditions – the Egyptian and Babylonian – attained the *technical* level of the Euclidean arithmetic and plane geometry; but their methods were always presented in the form of the solution of specific problems – over the course of centuries (indeed, millenia), they showed no sign of developing a formal proof-technique by which their concrete methods might be converted into a theoretical structure. In the same way, there seems small reason to suppose the early Greek practical tradition should have spontaneously evolved into such a theoretical system. By contrast, the early development of Greek natural philosophy soon arrived at the stage where attention had to be paid to the forms of discourse.[6] Being a quest for the principles and causes of things in general, natural philosophy was fundamentally disputative. In the Eleatics, we see the deliberate application of logical forms in the presentation of criticisms of Milesian cosmology. Thus, we have the earliest uses of logical forms – most important, of the indirect proof-technique, "reduction to the impossible" – as well as clear motives for their introduction, in the context of the emerging dialectic of the early fifth century. Szabó maintains that when the mathematicians were for their own reasons required to apply such deductive forms – as in their studies in number theory and in the theory of incommensurable lines – these were already available through the earlier philosophical debates.

Prima facie this might appear to be an appropriate view. For surely, it is the business of mathematicians to solve mathematical problems, not to pronounce upon the general structure of argumentation. Conversely – if our impressions based on the later development of philosophy do not mislead us – precisely such a question is among the legitimate concerns of philosophers. Nevertheless, we are well reminded by an eminent historian of logic that mathematicians need not depend on prior philosophical studies as the source of the basic logical forms they employ.[7] It is thus important that we clarify the possible alternative modes by which such deductive forms may have entered into mathematical study.

First, what shall we intend by "deductive method"? Certainly, if we seek fully articulated axiomatic systems, we shall not expect to find them among the early pre-Euclidean studies. But if we define our quest as the presentation of propositions as true through the ordered sequence of other propositions, in which each passage from two or more propositions to another (i.e., their "conclusion") is effected according to set rules (i.e., "laws of logic") – then mathematics even in the far more ancient Oriental traditions was already deductive. Babylonian and Egyptian mathematicians had to begin their solutions of problems by first stating the givens, and then through a recognized sequence of steps produce a desired conclusion (the construction or computation of a required term).[8] The very nature of mathematics compels adherence to some such ordered sequence, in effect, to a logical method. Indeed even the method of indirect reasoning had to figure among the most ancient techniques. For most problems are accompanied by a "proof", or as we might say, a "check". While of course most of the surviving computations do indeed check out, surely countless computations did not; for otherwise, checking would not have become part of the usual computing procedure. Now, in the event that the "check" did not conform with the stated conditions of the problem, the computist would repeat his work until it did; that is, he would recognize the former solution to be incorrect. We thus see the germ of the method of indirect reasoning.

It is thus clear that the appearance of such forms in mathematical work need not signify their prior study by philosophers. Indeed, we might press this point to argue that the efforts by early mathematicians (say, the fifth-century Pythagoreans) to put in order the body of arithmetic and geometric techniques they were assimilating from the older traditions gave rise to the awareness among philosophers that the same ideas and patterns of reasoning might be applied in their cosmological speculations. Lending credibility to this view is the Pythagoreans' organization as a school, within which the instruction of mathematical subjects might well have occasioned such an effort of compilation. Moreover, it is said that Parmenides had studied the Pythagorean ideas before breaking away. In this way, the Eleatics could have derived their logical forms from the prior example of the mathematicians. Lacking any definitive documentation on this question, we shall not propose this view as more than a possibility. At any rate, Szabó's insistence that the dialecticians *could not* have derived their deductive methods from mathematics[9] is hardly compelling, and thus inadmissible as support for his thesis.

Second, Szabó characterizes Greek mathematics of the formal type as "anti-empirical" and as deliberately in avoidance of "graphical" methods of proof.[10] The latter, as a description of classical Greek geometry may appear surprising; for the construction of diagrams is always a characteristic vehicle for the organization of geometric proofs. In principle, however, the diagram might be dispensed with, since the argument can be effected entirely by means of *verbal* reference to the relevant prior principles. Certainly Euclid and those in the formal tradition never presume to prove a theorem via the actual measurement of the elements in a drawn diagram.[11] Szabó's point about the non-graphical nature of Greek geometry is thus apt. Indeed, the converse emphasis on the visual character of Greek mathematics has led some scholars into absurdities about the meaning and limitations of Greek methods in certain fields, particularly number theory.[12]

But is this avoidance of graphical methods and this recognition of the *generality* of the theorems in mathematics the same as *anti-empiricism*? The geometer knew that his draftsmanship was not a limiting factor on the truth of his theorems; these are about geometrical entities and are true – or, better, are *known* to be necessarily true – through the correctness of the logical sequences followed in their demonstrations. But what *are* these geometrical entities? Impressed by the mathematicians' ability to draw accurate general conclusions, despite the unavoidably specific and limited nature of any concomitants (e.g., drawn figures, specific numbers) of his arguments, Plato proposed that the reasoning was really founded upon an intuition of abstract entities (the *Ideas*, or *Forms*) existing apart from any physically perceived objects.[13] Now, if the early mathematicians did indeed consciously seek through formal methods to capture the essence of such entities, as Plato supposes, then we should have to acknowledge the anti-empirical aim of their work and its roots in the Eleatic doctrines.

But this account of the mathematicians' enterprise is not the only one possible. One might, for instance, adopt the view argued by Aristotle: that the things studied in mathematics are *abstracted* from sense-particulars; they do not exist apart from those particulars in actuality (although, in a sense, they might exist conceptually); their special character is their generality – an issue discussed in detail in the *Posterior Analytics*.[14] Under this view the geometer would conceive of the "triangle", for instance, not as some abstract ideal entity, but only as a generalization derived from his inductive experience of sense-perceived triangles. In this way the specific Eleatic tenor of his epistemology disappears altogether.

No one would deny that the Eleatics were a dominant influence on Plato's philosophy of mathematics. But we have no means of determining whether the mathematicians themselves subscribed to this theory, or to the Aristotelian – if, indeed, to any philosophical account at all. Thus, in asserting the anti-empirical character of Greek mathematics and then deducing from this evidence of the influence of the Eleatics, Szabó has lapsed into circular reasoning.

Third, Szabó's detailed investigations into the roots of axiomatic terminology – "postulate", "axiom", "hypothesis", and so on – and their dialectical connotations,[15] as well as his tracing the *content* of the Euclidean first-principles back to the Eleatic arguments on being, plurality, motion and the like,[16] are designed to show that the search for axiomatic foundations of mathematics had already begun in the fifth century. Thus, if we possessed clear signs of an axiomatic interest within this period of mathematics, we might have to admit some influence from the side of the Eleatics. But in the *mathematical* writings which have survived from the pre-Socratic period, however scant and fragmentary these might be, there is nothing to suggest a concern over such foundational issues. Thus, if the mathematicians themselves manifested no awareness in their work of these philosophical issues, the thesis of a pervasive Eleatic influence dissolves. I find it important, then, briefly to survey the principal mathematical fragments and testimonia to establish this point.

Oenopides of Chios. Proclus, doubtless depending on Eudemus, assigns to Oenopides credit for having "first examined the problem" of constructing the perpendicular to a given line through a given point (*Elements* I, 12).[17] Following Heath and others, Szabó maintains that Oenopides must have initiated the first *theoretical* examination of constructions via compass and straight-edge, in the manner presented by Euclid in Postulates 1–3; for it is hardly conceivable that such elementary results should have remained unknown, that is, through practical techniques, until as late as 440 B.C. or so.[18] The latter point is undoubtedly correct. Yet the former need not follow. For we note that Eudemus is referring here to a *text*: he goes on to say that Oenopides called the drawing of the perpendicular "gnomon-wise in the archaic manner" and that he found this construction useful in astronomy. We easily infer that Eudemus found Oenopides' construction among the materials in a work on astronomy, most likely on the construction of sundials. In the context of such a work, the geometric constructions were certainly secondary to the main interest, so that a deliberate axiomatic approach to them is hardly to be expected. Eudemus's naming Oenopides as the "first" to effect these constructions indicates only that he was unable to find a writing earlier than Oenopides in which the same had been done. Thus, this testimonium may be useful to assist us in gaining a sense of the degree to which mathematical knowledge was becoming public through writings at that time, as opposed to being transmitted through a personal teacher-student relationship which combined a written and oral tradition. But it hardly justifies assigning to Oenopides the axiomatic presentation of the plane constructions in Euclid's *Elements* I.

Hippocrates of Chios. The most important extant mathematical writing from the pre-Euclidean period is the fragment of Hippocrates's quadrature of lunules.[19] From the technical point of view, it reveals a full mastery of the materials presented in Euclid's Books I, III and VI. Eudemus's account indicates further that the work had a careful deductive structure. Not only are the constructions presented and their quadratures effected in fully systematic fashion, but preliminary theorems are proven on which these quadratures depend; among these are results on the areas of circles and their segments which reappear in Euclid and Archimedes. But even allowing that the form of this presentation is due to Hippocrates, rather than to Eudemus in his paraphrase of Hippocrates, we may still justifiably question whether this work is aptly described as axiomatic.

First, the style of the work is comparable to the geometric treatises of Archimedes:[20] the writing is devoted to the examination and solution of a set of related problems. To the extent that certain materials, notably *theorems*, are required for this, they are established in an introductory section. Many other results are assumed in the course of the work, so that this study entails the application and extension of more elementary work. Certainly, there is a strong sense here of what a proof ought to consist of: a precisely ordered deductive sequence starting from accepted and known results. But this is not the same as an *axiomatic* organization of this material.[21]

Second, where Hippocrates' writing touches closely the items which receive axiomatic treatment in the *Elements*, we find him least satisfactory. For instance, what is the definition of proportion by which he establishes that sectors are in the ratio of the angles which subtend them? This is accompanied by the remark (presumably due to Hippocrates) illustrating what is meant by 'same ratio' by reference to the half, the third, and related proper parts. If this indicates Hippocrates' handling of this theorem, we are still far from the technique of the *Elements* which develops all the needed steps right back to the fundamental definition of proportion (V, def. 5). As another example, how did Hippocrates show that circles are as the squares on their diameters? This appears as *Elements* XII, 2 and requires for its demonstration there the "method of exhaustion" (based on X, 1 and the foundation of proportion theory of Book V). Szabó berates Toeplitz for denying Hippocrates any such fully formalized proof and for proposing a looser alternative, applying a direct limiting argument.[22] Apparently, Szabó holds that an indirect argument of the Euclidean type was available to Hippocrates. The issue is of importance; for access to such a proof would indicate with Hippocrates a concern for just such formal questions of proof which motivate the axiomatic foundation of the *Elements*. But, contrary to Szabó's claim, Toeplitz's reconstruction is not without support. Precisely such a limiting argument as he presents survives in Antiphon's quadrature of the circle; and the grounds for denying an early dating of the "exhaustion" technique rest on Archimedes' ascription to *Eudoxus* of the *first* proper demonstrations of such theorems as the volumes of the cone and pyramid which require it.[23] In view of these things, the usually accepted view of assigning to Eudoxus the technique of *Elements* XII appears preferable to the arbitrary dating of these back to Hippocrates some 50 years earlier.

Third, Hippocrates is obscure on just that question about his work which is most interesting from the dialectical point of view: did he maintain that his quadratures amounted in any sense to a quadrature of the circle? Aristotle and the commentators suspect he did, but modern commentators are properly reluctant to charge him with the gross logical error thereby entailed.[24] It seems more probable that Hippocrates established those results which were within his mathematical powers, and left to others the controversial matters. This schism between the mathematicians and sophists will arise again in our discussion of Theodorus; it speaks poorly for any view which highlights the positive influence of dialectic on mathematics at this time.

We may mention two other achievements due to Hippocrates. First, he *reduced* the problem of duplicating the cube to that of constructing two mean proportionals between two given lines.[25] Now, the logical method of "reduction" is known to Aristotle and may be viewed as a precursor of the method of "analysis" (or "analysis and synthesis").[26] Both these methods may be viewed as the mathematicians' technique, a powerful tool in the investigation of advanced problems from the fourth century onward. There is no reason to view its appearance with Hippocrates as a mark of dialectical influence.[27] Indeed, when Plato introduces the method of reasoning by "hypothesis" in the *Meno* (86), he does so with reference to a relatively advanced geometric problem. This appears more to suggest the initiation of the method in mathematics before its adoption in philosophy. I could imagine one's viewing the method of analogous reasoning as a precursor of reduction. But this would certainly be an imprecise comparison. Even Aristotle gives a mathematical example (i.e., Hippocrates' quadrature of lunules) to illustrate this method.[28]

The other work ascribed to Hippocrates was his contribution to the "Elements". Proclus, following Eudemus, says Hippocrates was the first to write a work of this type.[29] In the absence of the work, we cannot pronounce on the thoroughness of its axiomatization, or indeed on whether it was axiomatic at all, properly speaking. It seems to me reasonable to suppose that at Hippocrates' time, with the emergence of the professional teacher, a textbook collecting the materials he was prepared to teach would be a useful item for any mathematician to have at his disposal. The production of such a work, organized from simple materials toward more complex, would receive prescriptions for its format from the pedagogical context in which it would be used. There need thus be no more sophisticated influence derived from philosophical debates, such as those spawned by the Eleatic paradoxes; no self-conscious search for the fundamental unprovable first-principles from which the whole mathematical structure might be developed. Indeed, Plato's efforts to articulate the nature of this search and its importance both for mathematics and dialectic (in the *Republic*) would suggest that in that first effort, Hippocrates' "Elements", the search was barely, if at all, begun.

Theodorus of Cyrene. In my book on pre-Euclidean geometry I argue a major role for Theodorus in the formalization of geometry, particularly of those portions of Books I and II on the "application of areas".[30] From Plato we know Theodorus contributed to the early study of incommensurable magnitudes and earned praise for his precision in proof. Now, the theory of incommensurables represents a special problem for the geometer: its subject matter cannot be articulated nor its claims established by concrete methods, such as those typifying the older practical metrical tradition to which it is related. One might, for instance, recognize computational difficulties in the treatment of quantities like the square root of 2; but the assertion and proof of its *irrationality* is a matter of quite a different order. Indeed, attaining the level even of *defining* what incommensurability is, not to mention probing the various constructions to establish which of these produce incommensurable lines, cannot be possible without a fine control of formal proof methods, especially of the method of indirect proof, as well as a recognition that some very basic concepts require special, non-obvious, definitions.[31] For example, how are we to treat proportions among incommensurable magnitudes? Not at once seeing the way to do this, a geometer like Theodorus might well seek an alternative treatment, as in *Elements* II, dispensing with the techniques of proportions. Subsequently, mathematicians like Theaetetus and Eudoxus and their followers could provide suitable definitions by which these techniques might again be admissible under the more stringent logical conditions operating in proofs about irrationals.

In view of this special character of the study of incommensurables, I set the roots of the formal and axiomatic aspect of the *Elements* in Theodorus's work. I will develop this thesis further in the next section. But it is here important to recognize that Theodorus's own inspiration is being linked with the intrinsic nature of the mathematical subject he studied; it is not the response to external strictures of the dialectical philosophy. Plato provides a portrait of Theodorus in the *Theaetetus* which, if not mere dramatic license on his part, bears on Theodorus's debt to philosophy.[32] In his youth, Theodorus was exposed to the principles of Protagoras's philosophy. The explicitly relativist outlook of this school, prescribing the sharpening of dialectical and rhetorical skills so that one might argue his cause successfully, whether weak or strong, must have been uncomfortable for someone like Theodorus, whose talents lay in mathematics. He soon discontinued these studies. Plato describes him as still uneasy in debate, a teacher given to the lecture-method (rather than the question-and-answer technique so favored by Socrates), impatient with those who presume to set their own standards over against his ("they give me all the trouble in the world"), yet one whose standards of rigor in geometric proof were impeccable. How reasonable is it for us to presume strong and direct influences of dialectic on his methods and geometric outlook? To be sure, he must have acquired no small competence in logical technique through his training under Protagoras, and thus indirectly from the Eleatics. But in the highly contentious political and intellectual environment of mid-fifth-century Athens, this would be unavoidable in any course of higher education.

If this is all Szabó intends by attributing the "transformation of geometry into deductive science" to the influence of the Eleatics, then one could hardly object. But I suspect the above account is decidedly weaker than the thesis he advocates. Certainly, it would be odd to view Theodorus's work – which could not but be highly theoretical and deductive, marked by an incipient awareness of the axiomatic approach to geometry – as directly indebted to Eleatic methods, when Theodorus himself seems to have fled the arena of dialectical competition through distress over the absence of firm bases of thought and opinion in that field.

Archytas of Tarentum. Perhaps Archytas' greatest interest for us here is his affiliation with the Pythagorean school.[33] In the modern discussion of pre-Socratic philosophy and mathematics, no move is more common than the assignment to the Pythagoreans of remarkable feats of scientific and mathematical reasoning. Some view the elaboration of fully developed theories in arithmetic and geometry, indeed, even the authorship of whole books of the Euclidean corpus as now extant, as early Pythagorean accomplishments, despite the utter absence of documentary evidence from that period.[34] Thus, it is instructive to consider the work of Archytas, stemming from an advanced stage of the history of that school.

On just such points where Szabó's view would lead us to expect the dialectical influence, Archytas is least satisfactory. In the fragment on epimoric ratios the use of indirect arguments is awkward.[35] Indeed, the last step of the proof, which should properly be cast as an indirect argument, is actually phrased in direct form. Further, a comparison of Archytas' proof, in the Boethian version, with the same theorem proved in the Euclidean *Sectio Canonis* indicates that the number theory available to Archytas contained materials on the least terms of ratios, but not on the more sophisticated notion of relative primes – that is, the content of the first part of *Elements* VII, but not the rest. It is difficult to conceive how formally adequate proofs, let alone a complete axiomatic treatment of number theory, could be assigned to Archytas. The difficulty is of course all the greater in supposing any such theory among the earlier Pythagoreans.

One surviving fragment contains Archytas' definitions of the three means: arithmetic, geometric and harmonic.[36] These reveal a symmetry of phrasing which might be termed aesthetic; but from the viewpoint of any systematic inquiry into the properties of these means they are framed poorly, and later arithmetic authors prescribe the investigation and calculation of means according to alternative formulas. The results of Archytas' division of the canon have been preserved; but the principles under which the numerical ratios for the various intervals were determined seem not at all clear.

Archytas' solution to the problem of duplicating the cube is extant in a paraphrase by Eutocius of a version from Eudemus.[37] The geometrical insight is uncanny and reflects a persistent strength of the ancient Greek geometric tradition: the profound visual intuition of the relationships of plane and solid figures. But Archytas depends upon the conception of the interpenetration of surfaces of solids generated through the *motion* of planes. By contrast, one of Szabó's critical arguments links the *avoidance* of geometric motions (as in Euclid's definitions) to a sensitivity to the Eleatic rejection of motion and change as logically inconsistent.[38] It would thus appear that this level of awareness had not yet been attained by Archytas. These abstract features of Euclid's definitions and methods appear thus to have entered the tradition of *Elements* later, for instance, in the mid-fourth century as a response to the deutero-Eleaticism of Plato in the *Republic* and the *Parmenides*.

Concerning the alleged formalism of early Pythagorean mathematics, a word on the discovery of the irrational is appropriate.[39] As indicated in our remarks above on Theodorus, the notion of incommensurability and the confirmation that given constructed lines are incommensurable are *intrinsically* theoretical and demand deductive arguments of the indirect kind; no concrete or practical procedure can do more than suggest that such lines evade description in terms of ratios of whole numbers. Hence, if the discovery and verification of incommensurability, even in a single case like the side and diagonal of the square, had been achieved by Pythagoreans toward the beginning of the fifth century, we should have to admit some more or less direct debt to the dialectical forms of the Eleatics. Now, dating this discovery is complicated by the absence of any relevant and trustworthy documentation. While some scholars advocate an early date, there is no direct evidence to support their claim; indeed, the silence of the pre-Socratic literature on this notion until the close of the fifth century would appear to argue against them. Hoping not to be guilty of circularity myself, I find the later dating for this discovery both preferable and readily understood. Being by nature a theoretical result, the incommensurable requires a preparation in the skills of deductive reasoning before the anomalous facts emerging from computations and constructions could be articulated in the form of concepts like incommensurability and commensurability and of theorems asserting the incommensurability of specified lines. Those anomalies do not of themselves lead to the conclusion of incommensurability; for instance, the ancient Babylonians and Hindus knew of the computational problems in dealing with certain square roots, yet they appear never to have taken this as more than a practical difficulty, that is, to have recognized that such quantities were *in principle* inexpressible in rational terms. We do well not to underestimate the gap which separates these two stages of awareness. Believing the early Pythagoreans had already bridged that gap, one must accept the implications concerning the formal level of their mathematical tradition at that time.

Szabó rightly chides Becker for inconsistency in maintaining the thesis of "Platonic reform" of mathematics on the one hand, while proposing reconstructions of a deductive theory of the odd and even among the early fifth-century Pythagoreans on the other.[40] But I believe, on the basis of the materials presented in this section, that Szabó has affirmed the wrong side of this contradiction. He has proposed the Eleatic logical methods and idealist outlook had already abetted the conversion of mathematics toward deductive, axiomatic theory in the fifth century, so that no major contribution was left to be made by the Platonic philosophy in the fourth century. By contrast, I argue that the mathematical evidence discourages assigning to fifth-century mathematicians any efforts to axiomatize fields in arithmetic or geometry. The signs of such an interest are first seen in association with Theodorus at the threshold of the fourth century.

We thus see that Szabó's thesis of the Eleatic impact on fifth-century mathematics both reinforces and depends upon a distortion of the nature of that tradition, in imputing to it a formal character it did not have. Similarly, Szabó's view requires us to dismiss the explicit testimonies of Plato and Aristotle on the relation of mathematics and philosophy. It is a remarkable feature of the logical and epistemological writings of both philosophers that they turn again and again to mathematics for examples of concepts and methods. In Plato's *Phaedo, Meno, Republic, Theaetetus* and *Philebus* and in Aristotle's *Analytics, Physics* and *Metaphysics* one is reminded of a consistent theme: that the project of philosophy is to obtain the same kind of certainty, the same degree of rigor which mathematics has achieved – or in principle *can* achieve. We sense how strongly impressed both were at the success of the geometers of their time in their efforts toward this end. How is it possible that Plato and Aristotle could have been so mistaken in their view of the relation of mathematics and philosophy, if, as Szabó maintains, it was the Eleatic dialectic which had initiated the adoption of deductive method by the mathematicians? Szabó recognizes difficulty in his position, but offers an astonishing explanation: that in the half-century between Zeno and Plato the truth in the matter was simply forgotten.[41] In this way, the substantial corpus of Platonic and Aristotelian writings become irrelevant for understanding the pre-Socratic issues. In effect, then, Szabó can support his thesis only by inventing a suitably formal pre-Socratic Pythagoreanism and then by directly repudiating the documentary evidence of the fourth-century philosophers. As a method for studying the history of mathematics and philosophy, this is one which few should wish to adopt.

While thus dismissing the writers of the mid-fourth century, Szabó turns to Euclid at the very end of the century for evidences of Eleatic influence. He views the definitions, postulates and axioms ("common notions") which preface Book I of the *Elements* as a direct sign of such influence.[42] Now, even if we *could* detect in Euclid such a sensitivity to the issues which had preoccupied the Eleatics, this need not indicate a *direct* response to Eleaticism, that is, a mathematical reaction dating from the mid-fifth century and then transmitted through some continuous line to Euclid. For the Eleatic problems received a lively examination in both Platonic and Aristotelian circles in the latter part of the fourth century, so that any dialectical influence could well have been exerted at this later time. I will return to this view and its implications in the second part of the paper. But I should like now to consider how useful the hypothesis of Eleatic influence really is for the interpretation of Euclid's procedure.

Szabó has argued that Euclid so framed the basic principles of the *Elements* as to avoid just those difficulties to which the Eleatics had called attention. For instance, the definitions of point, line and so on seem deliberately to avoid reference to sensible things or operations. The "common notions" (known to Aristotle and Proclus as "axioms") articulate such evident principles about equality that one is surprised at the need to state them. Why, for instance, must one state that "the whole is greater than its part"? Wouldn't that be sufficiently obvious to be admissible in any proof without special comment? (In fact, many less obvious principles *are* introduced into proofs. Why should the truly unquestionable principles require such statement?) Now, it happens that precisely these definitions and axioms have a prior appearance in the history of dialectic. Ultimately, the parts-whole relationship and the essential nature of equality, being and unity were the center of interest for the Eleatics. Szabó would thus maintain that the statements in Euclid follow from the earlier examination of these concepts and principles by the Eleatics.

Let us take a specific example: Euclid adopts a statical conception of the circle – as the plane figure bound by a line such that all the lines drawn from a point within it (i.e., its center) to its edge are equal (Book I, Def. 15); he does not use the intuitively clearer conception of the circle as the figure generated by a finite line segment which rotates about one of its fixed end-points – a dynamical conception preferred by Proclus. Can we fathom Euclid's preference? Might it not be that Euclid knew of the Eleatics' refutation of the possibility of motion and change in true being? Mathematical entities – lines, figures and so on – are such types of abstract being; hence, how can we conceive of their motion?[43] Thus, if we follow Szabó's view, Euclid adopted the static conception as a resolution of this puzzle. Now, this bit of dialectic is not hypothetical. For Proclus presents just such an argument against the motion of mathematical entities and has no small difficulty attempting to reconcile his Platonic notion of geometric being with his dynamical conceptions of lines and figures.[44] Let us go further: among the "common notions" there is one which asserts the equality of figures which can be exactly superimposed one on the other. This is a principle of which Euclid makes remarkably little use. Given that it in essence presumes a physical operation – the motion of one figure until it comes in position directly over the other – might not Euclid have wished to minimize appeal to it, for the same reason of the immobility of geometric being?

Ostensibly plausible,[45] this view is revealed upon closer examination to be a fully inaccurate account of Euclid's procedure. First, on the acceptability of motions for geometric objects, we observe that Euclid *does* introduce and use dynamical conceptions – such solids as the cone, cylinder and sphere are defined in terms of generation by plane figures rotating about a fixed axis (Book XI); why not merely extend the statical mode already used for the circle, so to avoid the appeal to motion? Again, the "postulates" (1, 2, 3 and 5) all use dynamical operations: the *drawing* of a line segment or of a circle, and the indefinite *extension* of a given line segment – why not phrase these in terms of the mere *existence* of geometrical entities (lines, circles) which possess certain specified properties (a given center and radius; or two given points)? We note that neither before nor after Euclid was there any reservation about the admissibility of motions into purely geometric theorems – at least, as far as the mathematicians were concerned: Archytas solved the duplication of the cube by means of lines and surfaces generated by rotating plane figures; the followers of Eudoxus constructed mechanical devices to solve the same problem; Archimedes defined the spirals in terms of a double motion; later writers recognize several varieties of motion-generated curves and surfaces.[46] Plato and the Eleatics to the contrary notwithstanding, the geometers had no qualms at all about such conceptions.

As for the principle of superposition, were Euclid seeking dialectical approval, he would have been compelled to adopt a rather different mode – for if superposition is *in principle* unacceptable, then even *one* use of it is too many.[47] But Euclid uses it twice in Book I and again in Book III (24) to establish theorems on the congruence of triangles and circular segments, respectively. Might he be making a philosophically significant point: that some resort to this principle, itself based on the conception of geometric motion, is indispensable for the working out of his geometric system? Indeed, invariance of area under the so called "Euclidean" motions of translation and rotation can be adopted as the defining property of the Euclidean metric (for in the non-Euclidean geometries figures do not so remain invariant). But Euclid is not troubled by such concerns. Otherwise, his procedure should have been this: eliminate the superposition axiom altogether; adopt as a *postulate* one of the theorems which require it, for instance, the congruence of triangles of which two sides and the included angle are respectively equal;[48] then prove the others by means of this postulate.[49] Euclid does none of this. Nor do his successors reveal uneasiness over superposition: for instance, Archimedes employs it several times, in his study of equilibrium of planes and his proof that the ellipsoid is bisected by any plane passing through its center (note that Euclid asserts the analogue for circles as part of his *definition* of the circle).[50] Proclus provides alternative proofs of some of Euclid's theorems using the superposition principle where Euclid does not, and even gives a proof of the fourth postulate (that all right angles are equal) by means of this same principle.[51] In sum, then, the ancient geometers felt quite free to adopt the conceptions of motion and superposition in their proofs, despite the dialectical difficulties therein. However much the Eleatics and Plato after them might shout disapproval, the geometers remained remarkably deaf to their objections here.

There still remains a question about Euclid's procedure: why, then, *does* he appear to reduce to a minimum appeal to the principle of superposition? For it is clear that several of his theorems admit of far simpler proofs than he gives them if the principle is introduced. I believe the answer may lie in Euclid's conception of his geometric system. Theorems are not isolated propositions, to be proved in whatever way. They fit into an *ordered* sequence, and there are considerations under which certain choices of order might be viewed as preferable to others. Aristotle devotes no small attention to this very issue in the *Posterior Analytics*.[52] In proving a theorem, although any number of prior assumptions might serve, there will be one choice which affords us knowledge of the *cause* of the truth of what is asserted, and this is determined by the essence of the thing examined. In effect, geometric proofs develop in the order which reverses the order of the familiarity of these things to us, that is, of our inductive discovery of these facts.

If we may consider Euclid's procedure in this light, the manner of his introduction of the primary undemonstrated premisses becomes clear. On the one hand, if a postulate has been used to demonstrate a theorem, that theorem now becomes admissible as an assumption in the proofs of subsequent theorems; as in the order of the increase of our knowledge of things, each thing learned becomes the basis for learning new things. Hence, we do better to demonstrate by means of what we have just proved, if that is possible, rather than by continually referring back to the primary assumptions. I believe this accounts for Euclid's handling of superposition. On the other hand, if it is possible to prove a theorem without recourse to some basic premiss, one should avoid giving a proof which in fact appeals to that premiss. Here, the finest example lies in Euclid's use of the parallel-postulate. For he introduces it precisely at that point in Book I where its use is necessary (I, 29). The earlier theorem (I, 16) – that the exterior angle of a triangle exceeds each of the non-adjacent interior angles – might easily be viewed as a corollary of I, 32. But the latter theorem requires Postulate 5, whereas the former does not.

One notes that Euclid's commitment to such considerations of deductive order varies in different parts of the *Elements*: Books I, V and VII, for instance, reveal a strong sense of the sequential ordering of results. By contrast, Books II and IV generally lack a cumulative character, amounting to collections of independently established results. One may interpret this as a sign of the earlier provenance of the two latter books, a view confirmed by other aspects of content and method.

Such considerations, then, discourage acceptance of the thesis of a direct Eleatic responsibility for the rise of deductive mathematics: in particular, it misrepresents the pre-Socratic work in geometry and number theory and it proves a questionable instrument for the understanding of Euclid. We shall instead now seek to portray the rise of the axiomatic approach in geometry as in part a reflection of the dialectical interests of the mid-fourth century.

II. FORMAL ASPECTS OF THE WORKS OF EUDOXUS, EUCLID AND ARCHIMEDES

In Plato's *Republic* the acquisition of knowledge, both in mathematics and in philosophy, is described as a quest for the appropriate first-principle from which the truths of things may be derived. Until such universal insight is achieved, one must adopt "hypotheses" as the basis of reasoning, a provisional expedient which can also assist in the discovery of more basic principles. Plato's characterization is imprecise, and may perhaps have been overestimated as a prescription for the systematization of knowledge. But there appears to be here a sense of the project of axiomatization, in particular of the fields of mathematics, and one may easily suspect that work of this type was then being encouraged in the Academy.

This impression is confirmed by Proclus' summary of the pre-Euclidean work which culminated in the *Elements*. Beginning with Archytas, Leodamus and Theaetetus, thirteen contributors are named, the most notable being Theaetetus and Eudoxus, all having an association of some sort with Plato and the Academy. Moreover, a number of Plato's arguments, as in the *Parmenides*, revive difficulties which the Eleatics had introduced on the theoretical understanding of basic mathematical concepts (e.g., "unit", "number", "point", "line", "construction", "generation through motion", and the like). Somewhat later, in the *Physics* and other writings, Aristotle examined these issues; he criticized the notion of indivisibles, for instance, and provided synopses and commentaries on Zeno's arguments. The Eleatic methods and viewpoints so remained a topic of debate throughout this period. In view of these facts, together with the absence of comparable work in the fifth century, we may place the major contributions to the organization and axiomatization of mathematics among these fourth-century figures, whose treatises were collected and edited by Euclid at the very beginning of the third century.[53]

To what extent does Euclid's work actually reveal a concern over dialectical difficulties? One should note first that those parts of the *Elements* which appear most sensitive to such difficulties have the least bearing on the mathematics developed in the several Books. For instance, many of the definitions are but formal tokens, with no operational value for the proofs of theorems. The definitions of "point" and "line" before Book I, of "ratio" in Book V and of "unit" and "number" in Book VII have no use for the investigation of theorems; indeed, defining "point" as "that which has no part" is unsatisfactory even from the dialectical side, since both Plato and Aristotle disapprove of defining terms by means of negation. On the other hand, some terms which *are* of importance do not receive definition: for instance, the notion of "measure" is employed in Book VII, but is not there defined. Further, many of the definitions are straightforward and directly applied in the proofs; there is no indication in these instances that dialectical debates affected the introduction or choice of wording for these. As for the completeness of Euclid's postulates and axioms, the modern studies of foundations, such as those by Hilbert, have revealed how many principles are tacitly assumed by Euclid and in how many ways their formulation might be improved. For example, the "Archimedean axiom" appears as a *definition* in Book V, rather than in the form of a *postulate* as required for its application in Books V, X and XII. In view of such shortcomings, despite the long series of studies on the foundations of the "Elements" in the century before Euclid, one critic has recommended viewing Euclid's work not as an axiomatic effort at all, but rather solely as a mathematical treatment of plane constructions.[54]

But given the intensity of interest in the principles of scientific knowledge among the mid-fourth-century philosophers and the simultaneous effort among geometers to organize geometry in an axiomatic fashion, one can hardly suppose that these two enterprises proceeded in complete isolation from each other. By the end of this period both areas of inquiry were ripe for synthesis: the former in Aristotle's *Posterior Analytics*, the latter in Euclid's *Elements*. I wish briefly to explore the relation between these two works in two examples: the classification of the "first-principles" and the meaning of postulates of mathematical existence.[55]

Euclid divides the "first-principles" into three classes: definitions, postulates and axioms (that is, "common notions"). Proclus offers three rationales for this distinction between postulates and axioms, none of which adequately covers all five postulates and all nine axioms, as given by Euclid.[56] It would appear that through inadvertence, Euclid, or some earlier editor of the "Elements", created a puzzle for the dialecticians. Now, Proclus notes that several writers made no distinction between postulates and axioms. One might wish to take this view as a mark of the influence of dialectics, for Euclid would seem to be requiring only the acquiescence to the stated principles, and the essentially *open* character of such acquiescence (especially when one recalls the Eleatic challenges to the conceptions stated in the axioms) is noteworthy.[57] However, an important feature separates mathematical from philosophical writers on this point: the former, as cited by Proclus, employ only the term "postulate".[58] Proclus remarks further that writers similarly conflate the terms "theorem" and "problem"; "problems" and "postulates" alike address *construction* of geometric entities, whereas "theorems" and "axioms" entail the *statement* of their properties and relations. In the pre-Euclidean period it was the philosopher Speusippus who classed all propositions as "theorems", while it was the mathematician Menaechmus who classed them all as "problems".[59] This, I propose, underscores a persistent difference in outlook between the mathematicians and the philosophers: the former were primarily occupied with the constructive activity involved in geometric research, while the latter were most interested in the formal description of proofs. Euclid's adoption of the term "postulate" for those assumptions most directly concerned with the propositions on geometric construction can be understood as a derivation from his mathematical sources. The other principles may then have been formulated by way of satisfying more philosophical considerations and given the label "common notions" to signify their general applicability to *all* fields of mathematics.[60] But as Aristotle used the term "axiom" to designate these same "common" principles, the transferral of this term to Euclid's "common notions" in the later discussion of axiomatics is easily explained. One must admit, however, that the diversity of opinions on the subdivision of the "first-principles", from the time before Euclid right through to the time of Proclus, attest that Euclid's own choices in the *Elements* were neither fully clear nor definitive.

One of the classifications criticized by Proclus was that of Aristotle. Many scholars have since attempted to establish some sort of conformity between the Euclidean and the Aristotelian approaches, but without notable success.[61] I should like here to show how this issue reveals an interaction between pre-Euclidean mathematics and philosophy of a kind different from simple unilateral influence.

Aristotle's division of the principles in the *Posterior Analytics* (I, 2) separates the immediate first-premisses of apodictic syllogisms into (a) *theses* and (b) *axioms* – the former need not be proved or held in the prior knowledge of the learner; but the axioms, as it happens, necessarily are already known. The "theses" are then subdivided into *hypotheses* and *definitions* – the former assert one or the other of the two terms in a logical disjunction; definitions do not do this. Later (I, 10) Aristotle expounds this in a somewhat different manner: among the first-principles one must first assume (c) the *meaning* of the terms used and then (d) the *existence* of the entities so defined; in some instances this existence is merely *assumed*, but in others it is *proved*. Next there are (e) propositions which are *proper* to the subject-matter of each particular science, while there are also (f) propositions which are *common* to many or all of them.[62] As examples of such "common" propositions Aristotle cites the logical principle of the excluded middle and the condition that the subtraction of equals from equals leaves equal remainders. Now, this last class is variously called by him "the common things" (*ta koina*), "common opinions" and "axioms". As Aristotle's second example appears among Euclid's "common notions", one sees that Aristotle's "axioms" and Euclid's "common notions" refer to the same class of principles. In this same passage (I, 10) Aristotle next introduces the terms *hypothesis* and *postulate* which here seem to explicate (e). These may or may not be susceptible of proof, but in the teacher-student context they happen not to be proved. A "hypothesis" appears to be true to the learner – the term is thus not an absolute, but a relative term. When the student has no opinion of the truth of the assertion, or even doubts it, it is called a "postulate".

For Aristotle the terms "hypothesis" and "postulate" are thus embedded in a dialectical context. We shall consider later whether a similar attitude is taken by Euclid in his separation of the "postulates" from the "common notions".[63] But now, to help associate these schemes of first-principles, let us note the striking prominence Aristotle seems to accord to assumptions of *existence*, so to separate "definitions" from "hypotheses". The puzzle is that one fails to find in the mathematical literature from antiquity any explicit analogue to such assumptions and proofs of existence.[64] But this puzzle vanishes if we examine more closely Aristotle's intent. In these passages he is not really interested in prescribing the classes of assumptions preliminary to the construction of a formal axiomatic *system*. Rather, he is setting out what things one must do in order to prove a *theorem* scientifically; what things one must know or admit in order to understand a proof and recognize that it is a proof.

In this light let us regard the structure of any theorem in Euclid's *Elements*.[65] It begins with an assertion – the thing to be proved or to be constructed; to understand this much we already must know the meaning of certain terms, the relevant definitions. In the proof premisses are introduced: some of these are definitions; others are theorems which have already been proved or can be accepted as provable (perhaps deferred for separate proof as lemmas); still others are premisses which appear self-evident, so that they may be admitted without proof.[66] We have thus covered the appearance of definitions, postulates, axioms and prior theorems in the course of a proof, and on these items Aristotle's discussion and Euclid's practice are in agreement. But there is another kind of assumption made by Euclid: his proofs always provide an "exposition" (*ekthesis*) in which the theorem, having been stated in general terms, is now actualized as a particular configuration for the purposes of inquiry.[67] This exposition typically includes expressions like "let *ABC* be a triangle ...", "let the line *AB* join the points ...", and so on. In effect, one here asserts the *existence* of the entities discussed in the theorem; that is, they are brought into existence by a sort of intellectual *fiat*, a procedure indispensable for the successful management of the proof. Surely, it is this "exposition" which Aristotle had in mind when he spoke of the mathematicians' *assumption* ("hypothesis") that "there are such lines, units, etc." (*not* that "lines, units, etc., exist").[68] In this way Aristotle's discussion of the first-principles of proof becomes a straightforward commentary on the actual procedures of proof employed by Euclid's immediate precursors.

It is often maintained that the first three Euclidean postulates[69] are assumptions of geometric existence and that, in general, constructions serve the function of proofs of existence.[70] But I hold such a view to be anachronistic, assigning to the *Elements* a dialectical motive which it just does not have. First, where the existence of a solution is at issue, Euclid and the other Greek geometers established the result not by a construction, but by a *diorismus* – a theorem specifying the conditions under which the construction is possible. Second, the postulates and other alleged theorems of existence never employ a *language* of existence. For instance, in *Elements* I, 1 Euclid does not propose to prove the existence of equilateral triangles, but rather to show how the equilateral triangle may be constructed whose side is a given line segment.[71] The postulates are stated not as assertions of the *existence* of lines, points and circles under specified conditions, but rather of the *constructibility* of those lines, points and circles. Third, in the ancient geometry those situations comparable to cases in modern foundations of analysis and set theory where the question of existence is most interesting, assume the existence without comment: for instance, the existence of the fourth proportional is assumed throughout *Elements* XII, even though constructible alternatives could have been offered; the existence of intersections of given curves is assumed without a statement of the principle of continuity required.[72] It seems improbable that the geometric constructions were intended to establish existence when the issue of existence earns no place among the axioms.

Thus, there appears to be no dialectical motive behind Euclid's statement of the postulates or his presentation of geometric constructions. Moreover, in setting out the postulates, he did not aim to restrict the whole field of research on constructions to those which can be effected on these assumptions (in practice, to those constructible by means of compass and straightedge alone), nor even to suggest that these means were somehow privileged among the variety of constructing devices possible. We know of many efforts to solve the problems of angle-trisection, cube-duplication and circle-quadrature by means of constructions and curves effected by other than "Euclidean" means.[73] Indeed, sometimes such means were employed even when a Euclidean construction is possible.[74] Only late in the third century do we meet efforts to classify constructions and to establish an ordering of priority according to the means employed (i.e., "plane", "solid", and "linear"). Studies by Apollonius and Nicomedes, for instance, showed how certain types of constructions could be effected on the Euclidean assumptions, while others required conic sections or special motion-generated curves. Apparently this gave rise to a formal recommendation that only the simplest possible devices be employed, if a given construction is to be judged formally proper. Accordingly, both Apollonius and Archimedes came under criticism for their unnecessary assumption of "solid" means in certain constructions.[75]

In these third-century developments we perceive a way in which formal philosophical prescriptions might intrude on the geometer's research. To be sure, a formal commitment on the "priority" of assumptions might contribute to the tightening of the axiomatic structure of geometry.[76] But one can recognize that such conditions might impose an extraneous burden on the geometer in his search for the solutions of problems. Is the discovery of how to trisect an angle by means of the conchoid or the Archimedean spiral any the less useful for its incorrect application of "linear", rather than "solid" techniques? A related example is Archimedes' "mechanical" method: as geometry is "prior" to mechanics (according to Aristotle), mechanical principles are formally inadmissible into proofs about the purely geometric properties of figures. But does this diminish the heuristic power of the method? More to the point, does one impede the search for new discoveries in geometry by discouraging the application of such methods? Indeed, adhering strictly to such orderings of principles prevented the Greeks from perceiving many of the more general structural relations which connected ostensibly disparate problems. For instance, the area of the parabolic segment, the area of the plane spiral, the volume of the cone, the volume of an oblique section of a cylinder, the center of gravity of the paraboloid – these problems, all examined by Archimedes, would appear to be quite distinct each from the others; yet seventeenth-century geometers came to recognize that they could be reduced to the same procedure, the summation of consecutive square numbers, as signified by the expression $\int x^2\, dx$.[76a] Certainly, the axiomatic concerns of the ancients offered no assistance toward the recognition of such general relationships of structure linking diverse portions of the whole field of mathematics.

Let us return to Euclid's postulates. Proclus notes that the remaining two – the fourth asserting the equality of all right angles and the fifth providing a condition satisfied by parallel lines – differ from the former three in that they seem to have more the "theoretical" character of axioms than the "problematical" character of postulates. Indeed, following upon Aristotle's observation on the first-principles, that axioms ought to be indemonstrable, later writers on axiomatics, like Geminus, contended that these "postulates" were actually *theorems* whose proofs should be possible on the basis of the other assumptions. A number of proofs for both postulates have been preserved in Proclus' commentary. It is well known how diligently the quest for such proofs, especially of the parallel-postulate, was pursued until the status of this principle as a true postulate was established by nineteenth-century geometers. I. Toth has detected a variety of passages in the Aristotelian corpus which suggest that the search for such a proof was an active concern of geometers in the mid-fourth century.[77] Indeed, the careful organization of *Elements* I, in which the introduction of this postulate is deferred until precisely that point where its use is unavoidable, indicates the intensity of these studies. Certainly that research failed of its purpose, however, so that the principle retained its place as an unproven postulate within Euclid's theory of parallels.

If research on the "Elements" was strongly influenced by dialectical attitudes, to the extent that postulates and axioms were introduced only as provisional and debatable assumptions, we should expect to find the serious consideration of the postulates alternative to the parallel-postulate as the basis for a different geometric system. In this way, the Greeks should have advanced toward the elaboration of "non-Euclidean" systems, in particular the hyperbolic geometry. Indeed, Toth has argued that many of the theorems of "non-Euclidean" geometry were worked out in Aristotle's time, if only along the way toward a hoped-for indirect proof of the parallel-postulate. Unfortunately, no such systematic treatments have survived; indeed, save for the few suggestive, but highly problematic passages from Aristotle, there remain no fragments or even references to titles to indicate that such studies had ever been completed. It thus appears that one avenue of research which a geometer with strong dialectical inclinations might have discerned was never in fact followed out.

An explanation for this failure may be seen from another condition which Aristotle imposes on the formulation of axioms: they must be self-evident – that is, true as statements abstracted from our physical experience. Proclus remarks that one might dispute an axiom for the sake of argument, but he appears to accept Aristotle's condition and names no philosopher or mathematician who advocated that assumptions contravening experience might form the basis of a complete mathematical system.[78]

These considerations have thus revealed how the philosophical discussion of the principles of scientific knowledge drew from mathematical work, but that even the axiomatic aspects of such works as Euclid's *Elements* resulted primarily from mathematical concerns. Dialectical motives played at best but a limited role in the production of mathematical works. Nevertheless, the contributions of several geometers, notably Eudoxus and Archimedes, manifested a keen sensitivity to matters of formal precision. It is thus appropriate to review their work for possible signs of philosophical influence.

Eudoxus, an esteemed colleague of Plato in the Academy of the mid-fourth century, made remarkable contributions to geometry and mathematical astronomy. In the former area he is most noted for his theory of proportion and his method of limits (usually called the method of "exhaustion") used in the measurement of non-rectilinear surfaces and solids; both techniques can be associated with the solution of important difficulties in the foundations of geometry.

Euclid presents two forms of the theory of proportion in the *Elements*. The version in Book VII applies only to ratios of integers, but can be extended without difficulty to cover ratios of commensurable magnitudes. The one in Book V is general, applying both to commensurable and incommensurable magnitudes. While the former theory is sometimes ascribed to early Pythagoreans, I have already called into question the attribution of significant formal work to such fifth-century arithmeticians; rather, I have argued that this Euclidean theory was initiated with the work of Theaetetus after the beginning of the fourth century.[79] The theory in Book V is generally assigned to

Eudoxus; but for reasons which I have recently discovered and presented in detail elsewhere, there are good reasons for doubting this. For an alternative general technique of proportions, based on the same principle of convergence by consecutive bisection which typifies Eudoxus' "exhaustion" method, is applied in works of Archimedes and others and can readily be employed as the basis of a full theory of proportions. In addition to these approaches is a fourth, framed around the so-called "Euclidean" division algorithm (*anthyphairesis*), and attributable to Theaetetus.[80]

A comparison of these theories enables one to appreciate the formal motive which gave rise to them. The inadequacy of the numerical theory to cover incommensurable magnitudes led to the search for a technique which might apply rigorously to these. As the division algorithm was already basic to the number theory within which the study of incommensurables was conducted, it was natural to attempt an extension of the theory of proportion on the basis of this same technique. One can see how problems with the proofs of specific theorems encouraged the search for alternative approaches, first that utilizing the bisection-method (the one I have attributed to Eudoxus), and ultimately that based on equimultiples, the theory in Euclid's Book V. In this whole development, technical considerations appear to dominate. Indeed, the formal character of the theory of incommensurable magnitudes, where the theorems on proportion were to be applied, establishes from the start a thoroughly mathematical context for these formal problems, without reference to such external stimuli as dialectical paradoxes. It is as likely that Plato and Aristotle were spectators of the progress of the mathematical researches, as that they were instigators of them. On the other hand, the reasons for passing from the Eudoxean bisection version to the Euclidean equimultiple version of the theory seem more strictly formal than technical; in fact, the resultant theory, while formally elegant, tends often to obscure the mathematical ideas in a way the bisection theory does not, and in at least one theorem (VI, 33) actually introduces a logical error absent from its predecessor.

The roots of the "exhaustion" technique are in the pre-Euclidean efforts relating to the quadrature of the circle. One recognizes that among geometers and philosophers alike there were disagreements on the admissibility of certain techniques and concepts. In some mathematical work, procedures appear involving limits, indivisible magnitudes, assumptions and intuitions justified through sense-experience, and the like.[81] These were drawn into dialectical controversies over issues like the relation of perception to abstract thought and the necessary forms of logical reasoning. To claim that a mathematical technique like the method of "exhaustion" was designed to circumvent dialectical difficulties like the Eleatics' paradoxes on the divisibility and mutability of real being surely oversimplifies, if not entirely misrepresents, the context of the mathematical studies. Moreover, it is by no means clear in the fourth century whether the resolutions adopted in the Euclidean *Elements* yet commanded unanimous acceptance among mathematicians. For, as we shall see later, techniques whose formal inadmissibility was unequivocal nevertheless arise in third-century work.

This discussion has intended to show the nature and extent of the influence which dialectical controversies brought to bear on mathematical work, especially in the area of axiomatic studies. The evidence indicates that whatever that influence was, it was exerted in the environment of the fourth-century Academy, rather than earlier; that it affected the expression, rather than the substance, of mathematical concepts and techniques; that the most interesting axiomatic works can be understood as primarily an internal evolution of mathematical thought; and that the mathematicians' work was as much the source for dialectical inquiries as the humble recipient of the strictures of the dialecticians.

The relation of mathematical and philosophical studies in the period following Euclid appears comparably complex. Archimedes produced his major works after the mid-third century, and his earliest contributions dated from no less than three decades after Euclid. Despite this, one perceives that his early works rely on Eudoxean technical models, rather than the Euclidean versions.[82] Among these is *Plane Equilibria I*, the only one of Archimedes' extant works which qualifies as an effort of the axiomatic kind. Unfortunately, its success *as* an axiomatic effort is questionable; as many scholars have argued, the handling of the primary notions – equilibrium and center of gravity – is not uniform and appears to require additional assumptions on the physical properties of balance.[83] Of course, the work is not complete as far as its geometric technique is concerned, requiring many results from plane geometry and the theory of proportions, as well as the technique of "exhaustion".

The formal precision of Archimedes' proofs is beyond question; these served as a model both in his own time and afterward, well into the period of early modern mathematics, of the rigorous style of geometric demonstration. Nevertheless, he makes no claim of an axiomatic intent in his works, and we should avoid imputing any such motive to him. The structure of his treatises is ideally suited for the accurate presentation of specified geometric results, representing the product of original research. Typically, each develops one or two major results entailing the demonstration of a set of theorems, sometimes half a dozen or more depending on the number of cases involved. Leading up to this will be a sequence of theorems establishing results necessary for those theorems. Opening the work will be a few basic results not as specific for the principal theorems of the work, but useful in the proofs. Such lemmas often are stated only, the proofs being assumed from more elementary works. In similar fashion, steps are frequently elided in the course of proofs of the main body of theorems, sometimes with references to works in which the missing proof can be found; again, a lengthy demonstration may be abridged when a portion of it is strictly analogous to an argument already presented.

This makes clear that Archimedes has no aim to set forth exhaustively *all* the materials required; in such advanced studies he can assume much from the elementary literature. In another variation on Euclidean practice, he may *defer* the proof of a step until a later point in the writing, as in an appendix. Archimedes does provide definitions of terms, but only where these are necessary for the work and are presumed novel or unfamiliar.[84] Prefacing *Sphere and Cylinder I* are two interesting series of "lemmas" (*axiomata* and *lambanomena*) assumed in the theorems: the former comparable to "definitions" in Euclid, the latter to "postulates". The discrepancy in terminology would appear to indicate either that Archimedes was not familiar with Euclid's edition of the "Elements", or that he was not particularly disposed toward imitating its formal usage. By including among the *lambanomena* postulates on the relative magnitude of convex arcs and surfaces, as well as the "Archimedean axiom" on continuity, Archimedes reveals a profound insight into the formal treatment of this subject matter. But as there appears to have been no precedent for the former postulates in the earlier literature on axiomatics, mathematical or philosophical, we may be assured that his own experience in working out proofs led him to recognize the appropriate assumptions to make.

Among Archimedes' works two differ from the above pattern in an interesting way. Both the *Quadrature of the Parabola* and the *Method* are devoted in major part to the exposition of a "mechanical method" of research, lacking the full force of geometric demonstration. As noted above, the principal objection a formalist would have doubtless was the introduction of mechanical principles into the technically prior area of geometry. Archimedes justifies his use of this method in the preface to the *Method* addressed to Eratosthenes: by its use one may discover new results and gain insight useful toward their proof, even if the method itself does not amount to proof . It would appear that Archimedes let his guard down, as it were, in the interests of communicating his heuristic approach to a fellow mathematician. The situation of the *Quadrature of the Parabola* is comparable, in that Archimedes was here writing for the first time to an Alexandrian geometer, Dositheus, whom he did not know, but presumably might expect to have use for such a method. In the later writings to Dositheus, however, the formal standards were never again so relaxed; moreover, we receive several indications that Dositheus and his colleagues impressed Archimedes but little as far as their mathematical talents and capacity for creative research were concerned. In other words, the formal precision of these treatises appears to have been devised to command the approval of the Alexandrian professionals, a group whose interest in the formal details of proofs might be termed scholastic.

Pappus has preserved a number of alternative treatments of Archimedean theorems. As I argue elsewhere, one has reason to view these as based on early versions by Archimedes of studies on the sphere, the spiral and other figures which receive a more complete and formal treatment in the writings sent to Dositheus.[85] One may infer also that the earlier versions were addressed to Conon, a geometer much respected by Archimedes, as the prefaces to the formal treatises make clear. Considered then as communications by one geometer to another, the writings preserved by Pappus are of some interest with respect to their formal organization. Those aspects of looseness evidenced to some extent even in the formal treatises are much more prominent here. For instance, the "parts" of a proposition and its proof are not divided in the manner of the formal Euclidean and Archimedean works; assertions are made, not in general, but with reference to the particular elements of specific diagrams.[86] Proofs are severely abridged: major steps are assumed, whole theorems may be stated as "manifest", portions of proofs omitted as "similar" to arguments just presented; in the case of the theorems on spirals the entire convergence argument is absent, indeed is not even hinted at – although its lines could readily be supplied by one conversant with Archimedes' form of the "exhaustion" method. In the context of such a communication, the strictures on the formal organization of proofs would just get in the way, obscuring the essence of the mathematical ideas. In view of this, one might well suspect that the highly formal, detailed treatments of theorems in the treatises did not represent Archimedes' natural and preferred choice for the exposition of his findings; that even in the formal versions the omission of some steps need not be justified by the actual appearance of the step in a prior work (of course, sometimes it does), but may indicate only a sense of what is "obvious". In other words, Archimedes has not the intent to extend the axiomatic structure of mathematics. He is interested in communicating new discoveries in geometry in as clear and precise a manner as possible and according to a format which he deems will meet the approval of his professional colleagues.

Thus, far from abetting the progress of mathematical research, the emphasis on axiomatic issues appears to have distracted, perhaps even impeded, the geometers' work. Certainly, some geometers took this position. We have seen that aspects of Archimedes' work could not have satisfied certain formal criteria, as laid down by Aristotle. The *neuses* in *Spiral Lines* and the "mechanical method" in the *Quadrature of the Parabola* and the *Method* violated notions of formal priority of principles; one suspects that other treatments would face similar objections: the cubature of the cylindrical section in the *Method* (effected via reduction to the quadrature of the parabola) and the quadrature of the spiral in Pappus' version (reduced to the cubature of the cone). In his prefaces, Archimedes takes care to distinguish heuristic from demonstrative procedures, to articulate the assumptions he must make (most notably, the "lemma" on continuity), and the like; but these appear to be defensive gestures, aimed to convince his readers that his discoveries match those of Eudoxus and the other fine geometers of past generations.

In fact, Archimedes several times makes explicit his impatience with the formalists. In the prefaces to *Sphere and Cylinder I* and *Spiral Lines* he expresses regret at Conon's death, for only Conon, an extremely able geometer, had been qualified to judge Archimedes' work. That office was now occupied by Alexandrians who kept sending for details of proofs, but never contributed any new results of their own. This complaint is repeated by later geometers. From Apollonius (preface to *Conics*, Book IV) we know that Conon himself had received some abusive criticism from members of the Alexandrian community – although Apollonius, as a geometer, insisted on the merit even of these offending works. Again, Apollonius was criticized on the basis of some works he allowed to circulate in preliminary versions.[87] An echo of this division between the geometers and their more scholastically oriented colleagues is heard in a passage preserved by Pappus:[88] the geometer disclaims all competence or interest in the business of splitting hairs over first-principles; he wishes instead to gain praise through the utility of his findings. His case is made in the presentation of an ingenious method for evaluating areas and volumes by means of centers of gravity. A formalist might cringe. But how could a geometer not delight in this ancient presentation of the technique now known as "Guldin's rule"?

Were these complaints by the geometers merely a special plea? Did the strictures by formalists actually have any such impeding effects on their work? Certainly, Archimedes and Apollonius possessed techniques of their own useful for the discovery of results, even if these would not survive scrutiny in a formal work. But the justice of their case, I believe, must be decided with reference to the larger field of ancient mathematics. If we take into our view the subsequent development of geometry, using the materials presented by such late commentators as Pappus, Theon, Proclus and Eutocius as indicators, we perceive that among the studies which they take to be advanced work there appears to be not a single contribution which would overreach the abilities of a geometer in the late third century applying only those concepts and techniques then available. Almost all the results they mention could be assigned quite credibly to Archimedes or Apollonius or their immediate followers.[89] Such *stasis* is virtually incomprehensible in comparison with the development of mathematics since the Renaissance. Each interval of fifty or one hundred years has since embraced fundamental changes in the concepts, methods and problems studied by mathematicians. Surely no small part of the explanation for the stagnation in antiquity lies with the scholastic attitude toward mathematics, emphasizing the narrow investigation of purely formal questions. The impact of this emphasis must have been felt in geometry through the curriculum of higher education, by encouraging students toward this scholastic attitude and by limiting their exposure to heuristically useful methods.

III. SUMMARY

We have traced the relation of mathematics and philosophy in the fifth, fourth and third centuries with reference to the development of logical methods and axiomatics. Contrary to the view of Szabó, I have argued that the deductive procedures evident in fifth-century work can be understood as intrinsic to mathematical study, while any strong sense of axiomatic organization is there missing. Thus, the effect of the Eleatic dialectic leading to the rise of formal and axiomatic studies in arithmetic and geometry could only occur later, in the environment of the fourth-century Academy where such an interest in mathematics was combined with a renewal of the logical and philosophical views espoused by the Eleatics.

In this way, many of the features of the formal Euclidean system, which Szabó reads as evidence of a more or less direct influence by the Eleatics in the fifth century, are now argued to be the response to the inquiries into logic in the circles of Plato and Aristotle. But even here, the influence was largely superficial, affecting the form of the presentation of mathematical results, yet only little the problems and concepts and methods of examining them. Where a mathematical advance *was* closely tied to formal matters – most notably, in Eudoxus' contributions to proportion theory and the method of "exhaustion" – any view of one-sided dialectical influence grossly oversimplifies. The nature of these studies combined philosophical and mathematical problems; and in providing material for the continuing dialectical debates on related issues, the mathematicians were as active in shaping the development of philosophical views as responsive to philosophical recommendations concerning the appropriate methodology.

By the third century both philosophers and geometers had resolved the larger issues on the systematization of knowledge, as, respectively, the Aristotelian theory of science and the Euclidean compilation of geometry. The subsequent history of mathematics indicates that the success of this axiomatizing effort eventually served to discourage the creative forms of research which could have advanced mathematical knowledge. Earlier, the philosophical interest in mathematical questions and the exposure of mathematicians, in the course of their education, to the dialecticians' conscious interest in epistemology and the structure of knowledge, had the stimulating effect of making mathematicians aware of the autonomy of their field as an intellectual discipline, rather than as but an art for the practical solution of problems. Unfortunately, mathematics in later antiquity became subordinate to the objectives of the philosophical curriculum. Students might be trained in the subtleties of the foundations of elementary geometry, but they rarely acquired the techniques needed for pursuing researches in advanced fields.

In tracing the interaction of mathematics and philosophy into the later stages of formal studies in axiomatics, we have progressed far from the difficult question with which we opened: what was the manner of the first introduction of deductive methods into Greek geometry? While I have argued that Szabó's hypothesis of a specific influence by the Eleatics is unpersuasive, I have not yet presumed to offer a substitute. Even so, Szabó's view merely pushes our question further back. For we must still inquire into the infusion of rational and deductive modes into the earlier natural philosophy. It does little to assert that dialectics provided the logical model for mathematics, or conversely that mathematics performed this function for dialectics. For there still remains the more basic issue of how deductive methods came to be established at all as the appropriate basis of intellectual inquiry.

I believe that the answer to this question must be sought in the wider cultural context of sixth and fifth century Greece. The political, economic and social environment was then such as to encourage individual expression and thus to give rise to the problem of arbitrating among conflicting policies and opinions. The associated intellectual climate was critical in spirit, sometimes to excess, as in the doctrines of the Eleatics, of the Sophists and of the sceptics. This environment encouraged all thinkers, among these the mathematicians, to look to the coherence of their basic assumptions. In dialectics the grounds of knowledge became a central issue: what is true? how does one know what is true and distinguish it from the false? how does one communicate and teach? The importance of these questions for the thought of Socrates, Plato, Aristotle and others is clear. But surely this general environment had an equivalent impact on the mathematicians. Examining their arithmetic and geometric techniques, they began to seek justifications, such being especially important in problematic areas like the study of incommensurables. Once the quest for justification was underway, the nature of mathematics itself would lead to the implementation of deductive forms, as I have argued. Furthermore, the incentive to teach spurred efforts to organize large areas of mathematics into coherent systems. Hippocrates of Chios was but the first of more than a dozen compilers of "Elements" active within the century before Euclid.

We thus argue that the origins and elaboration of formal methods in mathematics, both in the pre-Euclidean period and afterward, are best understood as a development internal to the mathematical discipline. In response to the same intellectual climate, mathematics and philosophy advanced along parallel lines of development. But the areas of direct interaction between them appear to have had only a limited impact, at least on mathematical studies. Indeed, judging from the epistemological views of Plato and Aristotle, one cannot escape the conviction that the influence of mathematics on philosophy was far more significant than any influence in the converse direction.

BIBLIOGRAPHICAL NOTE

The following works are cited by abbreviations:

AGM = A. Szabó: 1969, *Anfänge der griechischen Mathematik*, Budapest and Munich/Vienna.

EEE = W. Knorr: 1975, *The Evolution of the Euclidean Elements*, Dordrecht.

HGM = T. L. Heath: 1921, *A History of Greek Mathematics*, 2 vol., Oxford.

NOTES

[1] A. Szabó: 'Transformation of mathematics into deductive science and the beginnings of its foundation on definitions and axioms', *Scripta Mathematica* **27** (1964), 27–48A, 113–139 (p.137).

[2] "I do not think we should assume that mathematicians cannot use a logically valid pattern of reasoning in their work until some philosopher has written about it and told them that it is valid. In fact we know that this is not the way in which the two studies, logic and mathematics, are related." – W. C. Kneale in his commentary to A. Szabó, 'Greek dialectic and Euclid's axiomatics' (in I. Lakatos (ed.), *Problems in the Philoso-*

phy of Mathematics, Amsterdam, 1967, pp. 1–27), p.9. While acknowledging that the Eleatics were the first to make conscious use of indirect arguments in dialectic, N. Bourbaki deems it most probable that the mathematicians of the same period had already availed themselves of the same method in their own work (*Éléments d'histoire des mathématiques*, Paris, 1969, p. 11). The "internalist" position is advocated strongly by A. Weil: "[the question is] what is and what is not a mathematical idea. As to this, the mathematician is hardly inclined to consult outsiders.... The views of Greek philosophers about the infinite may be of great interest as such; but are we really to believe that they had great influence on the work of Greek mathematicians?... Some universities have established chairs for "the history and philosophy of mathematics": it is hard for me to imagine what those two subjects can have in common." – 'History of mathematics: why and how', (pp. 6–7): lecture delivered at the International Congress of Mathematicians held in Helsinki, August 15–23, 1978.

[3] In addition to the essay cited in note 1, Szabó has expounded his views on the rise of deductive method in the following: '*Deiknymi* als mathematischer Terminus für *beweisen*', *Maia* **10** (N.S.) (1958), 106–131; 'Anfänge des euklidischen Axiomensystems', *Archive for History of Exact Sciences* **1**, (1960), 37–106; 'Der älteste Versuch einer definitorisch-axiomatischen Grundlegung der Mathematik', *Osiris* **14**, (1962), 308–369. These and other essays have been reworked as the basis of his *Anfänge der griechischen Mathematik*, Budapest/Munich, 1969 (esp. its third part, 'Der Aufbau der systematisch-deduktiven Mathematik'). It is this last-named work to which I will most frequently refer here (to be cited as *AGM*). The first and second parts of *AGM* are concerned with the pre-Euclidean study of incommensurability and the terminology of early proportion theory, respectively. These will not be discussed here, but many points have been examined by me in *The Evolution of the Euclidean Elements*, Dordrecht, 1975.

[4] P. Tannery, *Pour l'histoire de la science hellène*, Paris, 1887, pp. 259–260. His view was elaborated by F. M. Cornford (1939) and J. E. Raven (1948), but has been discounted by most scholars since; for references and discussion, see W. Burkert, *Lore and Science in Ancient Pythagoreanism*, Cambridge, Mass., 1972, pp. 41–52, 285–289 and my *EEE*, p. 43. On the pre-Euclidean "foundations crisis", see H. Hasse and H. Scholz, 'Die Grundlagenkrisis der griechischen Mathematik', *Kant-Studien* **33** (1928), 4–34. This view has been much modified and criticized, as by B. L. van der Waerden (1941) and H. Freudenthal (1963); see my *EEE*, pp. 306–313.

[5] *AGM*, III.1: 'Der Beweis in der griechischen Mathematik', esp. pp. 244–246.

[6] *AGM*, III.3: 'Der Ursprung des Anti-Empirismus und des indirekten Beweisverfahrens.'

[7] See the remark by W. C. Kneale, cited in note 2.

[8] For texts from the ancient Babylonian tradition, see O. Neugebauer, *Mathematical Cuneiform Texts*, New Haven, 1945. For the Egyptian tradition, see A. B. Chace *et al.*, *The Rhind Mathematical Papyrus*, Oberlin, Ohio, 1927–29.

[9] 'Transformation' (see note 1), pp. 45–48; *AGM*, pp. 292f.

[10] *AGM*, III.3: I render Szabo's *anschaulich* as "graphical"; other possibilities are "illustrative", "visual", "perceptual", or even "intuitive". I will understand him to refer to a kind of demonstration based on concrete perceptual acts, such as setting out numbered objects or constructing suitably illustrative diagrams.

[11] Of course, the ancient Babylonian geometers did not do this either. In this sense they too might be viewed as using "non-graphical" approaches in geometry. As far as the Greek classical tradition is concerned, we should beware pushing this point on the non-graphical or abstract nature of the discipline too far. For within it the production of the appropriate diagram was always an integral part of the proof. Proclus, for instance, includes "exposition" (*ekthesis*) and "construction" (*kataskeue*) as proper parts of any demonstration (see note 65 below) and the term for "diagram" (*diagramma*) could actually serve as a synonym for "theorem" (see my *EEE*, ch. III/II). Indeed, the availability of the diagram must have eased considerably the burden of the awkward geometric notation used by the Greeks and perhaps explains why they never saw fit to overhaul it.

[12] See my 'Problems in the interpretation of Greek number theory', *Studies in the History and Philosophy of Science* 7 (1976) 353–368.

[13] On this much-discussed aspect of Plato's epistemology, see, for instance, F. M. Cornford, 'Mathematics and dialectics in the *Republic* VI–VII', (1932), repr. in R. E. Allen (ed), *Studies in Plato's Metaphysics*, London, 1965.

[14] For a discussion of the relevant passages, see T. L. Heath, *Mathematics in Aristotle*, Oxford, 1949, ch. IV, esp. pp. 64–67 and J. Barnes, *Aristotle's Posterior Analytics*, Oxford, 1975, p. 161.

[15] *AGM*, III. 6–9, 13, 17, 21–23, 26.

[16] *AGM*, III. 18, 20, 24, 25.

[17] Proclus, *In Euclidem*, ed. G. Friedlein, Leipzig, 1873, p. 283.

[18] *AGM*, III. 19.

[19] The fragment is preserved by Simplicius, *In Aristotelis Physica*, ed. H. Diels, Berlin, 1882, pp. 60–68. See T. L. Heath, *A History of Greek Mathematics*, Oxford, 1921, I, pp. 182–202.

[20] Notably, *Quadrature of the Parabola* and *Sphere and Cylinder I*. The same format of exposition is followed in several of the discussions in Pappus' *Collection IV* and *V*.

[21] As we shall develop below, neither are the works of Archimedes, save for *Plane Equilibria I*, accurately viewed as efforts at axiomatization.

[22] *AGM*, pp. 450–452. Given the great importance of the Hippocrates-fragment, Szabó's discussion of it in *AGM* (part III) is remarkably slender, and what he does say inconsistent. On the one hand, Szabó insists on the deductive, even axiomatic, form of Hippocrates' work: the "Elements" attributed to him by Proclus must surely have had some sort of foundations (pp. 309f, 342). Szabó, of course, wishes to assert the nascent axiomatic form of early Greek geometry as a mark of Eleatic influence. Yet he elsewhere stresses the non-axiomatic nature of Hippocrates' study of the lunules; for instance, Hippocrates' "beginning premisses" are *lemmas*, whose proofs are given or assumed, *not* "first-principles" in the Aristotelian axiomatic sense (pp. 330f). Presumably, if the mathematicians had advanced too far in this direction so early, one might propose them as rivals to the Eleatics in the initiation of foundational inquiries. Similarly, Szabó points out that Hippocrates does not use indirect reasoning in the fragment (p. 331) – without, we should observe, recognizing the inappropriateness of such reasonings for this subject-matter; while he leaves the issue open, for want of explicit documentation, as to whether Hippocrates knew of this method, he clearly wishes to deny to the mathematicians priority in its use. Yet his criticism of Toeplitz'

reconstruction of a direct "inductive" proof of the area of the circle is based on the possibility that Hippocrates could have applied some form of the "exhaustion method". This latter technique, ascribed to Eudoxus and applied throughout *Elements* XII, is especially characterized by its use of indirect reasoning. Szabó's intent here is to attack Toeplitz' thesis of the "Platonic reform" of geometry, for Szabó views the adoption of formal methods in geometry as a direct response to the Eleatics much before Plato. It is clear that Szabo has shifted his positions on the interpretation of Hippocrates' work, without regard to consistency, in order to gain a favorable argumentative stance as context recommends.

[23] See Heath, *HGM* I, pp. 221f, 327–329.

[24] *Ibid.*, pp. 198f.

[25] Proclus, *In Euclidem*, p. 213.

[26] On the nature of this method and the extensive scholarship on it, see J. Hintikka and U. Remes, *The Method of Analysis*, Dordrecht, 1974, esp. ch. I.

[27] I do not here claim that Szabo presumes to maintain this.

[28] *Prior Analytics* II, 25.

[29] *In Euclidem*, p. 66.

[30] *Evolution of the Euclidean Elements*, ch. VI/IV.

[31] On the early history of the study of incommensurables, see my *EEE*, ch. II. As far as the earliest discoveries are concerned, I argue a dating not much before 420 B.C. and a method involving a form of "accidental" discovery rather than any deliberate formal proof (e.g., as in the indirect argument based on the odd and even). The possibility – indeed, the likelihood – that these early discoveries were accidental in some such way undercuts any attempt, such as Szabo's, to use the *formal* character of the study of incommensurables as a debt to the Eleatics (see *AGM*, I, 12; III, 2).

[32] *EEE*, ch. III/III–IV.

[33] On Archytas, see Heath, *HGM*, I, pp. 213–216.

[34] This view is prominent in B. L. van der Waerden, *Science Awakening*, Groningen, 1954, ch. 5, esp. p. 115.

[35] The fragment is preserved in Latin translation in Boethius, *De institutione musica*; see my discussion in *EEE*, ch. VII/I.

[36] Cited by Porphyry, *In Ptolemaei harmonica* (cf. H. Diels and W. Kranz, *Fragmente der Vorsokratiker*, 6. ed., Berlin, 1951, 47*B*2).

[37] Eutocius, *In Archimedem*, in the commentary to *Sphere and Cylinder* II, 2; cf. Archimedes, *Opera*, ed. J. L. Heiberg, III, Leipzig, 1915, pp. 84–88 and Heath, *HGM*, I, pp. 246–249.

[38] *AGM*, III. 20. We return to the use of motion in the Greek geometry below.

[39] See my *EEE*, ch. II and note 31 above.

[40] *AGM*, III. 30, esp. pp. 446f. Becker develops his view in 'Die Lehre vom Geraden und Ungeraden . . .', *Quellen und Studien*, 1936, 3:B, pp. 533–553. While here subscribing to Becker's reconstruction, Szabo had earlier dismissed it on the grounds that knowledge of certain *results* (such as the properties of odd and even numbers presented in *Elements* IX) does not in itself justify ascribing to the Pythagoreans comparable formal *proofs*. As elsewhere Szabó here adopts inconsistent positions as context demands (cf. note 22 above).

[41] *AGM*, III. 10, 13, esp. pp. 329, 341f.

[42] *AGM*, III. 17–20 (on the postulates), 21–25 (on the axioms). Heath covers this material in considerable detail in *Euclid's Elements*, 2. ed., Cambridge, 1926, I, ch. IX. We will take up the question of Euclid's relation to Aristotle on the classification of the "first-principles" in the second part of this paper.
[43] *AGM*, III. 20, 29.
[44] *In Euclidem*, pp. 185–187.
[45] Indeed, it is accepted by P. Bernays in his commentary to the paper by Szabó cited in note 2 above. S. Demidov also raises this view in his commentary on the present paper.
[46] Archytas: see note 37 above. Eudoxus *et al.*: Heath, *HGM*, I, p. 255 and I. Thomas, *History of Greek Mathematics*, London, 1939, I, pp. 262–266, 388. Archimedes, *Spiral Lines*, preface, definitions (preceding prop. 12), and prop. 1, 2, 12 and 14. Motion-generated curves (e.g., "quadratrix", "conchoid"): Heath, *op. cit.*, pp. 238–240, 260–262. Related to this is the solution of problems by means of *neuses* ("inclinations", that is, constructions involving a sliding ruler); see Heath, *op. cit.*, pp. 235–238, 240f, and my article, 'Archimedes' *Neusis*-Constructions in *Spiral Lines*', *Centaurus* **22** (1978), pp. 77–98.
[47] Heath, *Euclid*, I, pp. 224ff.
[48] Such a recasting of the proof was recommended by Russell; cf. Heath, *ibid.*, pp. 249ff.
[49] For instance, the congruence of circular segments subtending equal arcs in equal circles would follow from an "exhaustion" proof based on the triangular case.
[50] Archimedes: *Plane Equilibria I*, Axiom 4 and prop. 9, 10 (*aliter*); *Conoids and Spheroids*, prop. 18. Pappus employs superposition in his proof of the proportionality of arcs and sectors in equal circles (*Collection* V, 12).
[51] *In Euclidem*, pp. 188–190, 249f (citing Pappus).
[52] *Posterior Analytics*, I, 13.
[53] For a review of the *Elements* and its relation to the pre-Euclidean studies, see my *EEE*, ch. IX.
[54] A. Seidenberg, 'Did Euclid's *Elements*, Book I, Develop geometry axiomatically?' *Archive for History of Exact Sciences* **14** (1975), 263–295.
[55] The literature on the Euclidean and Aristotelian divisions of the first-principles is rather large. In addition to the studies by Heath and Szabó cited in note 42 above and to Barnes' notes on the *Posterior Analytics* (cf. note 14), the following may be considered: H. D. P. Lee, 'Geometrical method and Aristotle's account of first principles', *Classical Quarterly* **29** (1935), 113–124; B. Einarson, 'On certain mathematical terms in Aristotle's logic', *American Journal of Philology* **57** (1936), 33–54, 151–172; K. von Fritz, 'Die APXAI in der griechischen Mathematik', *Archiv für Begriffsgeschichte* **1** (1955), 13–103. J. Hintikka discusses this question in his paper in the present volume. Recent contributions include B. L. van der Waerden, 'Die Postulate und Konstruktionen der frühgriechischen Geometrie', *Archive for History of Exact Sciences* **18** (1978), 343–357. On constructions and existence-proofs, see note 70 below.
[56] *In Euclidem*, pp. 178–184.
[57] Cf. Szabó, *AGM*, III. 17, 20, cf. 21–26.
[58] For instance, opening *Plane Equilibria I*, Archimedes "postulates" properties of

equilibrium and centers of gravity. Proclus would prefer that he have used "axiom" in this context (*In Euclidem*, p. 181).

[59] Proclus, *In Euclidem*, pp. 77–78.

[60] Surely it is this wide applicability which accounts for the designation of these principles as "common", both by Aristotle and by Euclid; cf. the passages discussed by Heath, *Mathematics in Aristotle*, pp. 53–57, 201–203. Admittedly, Aristotle also describes the "axioms" as impossible to be mistaken about (*Metaphysics* 1005*b*11–20), but yet indemonstrable (*ibid.*, 1006*a*5–15). Apparently this has given rise to the alternative view that these "common notions" are "common *to all men*", as suggested in Heiberg's rendering of Euclid's *koinai ennoiai* as *communes animi conceptiones* (Szabó also subscribes to this view; cf. *AGM*, III. 25).

[61] See the contributions cited in note 55.

[62] See note 60.

[63] One may note that the attempt, as by Szabó, to assign Euclid a dialectical motive for both his "postulates" and his "common notions" would impute to him an even more extreme dialectical position than that suggested by Aristotle; for this would make Euclid's *reason* for articulating the "common notions" the desire to oblige those extremist critics who might challenge not only the "postulates" (some of which might, after all, appear to permit of proof) but also the "axioms" which seem self-evident.

[64] I will review below the familiar thesis of Zeuthen, that constructions served the role of existence proofs in the ancient geometry. One should note well that the modern conception of proofs of existence in mathematics and logic has been greatly influenced by developments in the fields of analysis and set theory in the late nineteenth century. There is thus a real danger of interpreting early geometry anachronistically in the case of such questions.

[65] Proclus gives an account of the formal subdivision of a theorem and its proof: *In Euclidem*, pp. 203–205; cf. Heath, *Euclid*, I, pp. 129–131.

[66] Indeed, there are many such steps in the ancient geometry which passed by without being then recognized as tacit assumptions; cf. the discussion by Becker cited in note 72.

[67] Typically, auxiliary elements may be introduced in the "construction" (*kataskeue*) later in the proof; cf. Hintikka's discussion of auxiliary constructions in the book cited in note 26.

[68] 76*a*31–35. I have since come upon the article by A. Gomez-Lobo, 'Aristotle's hypotheses and the Euclidean postulates', *Review of Metaphysics* **30** (1977), 430–439 in which the same view is argued in detail.

[69] These postulates assert (1) that the line segment connecting two given points may be drawn; (2) that a given line segment may be indefinitely extended in a straight line; and (3) that the circle of given center and radius may be drawn. The remaining two postulates do not involve construction as such: (4) that all right angles are equal; and (5) that if two lines are cut by a third such that the interior angles made on one side of the third are less than two right angles, then the given lines, extended sufficiently far, will meet on that side of the third.

[70] The primary statement of this thesis is by H. G. Zeuthen, 'Geometrische Konstruktion als Existenzbeweis . . .', *Mathematische Annalen* **47**, (1896), 272–278. It has since been elaborated by O. Becker, *Mathematische Existenz*, Halle a. d. S., 1927; A. D.

Steele, 'Ueber die Rolle von Zirkel und Lineal in der griechischen Mathematik', *Quellen und Studien* **3:B**, (1936), 288–369; and E. Niebel, '. . . die Bedeutung der geometrischen Konstruktion in der Antike', *Kant-Studien*, Ergänzungsheft, **76**, Cologne, 1959. I criticize Zeuthen's view in *EEE*, ch. III/II (cf. also note 64 above). Van der Waerden and Seidenberg also deny the existential sense of Euclid's postulates (cf. notes 54 and 55 above).

[71] Even proofs of incommensurability do not assume the form of negative existence theorems (i.e., "there exist no integers which have the same ratio as given magnitudes"). In this regard, the closest one comes to an existential expression is in *Elements* X, 5–8; e.g., "commensurable magnitudes have to each other the ratio which a number has to a number" (X, 5). Cf. also X, Def. 1: "magnitudes are said to be commensurable when they are measured by the same measure, but incommensurable if none can become (*genesthai*) their common measure". In such instances an existential statement is hardly to be avoided. Even so, it is remarkable how these formulations emphasize the *properties* of assumed magnitudes, rather than the *existence* of magnitudes having these properties.

[72] These issues are examined by O. Becker, 'Eudoxos-Studien I–IV', *Quellen und Studien* **B:1–3**, 1933–36 (esp. II and III).

[73] For a survey of these constructions, see Heath, *HGM*, I, ch. VII and the items cited in note 46 above.

[74] For instance, Hippocrates' third lunule (Heath, *ibid.*, p. 196) and Archimedes' construction in *Spiral Lines*, prop. 5, are both effected by *neusis*, even though a "Euclidean" construction is possible; this is discussed in my article, cited in note 46 above.

[75] Pappus, *Collection* IV, 36; cf. my article on *neusis* (note 46 above).

[76] Aristotle set out a doctrine of the "priority" of principles in the *Posterior Analytics* I, 6–7.

[76a] In the *Method* Archimedes reduces the problem of the volume of cylindrical section to that of the area of the parabolic segment. Doubtless, he knew of the equivalence of the problems of finding the area of the parabola and that of the spiral; but this notion is used in neither of the treatises devoted to these problems and does not seem to have appeared before Cavalieri's *Geometria Indivisibilibus* (1635). The reduction of the area of the spiral to the volume of the cone is employed in Pappus' *Collection* IV, 22 in a treatment which appears to stem from Archimedes (see my article cited in note 85). The center of gravity of the paraboloid was solved by Archimedes in a lost work *On Equilibria* and applied in the extant work *On Floating Bodies*; see my discussion in 'Archimedes' lost treatise on centers of gravity of solids', *Mathematical Intelligencer* **1**, (1978), 102–109. In this instance, it seems that Archimedes used an independent summation-procedure based on the same expressions proved in *Spiral Lines* and *Conoids and Spheroids*.

[77] I. Toth, 'Das Parallelproblem im Corpus Aristotelicum', *Archive for History of Exact Sciences* **3**, (1967), 249–422.

[78] It is not even clear whether among the Eleatics the arguments on the nature of being were intended as a possible basis for a cosmological system, or whether they were merely a negative device for the criticism of other cosmologies.

[79] See my *EEE*, ch. VII. Theaetetus' role in the foundation of number theory was

argued by H. G. Zeuthen, 'Sur les connaissances géométriques des Grecs avant la réforme platonicienne', *Oversigt Dansk. Videns. Sels. For.*, 1913, pp. 431–473.

[80] I present the reconstructed "Eudoxean" theory in 'Archimedes and the Pre-Euclidean Proportion Theory', *Archives internationales d'histoire des sciences*, **28** (1978), 183–244. The anthyphairetic theory was first proposed by Zeuthen and by Toeplitz, later elaborated by O. Becker, 'Eudoxos-Studien I' (see note 72). I review and modify his reconstruction of this theory in *EEE*, ch. VIII/II-III and Appendix B.

[81] See, for instance, the circle-quadratures of Antiphon and Bryson (Heath, *HGM*, I, pp. 220–226). Democritus conceived solids as somehow constituted of parallel indivisible plane sections (*ibid.*, pp. 179f). The latter has been construed (I believe, mistakenly) to be the basis of an ancient infinitesimal analysis; cf. S. Luria, 'Die Infinitesimaltheorie der antiken Atomisten', *Quellen und Studien* **2:B**, (1933), 106–185.

[82] On Archimedes' early works and their dependence on pre-Euclidean sources, see my 'Archimedes and the *Elements*', *Archive for History of Exact Sciences* **19**, (1978), 211–290 and the article cited in note 80.

[83] For a review of criticisms of this work, see E. J. Dijksterhuis, *Archimedes*, 1957, ch. IX and my articles cited in notes 80 and 82 above.

[84] See, for instance, the definitions given in *Spiral Lines* and *Conoids and Spheroids* (prefaces).

[85] See my article cited in note 82 above and also 'Archimedes and the spirals', *Historia Mathematica* **5**, (1978), 43–75.

[86] Among the Archimedean treatises only *Quadrature of the Parabola* adopts this usage.

[87] See Apollonius, *Conics* I, preface and Hypsicles, *Elements* XIV, preface.

[88] Pappus, *Collection* VII, ed. Hultsch, pp. 680–682.

[89] In certain fields outside the area of formal geometry notable advances *were* made in later antiquity: for instance, plane and spherical trigonometry, mathematical astronomy and number theory (in the Diophantine tradition). Each of these was associated with techniques of practical computation, rather than theoretical geometry, and none was systematized along the lines of Euclid, Archimedes and Apollonius. The non-axiomatic character of much ancient mathematics and mathematical science is examined by P. Suppes in his contribution to this volume and by F. Medvedev in his commentary.

FILIPPO FRANCIOSI

SOME REMARKS ON THE CONTROVERSY BETWEEN PROF. KNORR AND PROF. SZABÓ

As a premise to his theory of the origin of the axiomatic method, Szabó puts forward the abstractness of the Greek mathematics, which distinguishes it from the practical nature of Egyptian and Babylonian science. Moreover he maintains that Greek mathematics was abstractive and antiempirical right from the beginning, without any gradual process leading to these features. There was, on the contrary, Szabó supposes, a sudden revolution which affected only the Greek way of thinking, so that in consequence of this revolution, we have, not a transformation of earlier scientifical thought, but the birth of science itself.

The authors of this turn of helm were, according to Szabó's theory, the philosophers of Elea. They had the first idea of dialectical thinking and they formulated consciously the fundamental principles of logic. As a result of the antiempirical and antisensual character of Eleatic gnoseology, not only mathematics received the traits of a deductive science in the general sense of these words, but also basic elements of the deductive-axiomatic method appeared; e.g., the concepts of proposition, theorem, axiom etc., and the procedure of indirect demonstration.

It was Plato who received the heritage of the Eleatic school and developed it, giving to mathematical entities a distinguished place in his metaphysical system, and praising the reasoning of the mathematicians as a pattern for the dialectical thinkers.

Finally, Szabó distinguishes between the birth of the axiomatic method and the systematization we find in the *Elements* of Euclid. The former was, as we said above, a product of Eleatic philosophy; the second is the final act in a process that lasted more than two centuries, some traces of which may be discovered in the works of authors from Plato to Euclid himself.

There are two different approaches to the same subject in the theory of Professor Szabó on the origin of mathematical axiomatics and in Professor Knorr's comments on the latter, as they have different ideas of mathematics itself. One of them speaks of this science in a Platonic sense of the word, as γνῶσις τοῦ αἰεὶ ὄντος (*Rep.* 527b), that is,

J. Hintikka, D. Gruender, and E. Agazzi (eds.), Pisa Conference Proceedings, Vol. I, 187–191.

as a science "founded upon an intuition of abstract entities [1]", whereas Knorr claims that "it would be possible a different interpretation of the mathematician's work, as it could be found in Aristotle's view of the abstraction of general forms from the inductive consideration of particulars" [2].

Of course, I do not claim for myself the privilege of judging an ideology: if one is a Platonist, and the other an Aristotelian, each may be right. The question is only: Did ancient mathematics receive more influence from Platonism, or from Aristotelism? Does Greek mathematics reflect more Platonic or Aristotelian thinking?

I think the answer to these questions is rather easy. Many scholars both in antiquity and in modern times emphasized Plato's influence upon mathematics [3]. On the other hand, Euclid does not reflect any Aristotelian point of view. Certainly, there are some traces of Aristotelian thought in the ancient commentaries on Euclid [4], but hardly in Euclid's work itself. Even the terminology of mathematical axiomatics is independent on Aristotle's syllogistic logic.

There appears to be some difference between the two opponents on how they construe the fundamental concepts. Mr. Knorr speaks of axiomatization, but it is not quite clear—or it is at least for me not quite clear—how he wants this term (axiomatization) to be understood. One is tempted to think of Hilbert's approach to the problem. But did axiomatization in such a sense exist in antiquity? Szabó speaks on the other hand of a fundamental transformation of mathematics by the Greeks. That is, he has in mind the process I mentioned in the summary: not only mathematics became 'deductive' in the general sense of this word, but such fundamental new concepts appeared, as *proposition*, *theorem*, *demonstration*, and the *principles* of mathematics [4]. (I should like to emphasize here, speaking of the mathematical *principles* (*ἀρχαί*), not only the importance of *axioms* and *postulates*, but also that of the *definitions*, but which mathematics becomes speak independent of empirical reality: the proofs of propositions are not referred simply to empirical facts, but to the definitions of the same facts, i.e. how the facts were defined).

Therefore Szabó characterizes Greek mathematics as "antiempirical and consciously in avoidance of graphic methods of proof" [5]. Knorr seems to agree, at least to a certain degree, to accept such a point of view as applied only to *arithmetic* [6], but he is surprised that the same words of Szabó might refer also to Greek geometry because, as Knorr puts it, "the construction of diagrams is a characteristic vehicle for the organization of the proof" [7]. I wonder whether Mr. Knorr has not forgotten that Plato himself spoke of a kind of misleading visibility in geometry: the visible diagrams of the geometers should only remind us of things that are not visible at all [8]. Moreover Szabó was not alone in emphasizing the antiempirical traits of Euclid's geometry. As the mathematician Reidemeister wrote: "die Euklidischen Elemente dürfen nicht jenen anschaulichen Sachverhalten gleichgesetzt werden, die sie beweisen... (Mit den Pythagoreern) vollzieht sich etwas Ausserordentliches. Das anschauliche Quadrat... entschwindet beim näheren Zugriff der Untersuchung" etc. etc. [9].

There also are other divergences between the two scholars. Knorr accepts the Eleatic influence upon Greek mathematical thinking, but not the chronology, as Szabó maintains that all evidence of axiomatization should lead us to the fifth century and to the Eleatics. On the other hand Knorr speaks rather about "Plato's Eleatism". He believes also that it is unnecessary to dwell on the philological aspects of the problem [10] in order to decide the unsettled point. That is a pity, as Szabó's evidence is rather of philological kind: even Knorr begins his work with Oenopides of Chios, while on the contrary other historians, not only Szabó but also Becker, have shown by philological methods that earlier proved theorems of mathematics culminating in the proof of incommensurability of diagonal and side in a square [11]. Knorr seems to reject all these discoveries by Becker in order to begin his argument in much later period.

By combining this divergence on chronology with the contrasting views of Szabó and Knorr as to the relation of Greek mathematics to sense-experience, we come to the core of our subject matter: the origin of axiomatic foundation of mathematics. Knorr says that "Szabó's arguments were circular, in asserting the antiempirical character of Greek mathematics and then deducing from this the influence of the Eleatics" [12]. He maintains on the contrary that the axiomatizing work did not effectively begin until the early fourth century, stimulated by the above mentioned "revival of Eleatism in Plato's Academy", and that we could speak only in this sense of an Eleatic influence on the axiomatization. Knorr's opinion induces me to ask myself questions such as these: If Plato claims that a discussion of his could attain the rigor of mathematical reasoning [13], are we allowed to suppose that he would have expressed such a wish when the mathematical method of demonstrating was not yet available in its due rigor? If he writes that geometry is founded on the principles of *odd* and *even*, and on the three kinds of angles [14], are we entitled to think that the theory of *odd* and *even* had not yet become a mathematical one, and that deductive method and indirect demonstration were not employed at that time? In short, how could mathematics have such a distinguished place in Plato's system, if it had not yet grown to an exact and abstract science? Is it not strange that Plato himself should have stimulated this growth only in a late or, as Knorr puts it, in a "new Eleatic" period of his life and thought? Besides, if Hippocrates of Chios had already written a work called *Elements*, how could he begin his work without a list of fundamental assumptions?

I think the key issue is that we must not confound the origin and development of the foundation of mathematics on axioms and definitions with that systematization that Euclid presented – especially of geometry – in his *Elements*. The former one, the foundation of mathematics on axioms and definitions, or rather the striving for such a foundation, was the product of a powerful turn given to human thought by a few men, Pythagoras, Parmenides and Zeno. The first of them conceived the idea of considering and studying numbers as abstract entities, as Plato affirms [15]. The second lent an exclusive existence to the Being: as a consequence of this metaphysical principle the basic principles of logic were possible on which axiomatic science could be constructed. Zeno defended his master's doctrine, accepted the implications of the discovery of the irrational; and in consequence of his thought mathematicians resolved to state as postulates that it is always possible to consider straight line joining two given points, and that it is possible to divide a line as many times as it is necessary [16].

On the other hand, it is obvious, and proofs were adduced by several scholars, that Euclid made in his systematizating work a choice among older traditional mathematical principles [17]. I see indeed gradual evolution process of *organization* of geometry by the Greek. We even have historical evidence of this in the series of authors of *Elements* that Proclus refers to [18]. I am, however, of the opinion that the origin of axiomatizing tendency and the discovery of the basic principles of mathematics was not a gradual process.

Mr. Knorr cannot imagine an early and sudden beginning of theoretical science: all had to go on gradually, from concrete to abstract according to Aristotle's way of thinking. I hope, he won't take offence if I say that Plato would have included him among the "Sons of earth grasping at rocks and oaks", of whom he speaks in the *Sophistes* (246 a–b), referring, as some scholars now suppose [19], to his disciple Aristotle and his fellows in the Academy. Mr. Knorr should have grasped more at historical and philogical evidences and less at "rocks and oaks".

University of Padua

NOTES

1. Knorr, W. R.: 'On the Early History of Axiomatics', this volume pp. 145–186.
2. See note 1 above, especially pp. 149, 165–167.
3. See for instance Ž. Marković, 'Les mathematiques chez Platon et Aristote', *Bulletin International de l'Academie Yougoslave des Science et des Beaux-Arts* 32 (1939), p. 1–21; A. Frajese, *Platone e la matematica nel mondo antico*, Roma 1963; *Attraverso la storia della matematica*, Firenze 1969, pp. 64–86.
4. See for instance Procl. *In I Eucl.* p. 75 Friedlein.
5. Knorr, 'On the early History of Axiomatics'. Cf. A. Szabó, *Anfänge der griechischen Mathematik*, Budapest 1969, p. 252.
6. Plato himself thinks so: See *Rep.* 526a.
7. See note 1.
8. Plato, *Rep.* 510d–e; *Phaedon* 73a.
9. Reidemeister, K.: 1949, *Das exakte Denken der Antike*, Hamburg, p. 10–11.
10. See note 1.
11. Szabó, *Anfänge*, p. 263–287; O. Becher, 'Die Lehre vom Geraden und Ungeraden im Neunten Buch der Euklidischen Elemente,' in *Zur Geschichte der griechischen Mathematik* (Wege der Forschung XXXIII), Darmstadt 1965, p. 136–7.
12. See note 1.
13. Plato, *Phaedo* 100c; *Theaet.* 162e.
14. Plato, *Rep.* 510c.
15. Plato, *Rep.* 527b.
16. Euclid, *Elem. I*, Post. I; prop. 10. Szabó, *Anfänge*, p. 405–407.
17. Szabó, A. 'Anfänge des euklidischen Axiomensystems,' in Zur Gesch. d. gr. Math., p. 394; Frajese A., Maccioni L.: *Gli Elementi di Euclide*, Torino 1970, p. 65 ff.
18. Procl. In *I Eucl.* p. 66–67 Fr.
19. Diano, C.: *Studi e saggi di filosofia antica*, Padova 1973, p. 278.

WILBUR RICHARD KNORR

ON THE EARLY HISTORY OF AXIOMATICS: A REPLY TO SOME CRITICISMS

In my discussion above on the formal geometric tradition in the pre-Euclidean period, I have sought to sketch out what I view to be a plausible view of the interrelation of mathematical and philosophical studies which were then contributing to the elaboration of deductive methods. I have indicated my grounds for finding the alternative view proposed by Prof. Szabó implausible. In answer to the remarks made by Dr. Filippo Franciosi, I wish to emphasize these points:

In defense of Szabó's view of the Pythagorean mathematics as an outgrowth of Eleaticism, Dr. Franciosi has indicated that Plato knew of a highly developed – indeed, axiomatic – arithmetic theory, and that therefore Plato could hardly have influenced the origins and elaboration of such a theory. But we must consider closely the nature of that arithmetic theory: on what grounds do we suppose there to have been a formal theory, of the type of that in Euclid's Books VII–IX, as early as 400 B.C.? Here, Szabó and Franciosi refer to the study of O. Becker, who reconstructed a Pythagorean theory of odd and even numbers on the basis of a systematization of the representation of integers by means of pebbles.[1] To be sure, such a representation of numbers was known, at least as early as the mid-fifth century; again, one *can* in this way provide step-by-step deductive proofs of each of the theorems on odd and even which Euclid treats (under a different representation of numbers) in Book IX. Nevertheless, we have stronger reasons for doubting that any such formal theory of arithmetic was in fact produced by the early Pythagoreans. First, Plato criticizes the arithmeticians of his time for their inattention to deductive issues; for instance, he says, they "hypothesize" what the odd and even are, rather than give proper formal definitions.[2] Surely, then, the Pythagorean arithmetic he knew was deficient in this fundamental respect of formal method. Second, as we have mentioned in connection with Archytas, not even the whole *technical* apparatus – let alone the formal structure – of number theory as we now possess it in Euclid's arithmetic books was available at the beginning of the 4th century.[3] Hence, we may be confident in asserting that number theory was still far from the consciously formal stage of Euclid at the earlier

J. Hintikka, D. Gruender, and E. Agazzi (eds.), Pisa Conference Proceedings, Vol. I, 193–196.

time of Plato. Third, Becker's argument is a reconstruction of what Pythagorean arithmetic *may* have been; his reconstruction is based on the *assumption* that this theory was already axiomatic in form. One is thus guilty of reasoning in a vicious circle if one proposes to use Becker's reconstruction as a validation of the claim that the early Pythagorean arithmetic was in fact axiomatic.

Dr. Franciosi has raised the question of Euclid's philosophical affiliation: was he Platonist or Aristotelian? Presumably, if we can ascribe to him a Platonist conception of mathematical being, we can thereby identify an Eleatic influence on the development of geometry, for without question the Eleatic views on being had great importance for Plato's ontology. Now it is true that Proclus labels Euclid a Platonist.[4] But his evidence on this count is hardly persuasive. It consists merely of the fact that Euclid closes the *Elements* with the construction of the five regular solids (Book XIII) and these are known as the "Platonic solids" and important for the Platonic cosmology (as in the *Timaeus*). The fact that Proclus himself was prominent in the late Academy in Athens may suggest why he should wish his readers to believe that Euclid too was a Platonist.

Of course, it is irrelevant to my argument which, if either, of these philosophies Euclid inclined toward. *If* Euclid *was* a Platonist, there *might* be cause for reading into his work an explicit dialectical – as opposed to mathematical – motive and this *might* point to an influence by the Eleatics, if only indirect, as Szabó proposes. But it is quite equally *possible* that the mathematician could hold a different view – or perhaps none at all – on the philosophy of mathematics. We know that the Aristotelian view (mathematical concepts as *abstractions* from physical properties) is one such possible view. Hence, Szabó is again reasoning in a circle in presuming Euclid's Platonism and then using this to confirm the Eleatic roots of his method. Evidence for those Eleatic roots must be gotten elsewhere; and the argument of my paper has been that such evidence cannot in fact be found.

I have recommended de-emphasizing any conscious dialectical-philosophical motivations on the part of the author of the *Elements*. Euclid himself did not compose all the substance of the *Elements* – and perhaps not even very much of it. He relied on the work of prior geometers, many of whom we know at least by name, who collaborated in the environment of the Academy in the middle and later parts of the fourth century.[5] The school of Aristotle was fully aware of the Platonic approaches in logic and epistemology and a debate flourished between the schools over the issue of how to organize scientific knowledge. The books Euclid relied on – works stemming from Theaetetus, Eudoxus and their followers – were composed in this environment. Surely, we would not deny that the logical debates being conducted around them influenced somehow the decisions made by these writers on the proper form and content of the first principles of geometry. Euclid himself presumably trained in this same atmosphere in Athens. Thus, I maintain, we have the best cause for relating Euclid's *Elements* to this fourth-century context of debate. By contrast, we have little cause to invent some direct, but hidden, link between Euclid and the Pythagoreans who worked in the shadow of the fifth-century Eleatics.

Piecing together a coherent account of the origins of deductive method in antiquity is especially difficult, not only because of the fragmentary character of the evidence, but also because of the potential unreliability of our intellectual intuitions in the matter. We are looking for the beginning of the deductive sense within mathematics. But no mathematician trained in contemporary mathematics would conceive that mathematicians ever viewed their work in a way essentially different from the way he views his own – for instance, without a sense of the meaning and need for proof and abstraction. Yet it is just these features, so characteristic of the formal tradition of ancient geometry, which appear to be lacking in the traditions prior to the Greeks of the fifth century. Ultimately, we can hope for plausibility in the account of how this change occurred in the pre-Euclidean period. But there are are higher and lower degrees of plausibility. I am convinced that any account will be the stronger and more plausible when it seeks to relate the *Elements* to the thriving mathematical and philosophical activity of the fourth century, and that Szabó's approach hypothesizing the subordination of Pythagorean mathematics to Eleatic philosophy in the poorly documented and mathematically less active fifth century will hardly produce a convincing and plausible interpretation.

NOTES

[1] Becker, O.: 1936, 'Lehre vom Geraden und Ungeraden . . .', *Quellen und Studien* **3:B**, 533–553.

[2] Plato, *Republic*, 510–511.
[3] See my *Evolution of the Euclidean Elements*, ch. VII/I.
[4] Proclus: 1873, *In Euclidem*, ed. by G. Friedlein, Leipzig, p. 68.
[5] I present a view of Euclid's relation to his predecessors in *Evolution*, ch. IX.

PATRICK SUPPES

LIMITATIONS OF THE AXIOMATIC METHOD IN ANCIENT GREEK MATHEMATICAL SCIENCES

My thesis in this paper is that the admiration many of us have for the rigor and relentlessness of the axiomatic method in Greek geometry has given us a misleading view of the role of this method in the broader framework of ancient Greek mathematical sciences. By stressing the limitations of the axiomatic method or, more explicitly, by stressing the limitations of the role played by the axiomatic method in Greek mathematical science, I do not mean in any way to denigrate what is conceptually one of the most important and far-reaching aspects of Greek mathematical thinking. I do want to emphasize the point that the use of mathematics in the mathematical sciences and the use in foundational sciences, like astronomy, compare rather closely with the contemporary situation. It has been remarked by many people that modern physics is by and large scarcely a rigorous mathematical subject and, above all, certainly not one that proceeds primarily by extensive use of formal axiomatic methods. It is also often commented upon that the mathematical rigor of contemporary mathematical physics, in relation to the standards of rigor in pure mathematics today, is much lower than was characteristic of the 19th century. However, my point about the axiomatic method applies also to 19th-century physics. There is little evidence of rigorous use of axiomatic methods in that century either. This is true not only of the periodical literature but also of the great treatises. Three casual examples that come to mind are Laplace's *Celestial Mechanics*, his treatise on probability, and Maxwell's treatise on electricity and magnetism.

Three examples from ancient Greek mathematical sciences that I have chosen to comment on are Euclid's *Optics*, Archimedes' *On the Equilibrium of Planes*, and Ptolemy's *Almagest*.

I. EUCLID'S *OPTICS*

It is important to emphasize that Euclid's *Optics* is really a theory of vision and not a treatise on physical optics. A large number of the

J. Hintikka, D. Gruender, and E. Agazzi (eds.), Pisa Conference Proceedings, Vol. I, 197–213.

propositions are concerned with vision from the standpoint of perspective in monocular vision. Indeed, Euclid's *Optics* could be characterized as a treatise on perspective within Euclidean geometry. The tone of Euclid's treatise can be seen from quoting the initial part, which consists of seven 'definitions'.

1. Let it be assumed that lines drawn directly from the eye pass through a space of great extent;
2. and that the form of the space included within our vision is a cone, with its apex in the eye and its base at the limits of our vision;
3. and that those things upon which the vision falls are seen, and that those things upon which the vision does not fall are not seen;
4. and that those things seen within a larger angle appear larger, and those seen within a smaller angle appear smaller, and those seen within equal angles appear to be of the same size;
5. and that those things seen within the higher visual range appear higher, while those within the lower range appear lower;
6. and, similarly, that those seen within the visual range on the right appear on the right, while those within that on the left appear on the left;
7. but that things seen within several angles appear to be more clear.

(The translation is taken from that given by Burton in 1945.)

The development of Euclid's *Optics* is mathematical in character, but it is not axiomatic in the same way that the *Elements* are. For example, Euclid later proves two propositions, "to know how great is a given elevation when the sun is shining" and "to know how great is a given elevation when the sun is not shining". As would be expected, there is no serious introduction of the concept of the sun or of shining but they are treated in an informal, commonsense, physical way with the essential thing for the proof being rays from the sun falling upon the end of a line. Visual space is of course treated by Euclid as Euclidean in character.

It might be objected that there are similar formal failings in Euclid's *Elements*, but it does not take much reflection to recognize the very great difference between the introduction of many sorts of physical terms in these definitions from the *Optics* and the very restrained use of language to be found in the *Elements*. Moreover, the proofs have a similar highly informal character. It seems to me that the formulation of fundamental assumptions in Euclid's *Optics* is very much in the spirit of what has come to be called, in our own time, physical axiomatics. There is no attempt at any sort of mathematical rigor but an effort to convey intuitively the underlying assumptions.[1]

II. ARCHIMEDES' *ON THE EQUILIBRIUM OF PLANES*

Because I want to discuss the Archimedean treatise in some detail, a review of the theory of conjoint measurement is needed. The mixture of highly explicit axioms of conjoint measurement (as we would call them) and very inexplicit axioms about centers of gravity make Archimedes' treatise a peculiarly interesting example.

Conjoint Measurement

In many kinds of experimental or observational environments, the measurement of a single magnitude of property is not feasible or theoretically interesting. What is of interest, however, is the joint measurement of several properties simultaneously. The intended representation is that we consider ordered pairs of objects or stimuli. The first members of the pairs are drawn from one set, say A_1, and consequently represent one kind of property or magnitude; the second members of the pairs are objects drawn from a second set, say A_2, and represent a different magnitude or property. Given the ordered pair structure, we shall only require judgments of whether or not one pair jointly has more of the 'conjoined' attribute than a second pair.

Examples of interpretations for this way of looking at ordered pairs are abundant. In Archimedes' case, we are dealing with the measurement of static moments of force, or torques, where the two properties that make up the conjoint attribute are mass (or weight) and distance from the fulcrum. Momentum is another familiar example of a conjoint attribute. Quite different examples may be drawn from psychology or economics. For instance, a pair (a, p) can represent a tone with intensity a and frequency p, and the problem is to judge which of the two tones sounds louder. Thus the individual judges $(a, p) \geq (b, q)$ if and only if tone (a, p) seems at least as loud as (b, q).

The axioms of conjoint measurement are stated in terms of a single binary relation defined on the Cartesian product $A_1 \times A_2$. All the axioms have an elementary character, except for the Archimedean axiom, which I shall not formulate explicitly along with the other axioms, but which I discuss below. In formulating the axioms, I use the usual equivalence relation $\approx$, which is defined in terms of $\geq$, i.e.,

$(a, p) \approx (b, q)$ if and only if $(a, p) \succsim (b, q)$ and $(b, q) \succsim (a, p)$. Later, we shall also use the strict ordering: $(a, p) > (b, q)$ if and only if $(a, p) \succsim (b, q)$ and not $(b, q) \succsim (a, p)$. The axioms are embodied in the following definition.

DEFINITION 1. *A structure* $\langle A_1, A_2, \succsim \rangle$ *is a* conjoint structure *if and only if the following axioms are satisfied for every a, b and c in* A_1 *and every p, q and r in* A_2:

Axiom 1. *If* $(a, p) \succsim (b, q)$ *and* $(b, q) \succsim (c, r)$ *then* $(a, p) \succsim (c, r)$;
Axiom 2. $(a, p) \succsim (b, q)$ *or* $(b, q) \succsim (a, p)$;
Axiom 3. *If* $(a, p) \succsim (b, p)$ *then* $(a, q) \succsim (b, q)$;
Axiom 4. *If* $(a, p) \succsim (a, q)$ *then* $(b, p) \succsim (b, q)$;
Axiom 5. *If* $(a, p) \succsim (b, q)$ *and* $(b, r) \succsim (c, p)$ *then* $(a, r) \succsim (c, q)$;
Axiom 6. *There is an s in* A_2 *such that* $(a, p) \approx (b, s)$;
Axiom 7. *There is a d in* A_1 *such that* $(a, p) \approx (d, q)$;
Axiom 8. *Archimedean axiom.*

The intuitive content of most of the axioms is apparent. Axiom 1 is merely the familiar requirement of transitivity and Axiom 2 that of strong connectivity. Axioms 3 and 4 express the independence of one component from the other. Axioms 3 and 4 actually follow from the other axioms, but in the treatment of Krantz, Luce, Suppes, and Tversky (1971), weaker solvability axioms are used than Axioms 6 and 7, and in that context, Axioms 3 and 4 are needed. In any case, they state an important conceptual property. Axiom 5 states a cancellation property. When it is formulated in terms of the equivalence relation $\approx$ instead of $\succsim$, it is called the Thomsen condition, especially in the theory of webs. As already remarked, Axioms 6 and 7 state simple solvability axioms. Finally, Axiom 8 must be some form of the Archimedean axiom. Of course, I mean not an axiom directly pertinent to the treatise we are discussing here, but the familiar Archimedean axiom which is usually attributed to Eudoxus and not to Archimedes. In its most familiar form, it says that if we are given two magnitudes and the first is less than the second, there is a finite multiple of the first that is larger than the second. To formulate the axiom in explicit mathematical form in the present context, with no concept of addition or multiplication directly given, is somewhat troublesome. Because it is not important for our present discussion, I shall leave the axiom in inexplicit form.

For subsequent discussion of the postulates stated in Archimedes' treatise, some elementary consequences of Axioms 1–4 of Definition 1 are useful.

THEOREM 1. *The relation* $\approx$ *is an equivalence relation on* $A_1 \times A_2$, *i.e., it is reflexive, symmetric and transitive on* $A_1 \times A_2$; *and the relation* $>$ *is irreflexive, asymmetric and transitive on* $A_1 \times A_2$.

It is also desirable to define corresponding relations for each component. Thus, for a and b in A_1, $a \geq_1 b$ if and only if for some p in A_2, $(a, p) \geq (b, p)$; and for p and q in A_2, $p \geq_2 q$ if and only if for some a in A_1, $(a, p) \geq (a, q)$. Then as before, we may define for $i = 1, 2$, $x \approx_i y$ if and only if $x \geq_i y$ and $y \geq_i x$; and $x >_i y$ if and only if $x \geq_i y$ and not $y \geq_i x$. Using especially Axioms 3 and 4, the independence axioms, we may easily prove the following theorem.

THEOREM 2. *For* $i = 1, 2$, *the relation* $\geq_i$ *is transitive and strongly connected on* A_i, *the relation* $\approx_i$ *is an equivalence relation on* A_i *and the relation* $>_i$ *is irreflexive, asymmetric and transitive on* A_i.

We can prove that any structure satisfying the axioms of Definition 1 can be given either an additive or a multiplicative representation in terms of real numbers. Because the multiplicative representation is most pertinent here, we shall state the basic representation theorem in that form. The reader is referred to Krantz et al. (1971, Chapter 6) for the proof of the theorem.

THEOREM 3. *Let* $\langle A_1, A_2, \geq \rangle$ *be a conjoint structure. Then there exist real-valued functions* φ_1 *and* φ_2 *on* A_1 *and* A_2, *respectively, such that for* a *and* b *in* A_1 *and* p *and* q *in* A_2

$$(a, p) \geq (b, q) \text{ if and only if } \varphi_1(a)\varphi_2(p) \geq \varphi_1(b)\varphi_2(q).$$

Moreover, if φ_1' *and* φ_2' *are any two other functions with the same property, then there exist real numbers* α, β_2, $\beta_2 > 0$ *such that*

$$\varphi_1 = \beta_1 \varphi_1'^{\alpha}$$

and

$$\varphi_2 = \beta_2 \varphi_2'^{\alpha},$$

provided there are elements a and b in A_1 *and p in* A_2 *such that* $(a, p) > (b, p)$, *and elements p and q in* A_2 *and c in* A_1 *such that* $(c, p) > (c, q)$.

More than the theory of conjoint measurement is needed to give a correct analysis of Archimedes' treatise, for he obviously assumes that weight and distance are extensive or additive magnitudes. (This point is documented in the later discussion.) It will therefore also be useful to have in front of us the modern theory of extensive magnitudes. A rather complete presentation of the theory is to be found in Krantz et al. (1971, Chapter 3). Because of their relative simplicity I shall state here the axioms of Suppes (1951). In this case the Archimedean axiom is easily stated explicitly. A binary operation $\circ$ on the set A of magnitudes, as well as a binary relation $\geq$, is introduced, and we define recursively $1x = x$ and $nx = (n-1)x \circ x$. As before, the relations $\approx$ and $>$ are defined as expected in terms of $\geq$.

DEFINITION 2. *A structure* $\langle A, \geq, \circ, \rangle$ *is a* structure of extensive magnitudes *if and only if the following axioms are satisfied for every a, b and c in A*:

Axiom 1. *If* $x \geq y$ *and* $y \geq z$ *then* $x \geq z$;
Axiom 2. $(x \circ y) \circ z \geq x \circ (y \circ z)$;
Axiom 3. *If* $x \geq y$ *then* $x \circ z \geq z \circ y$;
Axiom 4. *If* $x > y$ *then there is a z in A such that* $x \approx y \circ z$;
Axiom 5. $x \circ y > x$;
Axiom 6. *If* $x \geq y$ *then there is a natural number n such that* $y \geq nx$.

The six axioms of Definition 2 have an obvious content when A is a set of positive numbers closed under addition and subtraction of smaller numbers from larger ones, $\geq$ is the numerical weak inequality, and $\circ$ is the operation of addition. It should be noted that Axiom 3 combines monotonicity and commutativity. The numerical interpretation just given is itself the basis of the following representation theorem.

THEOREM 4. *Let* $\langle A, \geq, \circ \rangle$ *be a structure of extensive magnitudes. Then there exists a real-valued function* φ *on A such that for a and b in A*

$$a \geq b \text{ if and only if } \varphi(a) \geq \varphi(b),$$

and

$$\varphi(a \circ b) = \varphi(a) + \varphi(b).$$

Moreover, if φ' is any other such function then there is a real number $\alpha > 0$ such that $\varphi' = \alpha\varphi$.

Archimedes' Postulates

With the axioms of conjoint and extensive measurement given above as background, let us now turn to Archimedes' postulates at the beginning of Book I of *On the Equilibrium of Planes.* I cite the Heath translation.

> I postulate the following:
>
> 1. Equal weights at equal distances are in equilibrium, and equal weights at unequal distances are not in equilibrium but incline towards the weight which is at the greater distance.
> 2. If, when weights at certain distances are in equilibrium, something be added to one of the weights, they are not in equilibrium but incline towards that weight to which the addition was made.
> 3. Similarly, if anything be taken away from one of the weights, they are not in equilibrium but incline towards the weight from which nothing was taken.
> 4. When equal and similar plane figures coincide if applied to one another, their centres of gravity similarly coincide.
> 5. In figures which are unequal but similar the centres of gravity will be similarly situated. By points similarly situated in relation to similar figures I mean points such that, if straight lines be drawn from them to the equal angles, they made equal angles with the corresponding sides.
> 6. If magnitudes at certain distances be in equilibrium, (other) magnitudes equal to them will also be in equilibrium at the same distances.
> 7. In any figure whose perimeter is concave in (one and) the same direction the centre of gravity must be within the figure.

Looking at the postulates, it is clear that postulates 1, 2, 3 and 6 fall within the general conceptual framework of conjoint measurement, but the remaining postulates introduce geometrical ideas that go beyond the general theory of conjoint measurement. I shall have something more to say about these geometrical postulates later. For the moment I want to concentrate on what I have termed the *conjoint postulates.* The wording of Postulates 2 and 3 makes it clear that Archimedes treated weight as an extensive magnitude. We shall thus assume that $\mathscr{W} = \langle W, \succeq_1, \circ\rangle$ is a structure of extensive magnitudes, that $\langle W \times D, \succeq\rangle$ is a conjoint structure, and that $\succeq_1$ of $\mathscr{W}$ is the defined relation $\geq_1$ of the conjoint structure. Also, to formulate Postulate 3 explicitly we need a subtraction operation that is well defined for extensive structures: If $x > y$ then $x - y = z$ if and only if $x = y \circ z$.

The formulation of Postulates 1, 2, 3 and 6 then assumes the following elementary form, with subscripts of $\geq_1$ and $\geq_2$ dropped to simplify the notation.

1a. *If* $w_1 \approx w_2$ *and* $d_1 \approx d_2$ *then* $(w_1, d_1) \approx (w_2, d_2)$.
1b. *If* $w_1 \approx w_2$ *and* $d_1 > d_2$ *then* $(w_1, d_1) > (w_2, d_2)$.
2. *If* $(w_1, d_1) \approx (w_2, d_2)$ *then* $(w_1 \circ x, d_1) > (w_2, d_2)$.
3. *If* $(w_1, d_1 \approx (w_2, d_2)$ *and* $w_2 > x$ *then* $(w_1, d_1) > (w_2 - x, d_2)$.
6. *If* $(w_1, d_1) \approx (w_2, d_2)$, $w_3 \approx w_1$ *and* $w_4 \approx w_2$ *then* $(w_3, d_1) \approx (w_4, d_2)$.

The first three propositions of Book I can be proved from these purely conjoint postulates and the assumption that weight is an extensive magnitude. For detailed analysis I cite the Heath translation of the propositions and their proofs.

Proposition 1.

Weights which balance at equal distances are equal.

For, if they are unequal, take away from the greater the difference between the two. The remainders will then not balance [*Post.* 3]; which is absurd.

Therefore the weights cannot be unequal.

Proposition 2.

Unequal weights at equal distances will not balance but will incline towards the greater weight.

For take away from the greater the difference between the two. The equal remainders will therefore balance [*Post.* 1]. Hence, if we add the difference again, the weights will not balance but incline towards the greater [*Post.* 2].

Proposition 3.

Unequal weights will balance at unequal distances, the greater weight being at the lesser distance.

Let A, B be two unequal weights (of which A is the greater) balancing about C at distances AC, BC respectively.

Then shall AC be less than BC. For, if not, take away from A the weight $(A - B)$. The remainders will then incline towards B [*Post.* 3]. But this is impossible, for (1) if $AC = CB$, the equal remainders will balance, or (2) if $AC > CB$, they will incline towards A at the greater distance [*Post.* 1].

Hence $AC < CB$.

Conversely, if the weights balance, and $AC < CB$, then $A > B$.

My aim is to catch the spirit of Archimedes' formulation of these first three propositions *and* their proofs within the formalization I have given. To be as explicit as possible about my procedure, I use in the proofs elementary properties of extensive magnitudes that follow from the axioms of Definition 2, but only properties of conjoint structures that follow from Archimedes' postulates, not the full set of Definition 1.

PROPOSITION 1. *If* $(w_1, d_1) \approx (w_2, d_2)$ *and* $d_1 \approx d_2$ *then* $w_1 \approx w_2$.

Proof. Suppose $w_1 > w_2$. Let $z = w_1 - w_2$. Then $w_1 - z \approx w_2$. Then by Postulate 1a

(1) $\quad (w_1 - z, d_1) \approx (w_2, d_2)$,

but by Postulate 3 and the hypothesis of the theorem

(2) $\quad (w_2, d_2) > (w_1 - z, d_1)$,

and (1) and (2) are from the definitions of $>$ and $\approx$ jointly absurd.

PROPOSITION 2. *If* $w_1 > w_2$ *and* $d_1 \approx d_2$ *then* $(w_1, d_1) > (w_2, d_2)$.

Proof. Let $z = w_1 - w_2$. Then $w_1 - z \approx w_2$, and by Postulate 1a

$$(w_1 - z, d_1) \approx (w_2, d_2).$$

Therefore, by Postulate 2

$$((w_1 - z_1) \circ z_1, d_1) > (w_2, d_2),$$

and $(w_1 - z_1) \circ z_1 = w_1$, so

$$(w_1, d_1) > (w_2, d_2).$$

PROPOSITION 3. *If* $w_1 > w_2$ *and* $(w_1, d_1) \approx (w_2, d_2)$ *then* $d_2 > d_1$.

Proof. Suppose not $d_2 > d_1$. Let $z = w_1 - w_2$. Then by Postulate 3,

(1) $(w_2, d_2) > (w_1 - z, d_1)$,

but this we shall show is absurd. First if $d_1 \approx d_2$, then by Postulate 1a

(2) $(w_1 - z_1, d_1) \approx (w_2, d_2)$,

and (as in the proof of Prop. 1) (1) and (2) are jointly absurd. On the other hand, if $d_1 > d_2$, then by Postulate 1b

(3) $(w_1 - z_1, d_1) > (w_2, d_2)$,

and (1) and (3) are jointly absurd (from the asymmetry of $>$). Hence $d_2 > d_1$.

On one point my formalization is clearly not faithful to Archimedes. I have replaced his symmetrical relation *unequal* by the asymmetric $>$, but this is a trivial formal difference, easy to eliminate if desired.

The remaining propositions of Book I use the concept of center of gravity in either their formulations or proofs, and I defer the consideration of this much-disputed concept.

The postulates and propositions as I have reformulated them above are a part of the elementary theory of conjoint measurement on the assumption that the first component is a structure of extensive magnitudes as well. A casual perusal of modern textbooks on mechanics reveals quickly enough that postulates like the ones formulated here are not an explicit part of modern discussions of static moments of force. The reason is simple. Once a numerical representation is assumed, explicit conjoint axioms are not necessary. Take Postulate 1a, for instance, and use the multiplicative representation:

If $\varphi_1(w_1) = \varphi_1(w_2)$ and $\varphi_2(d_1) = \varphi(d_2)$ then

$$\varphi_1(w_1)\varphi_2(d_1) = \varphi_1(w_2)\varphi_2(d_2),$$

but this is just an elementary truth of arithmetic and consequently not necessary to assume.

The important historical fact is that the concept of a numerical representation was missing in Greek mathematics, and consequently explicit conjoint axioms were needed. There seems little doubt that Archimedes' statement of such axioms is historically the earliest instance of an explicit approach to conjoint measurement, certainly at least in terms of extant texts of Greek mathematics and science.

It has been noted by many modern commentators that Greek mathematicians were completely at ease in comparing ratios of different sorts of magnitudes, e.g., the ratio of two line segments to that of two areas. Given this tradition it is natural to query why Archimedes did not state the Postulates of Book I in terms of ratios. The answer it seems to me is clear. Proof that two weights balance at distances reciprocally proportional to their magnitudes, which is Propositions 6 and 7 of Book I, is the Greek equivalent of a numerical representation theorem in the theory of measurement. The conjoint postulates that Archimedes formulates provide a simple qualitative basis from which the Greek 'representation theorem' can be proved. (I shall have more to say later about this proof.)

I know of no other instance of conjoint concepts in Greek mathematics and science. Certainly modern examples like momentum were not considered, and no such concepts were needed in Archimedes' other physical work, *On Floating Bodies.* It is perhaps for this reason that the level of abstraction to be found, for example, in Book V of Euclid's *Elements* is not reached in *On the Equilibrium of Planes.*[2] A higher level of abstraction was superfluous because other pairs of magnitudes satisfying like postulates were not known.

Centers of Gravity

The most difficult conceptual problem of Archimedes' treatises concerns the status of the concept of center of gravity of a plane figure. This concept is essential to the formulation of Postulates 4, 5 and 7, but it is quite evident, on the other hand, that these postulates in themselves do not provide a complete characterization of the concept. By this I mean that if we knew nothing about centers of gravity except what is stated in Postulates 4, 5 and 7, we would not be able to derive the theorems in which Archimedes is interested, and which he does derive. As Dijksterhuis (1956) points out, it is possible to argue that the concept of center of gravity is being taken over by Archimedes from more elementary discussions and thus really has the same status as the geometrical concept of similarity in his treatise. On the face of it, this argument seems sounder than that of Toeplitz and Stein (published in Stein, 1930), who propose that the postulates are to be taken as implicitly defining centers of gravity once the postulates are enlarged by the obvious and natural assumptions.

It is also clear that a standard formalization of Archimedes' theory, in the sense of first-order logic, cannot be given in any simple or elegant way. It is possible to give the standard formalization of the part of the theory embodied in Postulates 1, 2, 3 and 6, as we have seen in the previous section.

Quite apart from the question of standard formalization, there are serious problems involved in giving a reconstruction in set-theoretical terms of Archimedes' postulates. In such a set-theoretical formulation, we can without difficulty use a geometrical notion like similarity. If we take over from prior developments a definition of center of gravity, then it would seem that Postulate 4, for example, would simply be a theorem from these earlier developments and would not need separate statement. Put another way, under this treatment of the concept of center of gravity, no primitive notion of Archimedes' theory would appear in Postulate 4 and thus it would clearly be an eliminable postulate. The same remarks apply to Postulates 5 and 7. It would seem that Archimedes has constructed a sort of halfway house; his postulates do not give a complete characterization of centers of gravity, but on the other hand, they cannot be said to depend upon a completely independent characterization of this concept.

Schmidt (1975) gives an interesting axiomatic reconstruction of Archimedes' theory, but his elegant postulates for centers of gravity are restricted to plane polygonal figures, whereas in Book II Archimedes is especially concerned with centers of gravity of parabolic segments. The 'reduction' of such segments to rectangles of equal area requires the results found in Archimedes' treatise *Quadrature of the Parabola.* (Schmidt's treatment of the 'conjoint' axioms discussed above does not use the standard modern results on conjoint measurement.)

It is worth noting that the fundamental pair of propositions (6 and 7) asserting the law of the lever, or what we may also term the law of static torque, does not really need any geometrical facts about centers of gravity, as do later propositions of Book I, and the whole of Book II. Archimedes could have used something like the following definition to get as far as Proposition 7: *The center of gravity of* (w_1, d_1) *and* (w_2, d_2) *is the distance* d_3 *such that* $(w_1, d_2 - d_3) \approx (w_2, d_3 - d_1)$. This definition assumes that distances are extensive magnitudes, but there is little difficulty about this assumption. It seems obvious to me why it is unlikely Archimedes even momentarily would have considered such a definition. The mathematically difficult and geometrically significant propositions all deal with the centers of gravity of geometric figures; in fact, the whole of Book II is concerned with finding the centers of gravity of parabolic segments, and for this purpose a geometric concept of center of gravity is necessity.[3]

From a purely axiomatic standpoint, therefore, Archimedes is no more satisfactory than a modern physical treatise with some mathematical pretensions. A good comparative example, perhaps, is von Neumann's book (1932/1955) on quantum mechanics, which contains a beautifully clear axiomatic development of the theory of Hilbert spaces, but not of quantum mechanics itself.

III. PTOLEMY'S *ALMAGEST*

The third and most important example I cite is Ptolemy's *Almagest.* It is significant because it is the most important scientific treatise of ancient times and because it does not contain any pretense of an axiomatic treatment.

It is to be emphasized that Ptolemy uses mathematical argument, and indeed mathematical proof, with great facility, but he uses the mathematics in an applied way. He does not introduce explicit axioms about the motion of stellar bodies, but reduces the study of their motion to geometrical propositions, including of course the important case of spherical trigonometry.

Near the beginning of the *Almagest,* Ptolemy illustrates very well in the following passage the spirit of the way in which assumptions are brought in:

> And so in general we have to state that the heavens are spherical and move spherically, that the earth in figure is also spherical to the senses when taken in all its parts; in position lies right in the middle of the heavens, like a geometrical center; and in magnitude and distance has the ratio of a point with respect to the sphere of the fixed stars, having no local motion itself at all. And we shall go through each of these points briefly to bring them to mind.

There then follows a longer and more detailed discussion of each of these matters, such as the proposition that the heavens move spherically. My point is that the discussion and the framework of discussion are very much in the spirit of what we think of as nonaxiomatic mathematical sciences today. There is not a hint of organizing these ideas in axiomatic fashion.

When Ptolemy gets down to details he has the following to say:

> But now we are going to begin the detailed proofs. And we think the first of these is that by means of which is calculated the length of the arc between the poles of the equator and the ecliptic, and which lies on the circle drawn through these poles. To this end we must first see expounded the method of computing the values of chords inscribed in a circle, which we are now going to prove geometrically, once for all, one by one.

The detailed discussion, then, on the size of chords inscribed in a circle emphasizes, above all, calculation and would make a modern physicist happy by its tone and results as well. This long and important analysis of computations is concluded with a numerical table of chords.

The thesis I am advancing is illustrated, in many ways even more strikingly, by the treatment of the motion of the moon in Book IV. Here Ptolemy is concerned to discuss in considerable detail the kind of observations that are appropriate for a study of the moon's motion and especially with the methodology of how a variety of observations are to be rectified and put into a single coherent theory.

Various hypotheses introduced in later books, e.g., the hypothesis of the moon's double anomaly in Book V, are in the spirit of modern astronomy or physics, not axiomatic mathematics. Moreover, throughout the *Almagest*, Ptolemy's free and effective use of geometrical theorems and proofs seems extraordinarily similar in spirit to the use of the differential and integral calculus and the theory of differential equations in a modern treatise on some area of mathematical physics.

IV. CONCLUDING REMARKS

In this analysis of the use of axiomatic methods and their absence in explicit form in ancient mathematical sciences such as optics and astronomy, I have not entered into a discussion of the philosophical analysis of the status of axioms, postulates and hypotheses. There is a substantial ancient literature on these matters running from Plato to Proclus. Perhaps the best and most serious extant discussion is to be found in Aristotle's *Posterior Analytics*. Aristotle explains in a very clear and persuasive way how geometrical proofs can be appropriately used in mechanics or optics (75b 14ff). But just as Aristotle does not really have any developed examples from optics, mechanics or astronomy, so it seems to me that the interesting distinctions he makes do not help us understand any better the viewpoint of Euclid toward the 'definitions' of his optics or the postulates of Archimedes about centers of gravity cited above.

Many of you know a great deal more than I do about the history of Greek mathematics and Greek mathematical sciences, but, all the same, I want to venture my own view of the situation I have been describing. I may be too much influenced by my views about contemporary science, but I find little difference between contemporary physics and the problems of Greek science I have been describing. Physicists of today no more conform to an exact canon of philosophical analysis in their setting forth of physical principles or ideas than did those ancient scientists and mathematicians who wrote about the subjects I have been discussing. There was certainly a sense of methodology deeply embedded in Euclid, Archimedes and Ptolemy, but it was not a sense of methodology that was completely explicit or totally worked out, just as Aristotle's own general principles are never exemplified in any detailed and complicated scientific examples of an extended sort. The gap between philosophical analysis, canons of axiomatic method, and actual working practice was about the same order of magnitude that it is today. What is surprising, I think, from a philosophical standpoint is that the gap seems, if anything, to have widened rather than narrowed over the past 2000 years.

NOTES

[1] Ptolemy's *Optics* is much more physical and experimental in character. A more mathematical example, without any explicit axioms at all, is Diocles' treatise *On Burning Mirrors* (Toomer, 1976). The detailed mathematical proofs are also interesting in Diocles' work because of the absence in most cases of reasons justifying the steps in the argument, but, as in a modern nonaxiomatic text, familiar mathematical facts and theorems are used without comment.

[2] The attitude toward abstraction is very clearly expressed by Aristotle in the *Posterior Analytics* (Book I, 5, 74a 17–25). "An instance of (2) would be the law that proportionals alternate. Alternation used to be demonstrated separately of numbers, lines, solids, and durations, though it could have been proved of them all by a single demonstration. Because there was no single name to denote that in which numbers, lengths, durations, and solids are identical, and because they differed specifically from one another, this property was proved of each of them separately. Today, however, the proof is commensurately universal, for they do not possess this attribute *qua* lines or *qua*

numbers, but *qua* manifesting this generic character which they are postulated as possessing universally". The reference to (2) is to one kind of error we can make in drawing a conclusion that is too specific or concrete. Errors of type (2) arise "when the subjects belong to different species and there is a higher universal, but it has no name" (74a 7).

[3] In closing this discussion, it is worth noting that Mach (1942), in his famous treatise on mechanics, seems to be badly confused on what Archimedes' work is all about. The focus of Mach's analysis is the famous Proposition 6 asserting that commensurable magnitudes are in equilibrium at distances reciprocally proportional to their weights. Mach is particularly exercised by the fact that "the entire deduction (of this proposition) contains the proposition to be demonstrated by assumption if not explicitly" (p. 20). A central point of Mach's confusion seems to be a complete misunderstanding as to the nature of the application of mathematics to physics. He seems to have no real conception of how mathematics is used to derive particular propositions from general assumptions, and what the relation of these general assumptions to the particular proposition is. He seems to think that any such proposition as the one just quoted must somehow be established directly from experience. His mistaken sentiments on these matters are clearly expressed in the following passage:

From the mere assumption of the equilibrium of equal weights at equal distances is derived the inverse proportionality of weight and lever arm! How is that possible? If we were unable philosophically and a priori to excogitate the simple fact of the dependence of equilibrium on weight and distance, but were obliged to go for *that* result to experience, in how much less a degree shall we be able, by speculative methods, to discover the form of this dependence, the proportionality! (p. 19)

This last quotation shows, it seems to me, the basic fact that is usually not explicitly admitted in discussing Mach's views on the foundations of mechanics. He simply had no coherent or reasonable conception of how mathematics can be used in science, and his wrong-headed analysis of Archimedes is but one of many instances that support this conclusion.

Stanford University

REFERENCES

Dijksterhuis, E.J.: 1956, *Archimedes*, Humanities Press, New York.

H.E. Burton, trans.: 1945, 'Euclid, *Optics*', *Journal of the Optical Society of America* **35**, 357–372.

Heath, T.L. (ed.): 1897, *The Works of Archimedes with the Method of Archimedes*, Dover, New York.

Krantz, D.H., Luce, R.D., Suppes, P., and Tversky, A.: 1971, *Foundations of Measurement*, Vol. 1, Academic Press, New York.

Mach, E.: 1942, *The Science of Mechanics* (5th English ed.; T.J. McCormack, trans.), Open Court, La Salle, Ill.

Schmidt, O.: 1975, 'A system of axioms for the Archimedean theory of equilibrium and centre of gravity', *Centaurus* **19**, 2–35.

Stein, W.: 1930, 'Der Begriff des Schwerpunktes bei Archimedes', *Quellen und Studien zur Geschichte der Mathematik, Physik und Astronomie* **1**, 221–224.

Suppes, P.: 1951, 'A set of independent axioms for extensive quantities,' *Portugaliae Mathematica* **10**, 163–172.

Toomer, G.J. (ed.): 1976, *Diocles on Burning Mirrors*, Springer–Verlag, Berlin.

von Neumann, J.: 1955, *Mathematical Foundations of Quantum Mechanics*, Princeton University Press, Princeton, N.J., (Originally published in 1932.)

S. S. DEMIDOV

ON AXIOMATIC AND GENETIC CONSTRUCTION OF MATHEMATICAL THEORIES

There are two methods of constructing mathematical theories: *genetic* and *axiomatic*. The first of them implies the construction of a theory through successive generalizations, on the basis of simple concepts established earlier. For instance, one can construct in this way the theory of real numbers proceeding from the basic unit: by applying a certain natural process (that of counting) one obtains from it the series of positive integers 1, 2, 3, 4, 5, ...; introduces the so called arithmetic operations for them; then, by making use of these operations, introduces the negative numbers and zero; then fractions as pairs of integers; and, finally, defines the real number as a section or fundamental sequence.

The term '*axiomatic* method' of construction of a theory will mean here the method of its organization in which the theory starts with the list of undefinable terms and unprovable axioms, including those terms from which the statements of the theory (theorems) should be deduced according to the rules of formal logic. If the last sentence is not explained (these rules being regarded as known), the informal axiomatic method is meant, otherwise one has in mind the formal axiomatic method [1]. Thus, D. Hilbert's axiomatic theory presented by him in his paper of 1900 [2] can serve as an example of the theory of real numbers organized according to (informal) axiomatic principles.

We come across both methods, genetic and axiomatic, even in ancient mathematics. The first of them is used by Ancient Greeks to construct arithmetic on the basis of the concept of unity, the second to construct geometry: through the first, second and fifth definitions of the first book of *Elements* Euclid introduces the concepts of point, line and surface, the relations between which are specified, to a considerable extent, by the third and sixth definitions. A natural question arises: why do ancient mathematicians use different methods for the construction of arithmetic and geometry? What, finally, prevents Euclid from limiting himself to the single basic concept of point, using motion to construct from it lines and then surfaces in

J. Hintikka, D. Gruender, and E. Agazzi (eds.), Pisa Conference Proceedings, Vol. I, 215–221.

the same way positive integers are constructed from unity? This question is all the more justified because ancient thought demonstrates a clear tendency to reduce its constructions to the minimal number of elements! It seems to us [3, 4] that these questions can be answered as follows: such construction in geometry encounters difficulties associated with the structure of the continuum as a set of points, exposed already by the Eleatics (Zeno's paradoxes – see [5, 6]). Apparently, the discovery of exactly these difficulties can explain the absence in Euclid's *Elements* of such sentences as "the interval is composed of points" and, in the final analysis, its axiomatic construction (for details, see [4]).

Euclid's *Elements* marks the first stage in the history of axiomatic method, called in the literature the period of *informal* (or *material*) axiomatization (see [7]). It is characterized first of all by the axiomatic theory being inseparably connected with a single interpretation as the expression of the most fundamental properties of the theory. Such an approach to the axiomatic theory is preserved up until the XIXth century, the end of which coincides with the end of the period of informal axiomatics.

During nearly the whole of the first period, which lasted more than two thousand years, no fundamental changes had been introduced into the structure of the axiomatic theory that were new in comparison with what was known in antiquity. Various attempts at modification of the system of initial assumptions in geometry or another method of expression of some of them did not, as a matter of fact, add, anything new. Numerous examples of axiomatic construction of really arithmetic theory, represented by Wolff's school, were fruitless.

It seems to us that I. Newton's attempt of the axiomatic construction of the theory of fluxions, recorded in his papers published recently by D. Whiteside [8] deserves, to be mentioned specially. This publication for the second time (earlier it was done by A. Witting in [9], which was published in 1911 and remained unnoticed) drew the attention of researchers (see the comments of D. Whiteside himself in [8], and also [9]) to the example of non-trivial use of the axiomatic construction by the great mathematician. The fragment which is of interest to us, and dates back to 1671, begins with the list of four axioms (they are cited according to [10, p. 167]; here and in what follows Newton assumes that A and B go zero simultaneously):

1) fl.$(A) =$ fl.(B), then $A = B$
2) fl.(A): fl.$(B) = k$, where k is a constant, then $A:B = k$
3) fl.$(A \pm B) =$ fl.$(A) \pm$ fl.(B)
4) simultaneous moments (i.e., infinitesimal increments) are proportional to their fluxions.

Proceeding from these axioms, Newton proves more than ten varied statements, the first of which is:

if $A:B = C:D$, then $A\cdot$ fl.$(D) + D\cdot$ fl.$(A) = B\cdot$ fl.$(C) + C\cdot$ fl.(A).

From this statement the rules for differentiation of product, ratio, power and root are deduced. In 1742 the axiomatic presentation of the fluxion method was given by C. Maclaurin. It is not known whether he was familiar with the above attempt of I. Newton, but his axioms differ from Newton's. Note, first of all, that some of the above axioms of I. Newton (for example, the third) represent statements which he could easily prove himself. We here encounter an example of *singling out* some of the statements of a theory as axioms, which is conditioned by the convenience of derivation of the main theorems of the theory from them.

If one takes into account the constant interest on the part of I. Newton (and also C. Maclaurin) in the foundation of the method of fluxions, as well as insurmountable difficulties facing them when they attempted to give it such a foundation, one can assume that in this case also the scientists turned to axiomatic method because of the impossibility of providing its genetic justification.

The second period in the history of axiomatic method, the beginning of which coincides with the end of the XIXth century (and which, in its turn, is divided into two periods: those of *semiformal* and *formal* axiomatization), is characterized, first of all, by the fact that the axiomatic theory is not associated with a single interpretation; any system of objects satisfying the axioms of a given theory can serve as its interpretation. The complete development of the corresponding views on axiomatic theory is found in D. Hilbert's *Grundlagen der Geometrie*, published in 1899.

The development of the new approach to the axiomatic method was stimulated by a considerable number of factors, among the most important of which were the revolutionary changes in geometry associated with the discovery of non-Euclidean geometries in the works of N. I. Lobachevsky, J. Bolyai, K.-F. Gauss, and B. Riemann.

The construction of new types of geometry indicated the possibility of obtaining them through simple changes of certain axioms of the traditional Euclidean system; this did not mean, however, the necessity of adopting a new point of view separating the axiomatic theory from its interpretation – all the more so as the possibility appeared to consider the new geometries in the frame of Riemannian theory as the possible types of geometries realized on the surfaces of constant curvature.

Another important factor promoting the formulation of the new approach to the nature of the axiomatic method was the rapid development of XIXth century algebra. The works of English algebraists of the second quarter of the XIXth century (A. de Morgan, G. Boole, and others) contain an abstract concept of the composition law which is applied to a whole number of new mathematical objects. The result of this was the algebra of quaternions (W. R. Hamilton), the algebra of matrices (A. Cayley), the algebra of logic (G. Boole), etc. The development of group theory and the theory of algebraic numbers leads to the axiomatic introduction of several new concepts which are independent of specific interpretations (for example, the definition of the abstract finite group, formulated by A. Cayley in 1854, Dedekind's formulation of the concepts of field, modulus, and ideal in 1871), and later, in the works of R. Dedekind and H. Weber, to the first steps in the construction of abstract algebra. The multiplicity of interpretations of constructed axiomatic theories, and their independent significance regardless of the connection with these interpretations, become clearly perceived by mathematicians. G. Boole said [11] in 1847:

They who are acquainted with the present state of the theory of Symbolical Algebra, are aware, that the validity of the processes of analysis does not depend upon the interpretation of the symbols which are employed, but solely upon the laws of their combination. Every system of interpretation which does not affect the truth of the relations supposed, is equally admissible . . .

He was one of the first who established the independence of axiomatic theories from their concrete interpretations.

An essential factor which increased the interest in the axiomatic method during the whole of the XIXth century was the general tendency of that age towards the rigorousness of construction of mathematical theories; first of all, of the integral and differential calculus, which represented a leading mathematical discipline of the century and at the same time was based on quite unsatisfactory foundations. When the integral and differential calculus was brought to the new level of rigorousness (B. Bolzano, N.-H. Abel, A. Cauchy, and K. Weierstrass), it was constructed on the foundation of the theory of real number which had been formulated in different ways in the seventies of the XIXth century by R. Dedekind, G. Cantor, Ch. Mérey, and K. Weierstrass through its reduction to the theory of positive integers, which at this time already required, in its turn, accurate construction. One of the best systems which provided the construction of such a theory was given by G. Peano in 1891. This system, containing five axioms, essentially formalized the genetic method of introduction of positive integers. Thus, the whole giant construction of the differential and integral calculus, real numbers, and positive integers turned out to be organized according to the *genetic* principle, starting with the initial number (in Peano's works it is zero). And since geometry also becomes subject to interpretation based on the theory of real numbers, the latter turns out to be the foundation practically of the whole mathematics of the XIXth century, which was thus organized according to *genetic* principle.

The works of G. Cantor, of the seventies of the XIXth century, mark the beginning of the development of set theory, to which the theory of positive integers and real numbers can be easily genetically reduced, and which starts to be regarded by the end of the century as the preferable foundation (in comparison with the theory of positive integers) of the whole of mathematics. During the construction of the set-theoretic foundation under the structure of mathematics, however, cracks in that foundation were discovered: the first paradoxes of the set theory were formulated, thus bringing back (although at a new level) the situation which mathematics encountered in the antiquity (the paradoxes of the infinity) and subjecting to doubt the validity of the whole construction. Speaking in 1904 about the theory of real numbers, Hilbert wrote [12] that it is found under the blows of the paradoxes of the set theory,

Die Vermeidung solcher Widersprüche und die Klärung jener Paradoxien ist vielmehr bei den Untersuchungen über den Zahlbegriff von vornherein als ein Hauptziel ins Auge zu fassen,

and further:

Ich bin der Meinung, das alle die berührten Schwierigkeiten sich überwinden lassen und dass man zu einer strengen und völlig befriedigenden Begründung des Zahlbegriffes gelangen kann, und zwar durch eine Methode, die ich die ***axiomatische*** nennen . . .

Thus, the axiomatic method was applied as a measure to lead the theory out of the impasse it was left in by the genetic method. As we have noted before, this situation most probably explains the axiomatic construction of geometry in Euclid's *Elements* and the attempts of I. Newton and C. Maclaurin at the axiomatic construction of the theory of fluxions.

As is shown by history, mathematicians in their constructions frequently prefer the genetic method, which is regarded as more natural, corresponding to the essence of mathematical objects, to counterbalance the axiomatic one which is more formal and intended mostly for the shaping of the already existing theories. According to J. Piaget, who studied this question from the psychological point of view [13], the genetic method is more fundamental in reality than the axiomatic one (see [14]). The efforts of intuitionists and constructivists are motivated by the desire to base mathematics exactly on the genetic principle. In spite of their constructions being interesting by themselves, however, they cannot in any sense cover a considerable part of, for instance, modern integral and differential calculus. On the other hand, it follows from the whole history of the axiomatic method and its remarkable success in this century that it is not only a form of presentation and the means of justification of mathematical theories but also a means for obtaining new mathematical results.

It seems to me (in opposition to the opinion of J. Piaget, who is supported, perhaps, by many mathematicians) that the *axiomatic method is a fundamental reality of our thinking no less than the genetic one. And if* (here we venture to formulate a hypothesis) *the genetic method corresponds to the constructive abilities of our mind, the axiomatic method represents a way of fixation of ideas produced by our intuition.* Sometimes such ideas can be later shaped into genetic constructions, sometimes, as when one has to perform operations that are *equivalent to construction of a continuum from a point, such construction fails,* and one has to formulate the theory axiometrically. W. Kuyk, in his interesting book on foundations of mathematics published recently [15], states that the 'discrete' and 'continuum' represent two ideas that cannot be constructively reduced to each other and stand in a peculiar relation of *complementarity* to each other which is in a certain way similar to complementarity of N. Bohr. *They cannot be connected genetically and their unification is possible only on the axiomatic principle.* The whole history of the axiomatic method is an illustration of this thesis.

REFERENCES

[1] Church, A.: 1956, *Introduction to Mathematical Logic*, Vol 1, Princeton.

[2] Hilbert, D.: 1900, *Über den Zahlbegriff*, Jahresbericht der Deutschen Math. Verein., Vol. 8.

[3] Demidov, S.: 1971, 'Sur l'histoire de la méthode axiomatique', Actes du XIIe Congress International d'Histoire des Sciences, t. IV, pp. 45–47, Paris.

[4] Demidov. S.: 1970, 'Évolution, extension et limites de la méthode axiomatique dans les sciences modernes sur l'example de la géométrie', *Archives int. d'hist. des sciences* **90–91**, 3–30.

[5] Bashmakova, I.: 1958, 'Lectures on the history of mathematics in Ancient Greece', *Istoriko-matem. issled.* **11**, , Moscow, 225–438.

[6] Szabó, A.: 1959, 'On the transformation of mathematics into a deductive science and the beginning of its justification', *Istoriko-matem. issled.* **12**, Moscow, 321–392.

[7] Kleene, S.K.: 1952, *Introduction to Metamathematics*, New York, Toronto.

[8] Whiteside, D.T. (ed.): 1967–1977, *The Mathematical Papers of Isaac Newton*, Vol. 3, 1670–1673, Cambridge Univ. Press. Cambridge, England.

[9] Witting, A.: 1911–1912, 'Zur Frage der Erfindung des Algorithmus der Newtonschen Fluxionsrechnung', *Biblioteca mathematica* (Folge 3) **12**, 56–60.

[10] Yushkevich, A.: 1977, 'On I. Newton's mathematical manuscripts', *Istoriko-matem. issled.* **22**, Moscow, 127–192.

[11] Boole, G.: 1952, *Studies in Logic and Probability*, London.

[12] Hilbert, D.: 1905, 'Uber die Grundlagen der Logik und Arithmetik', Verhandlungen des 3. internationalen Mathematiker-Kongresses in Heidelberg, Leipzig, S. 174–185.

[13] Piaget, J.: 1957, 'Méthode axiomatique et méthode operationnelle', *Synthese* **5**, 23–43.

[14] Sadovsky, V.: 1962, 'The axiomatic method of construction of scientific knowledge', *Filos. voprosy sovr. formal'n. logiki*, Moscow, pp. 215–262.

[15] Kuyk, W.: 1977, *Complementarity in Mathematics*, D. Reidel, Dordrecht-Boston.

F. A. MEDVEDEV

ON THE ROLE OF AXIOMATIC METHOD IN THE DEVELOPMENT OF ANCIENT MATHEMATICS

Scholars frequently consider the introduction of the axiomatic method and the deductive construction of mathematical theories in Ancient Greece as one of the greatest achievement of ancient thought. As is seen from the materials presented at the present conference, such an approach is viewed by the majority of participants as obvious, as a certain axiom, and the only points moot are various interpretations of the appearance and the early history of this method. The axiomatic method is regarded as typical of the general aspect of the scientific thinking of that time.

This thesis, however, may be subject to doubt, at least as far as the mathematics of the Ancient Greece is concerned. This doubt is based on the following grounds.

1. The axiomatic construction of mathematical theories, even in that limited sense in which it was understood by ancient scientists, was applied to a very narrow range of mathematical disciplines, essentially only to elementary geometry. One can hardly agree with Knorr's assertion [1, p. 145] to the effect that the Greeks undertook a conscious effort to axiomatize arithmetic. The established view that the first steps in the axiomatization of arithmetic were taken only by Grassman in 1861 and that it was only Dedekind in 1887 and Peano in 1889 who formulated a relatively complete system of arithmetical axioms [2, pp. 38–39] appears more plausible. In this connection, it is relevant to note that there even exists a study of the question why arithmetic was axiomatized so late [3]. The actual application of the axiomatic method in other scientific disciplines of that time (*Phenomena* and *Optics* by Euclid, *On Equilibrium of Plane Figures, or the Center of Gravity of Plane Figures* by Archimedes) turned out to be unsatisfactory. And one can probably agree with Knorr, if one discards his word "unfortunately", when he states that "only one of the extant works by Archimedes, *Plane Equilibria* I, qualifies as an effort of the axiomatic kind. Unfortunately, its success as an axiomatic effort is questionable" [1, p. 172].

J. Hintikka, D. Gruender, and E. Agazzi (eds.), Pisa Conference Proceedings, Vol. I, 223–225.

2. The axiomatized *Elements* of Euclid presented already available, existing knowledge. No essential scientific discovery has been witnessed to have been obtained axiomatically. Thus, it is possible to say that the ancient axiomatic method was not a tool of heuristics. Moreover, the attempts to apply it to a developing science not only did not allow mathematicians and scientists to obtain new results but prevented it. For example, the intension to extend it to algebra is described in *The History of Mathematics* in the following way: "The construction of algebra on the basis of geometry made it possible for the first time to justify in a general manner certain theorems and rules; during further development, however, geometrical vestments bounds, like a coat of mail, the living body of antique mathematics" [4, p. 78]. And although these words apply not so much to the axiomatic method as to the narrow character of purely geometrical conceptions of that time, this narrow character was associated with the axiomatic manner of presentation of Greek geometry.

3. After the formulation of axiomatized geometry its development as a science stopped for more than two thousand years, and only in the 19th century was this situation changed. Unlike the dormant geometry of this kind, the development of living science occurred outside the framework of the axiomatic method. For instance, one can hardly establish any association between the highest achievement of Greek mathematics – Diophantos' *Arithmetics* – and axiomatization. Moreover, in the course of subsequent presentations of elementary geometry ancients authors frequently gave up its Euclidean form, since it did not correspond to those practical needs that appeared in real life. Such was the case, for instance, with the *Metrics* by Heron of Alexandria and, according to Neugebauer, "the axiomatic school in mathematics exercised at this more elementary level as little influence as it exercises today on a land-surveyor" [5, p. 148].

4. The 19th and the first half of the 20th centuries were a period of an upward flight of axiomatic and deductive studies. An approach was developed that became extremely widely spread, to the effect that "verdient doch zur endgültige Darstellung und volligen ligischen Siecherung des Inhaltes unserer Erkenntnis die axiomatische Methode den Vorzug" [6, S.242] with Hilbert's words "verdient . . . den Vorzug" being frequently replaced by the words "is applicable only" and the like.

Recently, however, mainly in connection with Gödel's result

concerning the non-formalizability of a sufficiently rich mathematical theory, and partially independently of it, a circle of studies that pertain to the establishment of non-axiomatizability of one or another portion of mathematics have come to the foreground. This and the fact that there have always been enough opponents of the axiomatic method among mathematicians, and especially the fact that the great majority of them never did nor do apply the axiomatic method, causes certain doubts with respect to the common high estimate of this method in general and the achievement of Greeks in this field in particular.

If such doubts are recognized as justified to some extent, the increased interest to the axiomatic-deductive method of the ancient on the part of A. Szabó and his opponents causes one to prick up one's ears. This interest by itself is harmless and would have been useful if it did not damage other trends of research on the history of ancient mathematics, in particular, the study of its connections with the practical needs of that time.

Academy of Science of USSR

REFERENCES

1. Knorr, W.: 'On the early history of axiomatics,' above pp. 145–186.
2. Bourbaki, N.: 1969, *Eléments d'histoire des mathématiques*, Hermann, Paris.
3. Yanovskaya, S.A.: 1972, 'On the history of axiomatics', in: S.A. Yanovskaya, *Methodological Problems of Science*, Publ. House "Mysl", Moscow, pp. 150–180.
4. Neugebauer, O.: 1968, *The Exact Sciences in Antiquity*, Publ. House "Fizmatigiz", Moscow.
5. Yushkevich, A.P., (ed.): 1970, *The History of Mathematics* Vol. 1, Publishing House "Nauka", Moscow.
6. Hilbert, D.: 1930, Uber das Zahlbegriff', in *Grundlagender der Geometrie*, Teubner, Leipzig u. Berlin, pp. 241–46.

SECTION III

THE PHILOSOPHICAL PRESUPPOSITIONS AND SHIFTING INTERPRETATIONS OF GALILEO

MAURICE CLAVELIN

GALILÉE ET LA MÉCANISATION DU SYSTÈME DU MONDE

Tout historien des sciences doit résoudre, à un moment ou à un autre, une difficulté qui engage le sens même de sa recherche. D'une part, quand il étudie une oeuvre, il ne peut guère faire abstraction de ce qui l'a suivie, et qui, d'une certaine façon, représente sa finalité. Mais d'autre part, il est certain qu'en nous poussant inévitablement à distinguer dans toute oeuvre une partie vivante et une partie morte, une telle attitude risque aussi de détruire ce qui fait son originalité, et généralement lui a permis de jouer le rôle qui fut le sien. Mieux que tout autre, et parce qu'elle fut vraiment fondatrice, l'oeuvre de Galilée illustre cette difficulté. Nul ne peut nier, en effet, qu'elle ait son achèvement dans la mécanique céleste de Newton, et donc que le but auquel "tendaient" toutes les découvertes de Galilée – même si ce but ne lui était pas aussi évident qu'il l'est pour nous – est bien *la mécanisation du système du Monde.* Par où j'entends que lorsque Newton, pour reprendre une remarquable formule de Laplace, fait de l'astronomie "la solution d'un grand problème de mécanique, dont les éléments des mouvements célestes sont les constantes arbitraires"[1], il dégage, en l'actualisant, la fin que portait en elle l'oeuvre galiléenne. Mais déjà nous guette le péril que j'évoquais en commençant. S'il est vrai que la mécanisation newtonienne du système du Monde marque bien l'achèvement de l'entreprise inaugurée par Galilée, la meilleure façon d'évaluer sa contribution ne sera-t-elle pas de faire de ses propositions deux groupes: d'une part, celles que conservera Newton, d'autre part celles qui perdront toute signification une fois la science newtonienne mise en place? Retenant les premières, et écartant les secondes, n'aura-t-on pas ainsi toute chance de dégager le vrai visage de Galilée, tout en adoptant le point de vue le plus objectif?

Or il est aisé de montrer qu'il n'en est rien. Adopter une telle méthode, c'est oublier deux faits primordiaux. D'abord qu'au moment même où il jetait les fondements de la mécanique classique, Galilée ne pouvait – à la différence de ses successeurs – se borner à per-

J. Hintikka, D. Gruender, and E. Agazzi (eds.), Pisa Conference Proceedings, Vol. I, 229–251.

fectionner un système déjà existant; ensuite – et ce n'est pas moins important – que chacune de ses découvertes dut être imposée pas à pas contre une philosophie naturelle, dans laquelle lui-même avait grandi, et qui à tous les grands problèmes offrait des solutions, sans doute criticables, mais nullement absurdes. D'un mot, c'est oublier que l'oeuvre galiléenne ce fut avant toutes choses la définition d'une nouvelle attitude à l'égard de la nature, *le passage d'un univers conceptuel à un nouvel univers conceptuel.* Evaluer sa contribution sous forme d'inventaire – ce que Newton retiendra, ce qu'il ne retiendra pas – c'est donc bien, selon toute probabilité, manquer l'essentiel, et substituer un Galilée idéal, donc subjectif, au Galilée réel. Mais c'est dire du même coup que seule une méthode traitant les découvertes de Galilée comme autant d'*initiatives* par lesquelles il modifia radicalement les perspectives théoriques sur le monde, pourra nous révéler leur richesse, et par là même nous faire comprendre comment il aida, plus que personne, à la création de la mécanique classique. C'est à l'examen de quelques unes de ces initiatives, et en évitant toute technicité inutile, que sera consacrée cette étude.

UNE NOUVELLE IMAGE DU MONDE

La première – la plus ambitieuse aussi par son objet – est à coup sûr celle qui conduisit Galilée, après une incroyable série de découvertes astronomiques, à bouleverser à son niveau le plus général l'image du Monde. Riche d'une information dont nul avant lui n'avait disposé, il sut alors en quelques mois élaborer des idées qui équivalaient en fait à une nouvelle cosmologie. Une brève description du système traditionnel, hérité des Grecs, donnera les clefs indispensables pour cette analyse.

§1. De ce système, que je prendrai dans sa version aristotélicienne, l'essentiel tient sans doute en une phrase: avoir réussi à lier pour vingt siècles un fait constatable par tous – l'ordre rigoureux caractérisant les mouvements célestes – aux deux thèses philosophiques du géocentrisme, d'une part, d'une différence de nature entre la Terre et les corps célestes, de l'autre. Au départ de toute la construction, en effet, l'idée d'ordre affirmée comme une évidence première: le Monde est un tout ordonné, un cosmos. A cette idée, Aristote en ajoute une autre, pour lui non moins primitive: les corps naturels se meuvent spontanément de mouvement local; et comme entre toutes les trajectoires seules la ligne droite et la ligne circulaire ont la simplicité requise par la nature, il en tire sans attendre la conclusion qu'il existe dans la nature – constitutivement en quelque sorte – des mouvements naturels rectilignes et circulaires. Mais dire que le Monde est un tout ordonné, c'est dire – toujours selon Aristote – qu'il a un haut et un bas, c'est-à-dire un centre; si l'on rapproche alors cette dernière affirmation de la précédente – savoir l'existence de mouvements naturels rectilignes et circulaires – on aperçoit sans trop de peine pourquoi la cosmologie traditionnelle a pu admettre comme un dogme l'existence de trois mouvements naturels simples: un mouvement autour du centre (en fait celui des corps célestes ou supralunaires), un mouvement vers le centre, un mouvement à partir du centre (ou vers le haut)[2].

D'importantes conséquences vont aussitôt apparaître. Jouant le rôle de prémisses, les trois mouvements naturels vont servir en effet de fil conducteur pour la déduction des éléments, c'est-à-dire pour la mise en place des corps concrets dont est fait le Monde. A chaque mouvement naturel, raisonne Aristote, doit correspondre un élément; au mouvement vers le bas, par exemple, correspondra un élément tendant spontanément à s'approcher le plus possible du centre du Monde qui représentera son lieu naturel et où, par conséquent, il ne pourra que demeurer immobile: ce sera l'élément terre, constituant principal du globe terrestre, dont le centre coïncide donc, en vertu d'une intime nécessité, avec le centre du Monde. Ainsi la présence de la Terre au centre du Monde, de même que son immobilité, sont-elles déduites directement de l'idée si simple, si conforme à l'observation, que le monde est rigoureusement ordonné.

Mais suivons Aristote encore un instant. Au mouvement naturel vers le haut correspondra, de même qu'au mouvement naturel vers le bas, un élément bien différencié, en l'occurence le Feu, qui, lui, tendra spontanément à monter, c'est-à-dire à s'approcher au maximum de l'orbe lunaire. En revanche, la détermination de l'élément correspondant au mouvement circulaire naturel est nettement plus délicate. Un tel mouvement, argumente Aristote, étant équivalent en chacun de ses points, ne saurait avoir ni commencement ni fin; il sera donc éternel, et l'élément qui lui correspond sera soustrait à la génération et à la corruption; de même d'ailleurs qu'il sera dépourvu de gravité, puisqu'à aucun moment il ne s'approche du centre. Aussi est-il exclu qu'il soit l'un des quatre éléments terrestres (à la terre et au feu sont en effet venus s'ajouter l'air et l'eau), et la seule conclusion logique est d'admettre qu'il représente un cinquième élément – l'élément céleste ou éther – absolument distinct des quatre autres. C'est la thèse célèbre de l'hétérogénéité de la Terre et des corps célestes qui, comme le géocentrisme, paraît bien dériver en droite ligne du simple fait de l'ordre, et, comme lui encore, allait peser sur la cosmologie jusqu'à la fin du XVIe siècle.

Cette analyse ne prétend nullement être exhaustive. Elle suffit à montrer, néanmoins, et la grande cohérence du Système du Monde mis au point par les Grecs, et son accord exceptionnel avec les données de l'observation. Ce dernier point est sans doute le plus remarquable. Car la cosmologie traditionnelle, plus encore qu'une construction solidement raisonnée, c'est d'abord la mise en forme d'une expérience immémoriale, celle de l'homme observant le ciel de ses yeux nus. Conjuguant la régularité des mouvements célestes, l'aspect apparemment immuable des corps célestes, et pour finir l'absence, du moins à ce niveau, de toute preuve d'un mouvement de la Terre, le système traditionnel pouvait bien sembler l'expression nécessaire de l'évidence sensible. Seul un progrès radical de l'observation pouvait donc l'abolir et ainsi provoquer l'avènement d'une nouvelle image du Monde. A Galilée revint l'honneur d'accomplir ce progrès dont il sut aussi, sans perdre un instant, exploiter toutes les conséquences.

§2. Nous sommes en 1609. Galilée enseigne alors à l'Université de Padoue, sur le territoire de la République de Venise. Au cours de l'été, la nouvelle se répand avec insistance qu'un Flamand aurait mis au point une lunette grâce à laquelle des objets éloignés seraient vus aussi distinctement que des objets très proches. Galilée, qui entretient une vaste correspondance scientifique, en reçoit confirmation par l'un de ses amis fixé à Paris, Jacopo Badovere; puisant dans son expérience et ses connaissances, il entreprend aussitôt, pour son compte, la construction d'un tel instrument. Nous savons, par son propre récit, comment les choses se passèrent.[3] Prenant un tube de plomb, il adapte à ses extrémités deux lentilles, l'une plan-concave, l'autre plan-convexe; très vite, il s'aperçoit que si l'on prend la lentille plan-concave comme oculaire les objets, selon sa propre expression, sont "sensiblement agrandis et rapprochés". Les résultats sont d'abord modestes, puisque les objets sont seulement trois fois plus proches et neuf fois plus grands que vus à l'oeil nu; sa réelle habileté technique lui permet cependant d'obtenir bien vite un instrument très supérieur, grâce auquel, dit-il, "les corps sont agrandis mille fois et rendus trente ou quarante fois plus proches". Il est probable que Galilée exagère quelque peu, comme il est probable qu'il a dû bénéficier, plus qu'il ne l'avoue, de certaines informations, voire de certaines réalisations antérieures. A vrai dire, cela est de peu d'importance. Ce qui est capital, c'est que Galilée, à peine l'appareil mis au point, le tourne vers le ciel. Et le résultat est éclatant; en quelques semaines, au début de 1610, une incomparable série de découvertes – la plus importante qui ait jamais été accomplie en un temps aussi bref – bouleverse toutes les idées sur le Monde. Coup sur coup Galilée découvre les montagnes lunaires et les satellites de Jupiter, établit que la Voie lactée est constituée par un amas d'innombrables étoiles, donne l'explication correcte de la lumière cendrée de la Lune; un an plus tard environ, nouvelle série d'observations qui culmine dans la découverte des taches solaires et des phases de Vénus. Jamais on ne dira assez l'importance de ces quelques mois. Jusqu'au début de 1610, et quelle que fut l'ingéniosité des hommes, la science avait toujours pris corps dans les limites du monde naturellement perçu; en balayant ces limites et en faisant surgir du Ciel un univers insoupçonné, Galilée n'élargissait pas seulement de façon prodigieuse le champ des connaissances humaines: il transformait du tout au tout les conditions de la réflexion cosmologique.

§3. Première conséquence, premier bouleversement: l'élimination sans appel de l'hétérogénéité physique de la Terre et des corps célestes, et l'unification matérielle du Monde. Ouvrons par exemple le *Sidereus Nuncius* – le *Messager céleste* – ce court ouvrage qui en mai 1610 divulgua la première vague des grandes découvertes et consacra en quelques mois la gloire de Galilée.

Corps céleste le plus proche de la Terre, et donc le plus aisément observable, la Lune la première va fournir son témoignage. Observant notre satellite quatre jours après la conjonction, alors qu'il commence à s'éloigner du Soleil et présente la forme d'un croissant "aux cornes brillantes", Galilée remarque aussitôt un phénomène encore jamais perçu: loin d'avoir l'aspect d'une ligne unie, comme il le paraît à l'oeil nu, la limite qui sépare la partie éclairée de la Lune et la partie toujours sombre se caractérise au contraire par sa forme irrégulière et sinueuse. Cette première observation est déjà nettement incompatible avec la thèse traditionnelle selon laquelle les corps célestes ne pouvaient qu'avoir une surface sphérique parfaitement lisse. Mais la lunette révèle bien d'autres choses encore. Non seulement la frontière entre la zone claire et la zone enténébrée est irrégulière, mais des taches sombres, "nombreuses et complètement isolées de la région obscure, se trouvent presque partout répandues sur le secteur que baigne déjà la lumière du Soleil". Tournées vers celui-ci, ces taches sont dans la direction opposée comme couronnées de lumière, et c'est toujours par le côté le plus éloigné du Soleil qu'on les voit peu à peu disparaître au fur et à mesure que progresse l'illumination de la Lune. Or "ce spectacle, écrit Galilée, est absolument semblable à celui que nous offre la Terre au lever du Soleil quand les vallées n'étant pas encore inondées de lumière, nous voyons resplendir les monts qui les entourent; et tout comme l'ombre diminue dans les dépressions terrestres au fur et à mesure que monte le Soleil, ainsi ces taches lunaires perdent leur obscurité lorsque croît la zone lumineuse". Ce n'est pas tout. De même que des taches sombres subsistent assez longtemps dans la partie éclairée, de même on voit des points brillants surgir à l'intérieur de la partie obscure, très en-deça de la zone d'illumination. "Ces points augmentent peu à peu de grandeur, note Galilée, et après deux ou trois heures s'unissent à la partie lumineuse devenue plus large; cependant que de nouveaux points, jaillissant en foule ici ou là, s'allument dans la région obscure, s'agrandissent et finalement rejoignent à leur tour la surface illuminée qui s'est étendue jusqu'à eux". Or, "ne voit-on pas sur Terre de la même façon, continue Galilée, avant le lever du Soleil, alors que l'ombre occupe encore les plaines, les cimes des plus hautes montagnes briller sous les rayons solaires? Ne voit-on pas, après un court laps de temps, la lumière s'accroître tandis que des parties de plus en plus vastes de ces mêmes montagnes sont éclairées, et qu'enfin, une fois le Soleil levé, l'illumination des plaines rejoint celle des collines?" Arrêtons ici notre lecture. Si riche et si précise était l'argumentation que dès l'automne 1610 toute l'Europe savante avait accepté les conclusions de Galilée. Désormais, comme il l'écrivait avec fierté, toujours dans le *Sidereus Nuncius*, "chacun pouvait s'assurer de façon sensible que la surface de la Lune n'est ni lisse ni polie, mais accidentée et inégale, et à l'instar de celle de la Terre, partout recouverte de proéminences géantes, de dépressions et de gouffres profonds".

Cette première unification physique de la Terre et des corps célestes allait d'ailleurs recevoir confirmation, un peu plus tard, avec la découverte et l'observation minutieuse des taches solaires. L'existence des montagnes lunaires établissait que par l'aspect la Terre et la Lune ne sont en rien différentes. Les taches solaires ajoutaient, pour leur part, une précision décisive. Formations éphémères, dont l'observation permettait de suivre la naissance, les transformations et la disparition, elles prouvaient en effet que le Soleil, loin d'être formé d'une matière immuable, soustraite au changement, comme le voulait la cosmologie traditionnelle, était bel et bien le siège de phénomènes de génération et de corruption, exactement comme la Terre. C'est donc de tous côtés que l'idée d'une hétérogénéité physique de la Terre et des corps célestes était démentie par les faits.

§4. Mais les grandes découvertes astronomiques ne permettaient pas seulement d'imposer l'image d'un Monde matériellement semblable dans toute son étendue. Elles apportaient encore un extraordinaire faisceau d'arguments en faveur du copernicianisme, c'est-à-dire de la doctrine pour laquelle la Terre, loin de rester immobile au centre du Monde, se meut d'un double mouvement: mouvement de rotation sur elle-même en vingt-quatre heures, mouvement de révolution autour du Soleil en un an.

A vrai dire, l'unification physique, dont nous venons de parler, allait déjà puissamment en ce sens. En montrant que tous les corps étaient composés de la même matière, elle accordait aux corps célestes la *gravité* dont la cosmologie traditionnelle les avaient privés. Mais du coup un nouveau problème – celui de la *possibilité physique* du géocentrisme – était posé: comment comprendre en effet qu'un corps comme le Soleil, dont le poids ne peut qu'être immensément supérieur à celui de la Terre, puisse de quelque façon tourner autour d'elle? Par quelle action la Terre pourrait-elle bien, d'abord attirer cette masse énorme, ensuite équilibrer son effet? L'attraction magnétique – seul exemple dont disposait Galilée – incitait sans équivoque à une conclusion négative. Pour un esprit réaliste, comme l'était celui de Galilée, il y avait là comme une absurdité foncière, à laquelle seuls des hommes ayant perdu tout sens du concret pouvaient rester aveugles.

Mais les grandes découvertes apportaient encore en faveur de l'héliocentrisme nombre d'arguments que je qualifierai d'optiques, pour les distinguer du précédent. En voici deux parmi les plus importants. En octobre 1610, Galilée commence l'observation systématique de Vénus; Vénus est alors étoile du soir et à son maximum d'éloignement vis à vis de la Terre. A Galilée qui la suit de soir en soir, elle se présente d'abord sous la forme d'un disque rond et de petite dimension. Pendant plusieurs semaines, et tout en s'éloignant du Soleil, son aspect ne varie pas. Mais alors qu'elle approche de son élongation maxima (c'est-à-dire de son point le plus éloigné vis à vis du Soleil), son disque apparent perd sa forme circulaire, et au point de plus grand éloignement, se réduit à un demi-cercle. Quelques semaines plus tard, Vénus s'est à nouveau rapprochée du Soleil; loin de redevenir comme avant, sa surface visible a encore diminué, et prend l'aspect d'un croissant de plus en plus fin, jusqu'au moment où, passant en face du Soleil, elle devient invisible. Puis le même cycle, mais inversé, va se reproduire: simple croissant, alors qu'elle recommence à s'éloigner du Soleil, Vénus retrouve progressivement sa forme semi-circulaire, puis, après l'élongation maxima, sa forme circulaire. Or de telles variations, inexplicables si Vénus se borne à tourner autour de la Terre, deviennent aussitôt *normales* si l'on admet que Vénus tourne autour du Soleil. Et si Vénus tourne autour du Soleil, la probabilité n'est-elle pas considérable pour que la Terre, dont le volume comparé à celui du Soleil est à peine supérieur, en fasse autant?

Autre exemple, non moins impressionnant: les variations dans la grandeur apparente de Mars. En même temps qu'elle rapproche les corps célestes, la lunette astronomique a en effet le pouvoir de supprimer l'irradiation, c'est-à-dire ce phénomène par lequel la surface d'une source lumineuse est agrandie à l'oeil nu. Grâce à cette suppression, il devient alors possible de percevoir avec précision les variations survenant dans la grandeur apparente d'une telle source. Or, dans le cas de Mars, ces variations sont particulièrement importantes, puisque la planète apparaît environ cinquante fois plus grande lorsqu'elle est au plus près de la Terre que lorsqu'elle en est au plus loin. A nouveau l'héliocentrisme rend immédiatement compte de ces différences; car si la Terre, comme Mars, tourne bien autour du Soleil, il est parfaitement *normal* que Mars, vu de la Terre, soit nettement plus grand lorsqu'il est en opposition (c'est-à-dire dans la direction opposée au Soleil, donc au plus près de la Terre) que lors de la conjonction, c'est-à-dire pour nous, au delà du Soleil.

Naturellement, on ne saurait dire, en toute rigueur, que ces observations constituent une preuve formelle du copernicianisme. En fait, aucune preuve optique n'est décisive, puisqu'il est toujours possible, par des artifices *ad hoc*, de sauver les apparences dans une perspective géocentrique. Il reste que l'observation s'accordait spontanément avec l'hypothèse héliocentrique, et pouvait à chaque fois en être présentée comme une suite nécessaire. Tout se passait, en d'autres termes, comme si le ciel perçu était une conséquence analytique du système copernicien.

Dès lors nous percevons dans toute son ampleur cette première initiative par laquelle Galilée, prenant appui sur ses grandes découvertes astronomiques, sut modifier de fond en comble l'image cosmologique. A un monde partagé en deux régions distinctes que tout séparait, il est parvenu à substituer un monde unifié dont tous les corps, faits d'une même matière, relevaient désormais d'une même science; à une vision géocentrique, expression spontanée de l'expérience millénaire de l'humanité, il a su opposer avec succès une conception qui marquait de manière irréversible le primat de la raison sur les sens. Dans un cas comme dans l'autre – et chacun l'aura déjà compris – il venait de mettre en place deux des conditions préalables sans lesquelles jamais la mécanisation du système du Monde n'aurait pu devenir réalité.

VERS LE PRINCIPE D'INERTIE

§1. La deuxième initiative que je voudrais examiner maintenant est d'un tout autre type. Mécaniser le système du Monde, c'est montrer comment les mouvements réels de ses principaux corps peuvent se déduire, une fois déterminées certaines constantes, des équations d'une science générale nommée mécanique rationnelle. Pour construire une telle science une théorie mathématique du mouvement est de toute évidence nécessaire, encore que non suffisante; et cette théorie mathématique, comme chacun sait, repose d'abord sur l'idée que le mouvement rectiligne uniforme est un mouvement inertial – capable, en d'autres termes, de se conserver indéfiniment si aucune force nouvelle ne vient à se manifester. Que Galilée ait formulé ce principe de conservation du mouvement uniforme, puis l'ait utilisé pour construire une analyse mathématique du mouvement, la quatrième journée des *Discours et démonstrations mathématiques concernant deux sciences nouvelles* en apporte explicitement la preuve: "J'imagine", écrit Galilée en ouvrant sa théorie du mouvement des projectiles, "qu'un mobile a été lancé sur un plan horizontal d'où on a écarté tout obstacle: il est évident, d'après ce qui a été dit ailleurs et plus longuement, que son mouvement se perpétuera indéfiniment et de façon uniforme sur ce plan, pourvu qu'il soit prolongé à l'infini". Or la physique traditionnelle – prolongement de la cosmologie dont nous avons examiné plus haut les grandes lignes – affirmait exactement le contraire. Pour elle, tout corps possédait dans le Monde, en vertu de sa composition matérielle, un lieu naturel; tout corps grave, notamment, tendait à occuper un emplacement aussi proche que possible du centre du Monde, en l'espèce le centre de la Terre, ce qui lui permettait d'actualiser au maximum sa nature de corps grave. Un mouvement, quel qu'il soit, était donc d'abord le mouvement d'un corps déterminé, uni à un lieu déterminé, et cette conception excluait d'entrée, comme inintelligible, l'idée que le mouvement puisse se conserver de *lui-même* dans certaines conditions. Quand il énonce son principe de conservation, Galilée accomplit bien encore, et sur un point essentiel, ce passage d'un univers conceptuel à un nouvel univers conceptuel dont je parlais en commençant. Dans quelles conditions et sous quelles formes cette nouvelle mutation a-t-elle réussi à s'imposer?

§2. Au départ de cette recherche, une certaine conception de la gravité qu'il nous faut soigneusement définir. Responsable aussi bien du poids des corps que de leur tendance à se mouvoir vers le bas, la gravité possède pour Galilée deux propriétés capitales. D'abord – et cela est typiquement pré-newtonien – elle agit de l'*intérieur des corps*. Jamais Galilée n'a en effet admis l'idée d'attraction, dans laquelle il dénonce à plusieurs reprises l'une de ces "qualités occultes" dont la physique aristotélicienne faisait un usage inconsidéré[4]; dans toute notre analyse, la gravité désignera donc un attribut direct des corps, et en aucun cas l'effet d'une certaine action due à des corps extérieurs. Quant à sa deuxième, et non moins importante caractéristique, elle est d'agir en direction d'un centre, identifié avec le centre de la Terre, et sans qui on ne pourrait même pas la définir. De telles idées, au demeurant, n'ont rien d'original; ce sont celles que l'on trouve déjà chez les philosophes du XIVe siècle, ce sont celles que l'on trouve encore chez la plupart des contemporains de Galilée. Or, partant de cette base apparemment si pauvre, Galilée va parvenir à d'étonnants résultats.

Un dispositif élémentaire, auquel il a toujours accordé le plus vif intérêt, le *plan incliné*, est à l'origine de toute l'analyse. Soit en effet un plan incliné sur lequel, imagine-t-on, une boule peut rouler sans frottement. Abandonnée à elle-même, elle va manifester aussitôt sa tendance à descendre vers le bas, en même temps qu'elle offrira une certaine résistance à tout mouvement en sens contraire. Diminuons maintenant l'angle d'inclinaison: la tendance qui pousse la boule vers le bas va montrer une diminution corrélative, de même que la résistance à tout mouvement en sens contraire. Comment rendre compte de cette modification à la lumière de la conception galiléenne de la gravité? Dira-t-on que la gravité du mobile, sans autre précision, a subi une diminution? Mais le poids du mobile – poids absolu ou poids spécifique – n'est quant à lui en rien modifié; seule la tendance du grave à se diriger vers le bas, c'est-à-dire la *gravité en tant que force motrice*, a effectivement diminué. Une dissociation s'impose donc entre les deux fonctions qu'est censée exercer la gravité, c'est-à-dire d'une part sa fonction gravifique, d'autre part sa fonction motrice. L'expression "moment de descente" (*momento di discendere*), par laquelle Galilée désigne la tendance plus ou moins grande d'un mobile à se mouvoir vers le bas, va sanctionner précisément cette dissociation, faisant ainsi de la fonction motrice de la gravité un phénomène conceptuellement distinct, caractérisé par sa variabilité, et susceptible par conséquent d'être analysé en lui-même. Ce premier point acquis, transformons notre plan incliné en une surface parfaitement "horizontale", c'est-à-dire telle qu'elle n'éloigne ni ne rapproche notre mobile du centre commun des graves, soit en fait une surface sphérique; la gravité ne pouvant plus exercer sa fonction motrice, le mobile ne présentera plus aucun "moment de descente"; il sera, comme l'écrit Galilée, "indifférent entre la tendance et la résistance au mouvement", l'une et l'autre en somme étant neutralisées. Mais nous voici au seuil d'un problème capital, formulé pour la première fois: comment comprendre le comportement éventuel d'un mobile, dès lors qu'il n'y a plus en lui ni tendance ni résistance au mouvement?

Arrêtons-nous un instant. J'ai brièvement évoqué tout à l'heure la mécanique traditionnelle – pré-galiléenne – et rappelé notamment que pour elle le mouvement prenait toujours son sens par rapport à un corps déterminé, lié lui-même à une région bien définie du Monde. Instrument d'une remise en ordre, ou manifestation d'un désordre, le mouvement ne pouvait donc en aucun cas être un *état*, mais seulement un *processus* plus ou moins transitoire. Les mouvements naturels prenaient fin au moment même où les corps rejoignaient leurs lieux naturels; quant aux mouvements violents, leur durée traduisait uniquement le temps nécessaire à l'épuisement de leurs moteurs respectifs. Aussi était-il exclu *a priori* qu'un corps puisse jamais être indifférent entre le mouvement et le repos; expression privilégiée de l'ordre, le repos, pour citer A. Koyré, avait une dignité ontologique supérieure à celle du mouvement. Quand donc Galilée parvient à concevoir, et cela vers les années 1595–96, une situation où un mobile n'éprouve plus ni tendance ni résistance au mouvement, il fait bien plus que pousser un raisonnement jusqu'à son terme logique. En affirmant qu'il existe au moins un cas où un mobile peut être indifférent à l'égard du mouvement et à l'égard du repos, il introduit une idée entièrement neuve, amorce d'une compréhension inédite du Monde.

Revenons alors à notre plan horizontal – en fait concentrique au centre commun des graves – et sur lequel un mobile se trouve "indifférent et incertain entre le mouvement et le repos". Deux étapes vont conduire successivement Galilée au principe de conservation. Une première question se présente en effet à l'esprit: pour mettre en mouvement notre mobile, une force minima, c'est-à-dire en deça de laquelle il est impossible de descendre, n'est-elle pas nécessaire? Pour la mécanique traditionnelle, la réponse n'était pas douteuse: le repos étant seul compatible avec l'ordre, une force minima est dans tous les cas requise. Quelle position Galilée va-t-il adopter pour sa part? Sa réponse est sans la moindre équivoque: "Il est certain", écrit-il, "que sur une surface exactement équilibrée, la sphère demeure indifférente et incertaine entre le mouvement et le repos, de sorte que toute force, si petite soit-elle, suffira à la mouvoir, et qu'inversement toute résistance, pour faible qu'elle soit, celle de l'air ambiant par exemple, aura le pouvoir de la maintenir immobile".[5] Dans ces lignes, extraites d'un traité sur les machines simples, rédigé à Padoue entre 1595 et 1597, Galilée fait plus à nouveau que mener une analyse à son terme logique: il introduit à propos du comportement des corps une affirmation aussi positive que nouvelle: savoir que dans certaines conditions un corps peut n'opposer aucune résistance à un mouvement tendant à le déplacer latéralement. Ce n'est pas encore le principe de conservation; c'est l'énoncé d'une condition préliminaire, sans laquelle l'idée même de conservation n'aurait pu être conçue.

Car il reste à répondre – et ce sera la deuxième étape – à la question essentielle: étant admis qu'un mobile, sur un plan n'éloignant ni n'approchant du centre, peut être mû par une force si petite soit-elle, que fera-t-il une fois mis en mouvement? Ne va-t-il pas s'arrêter au bout d'un certain temps, comme on l'avait toujours pensé jusqu'alors? ou, au contraire, l'absence de toute résistance n'impose-t-elle pas une conclusion fort différente? Nous voici parvenus, en fait, à l'un des moments vraiment exceptionnels de l'histoire des sciences. Procédant avec une hardiesse qui, aujourd'hui encore, remplit le lecteur d'admiration, Galilée va formuler, pour la première fois, l'idée d'une conservation du mouvement acquis: *sur un plan horizontal, où toute cause d'accélération et de ralentissement a disparu, un mouvement une fois commencé se perpétuera indéfiniment.*[6] Le hasard veut que nous soyons en mesure de dater assez exactement cette affirmation fondamentale. Dans une lettre datée de 1607, Benedetto Castelli rapporte en effet, comme une doctrine alors professée par Galilée, que si un moteur est nécessaire pour commencer un mouvement, sa continuation indéfinie exige seulement qu'aucun obstacle ne survienne.[7] Dès 1607, par conséquent, Galilée avait opéré la mutation intellectuelle qui symbolise le mieux peut-être le passage du monde ancien au monde moderne: avoir compris que ce qui exige réellement explication, c'est non la continuation du mouvement une fois commencé, mais bien sa cessation. Certes, sous la forme où nous venons de le rapporter, le principe de conservation n'est pas encore suffisant pour mathématiser le mouvement des projectiles; Galilée devra même le modifier d'une façon non négligeable en remplaçant la surface sphérique, qui pourtant avait permis son élaboration, par une surface plane; ainsi sera-t-il en mesure d'assurer le caractère rectiligne du mouvement inertial, nécessaire à la genèse du mouvement des projectiles, et d'en déterminer aussitôt la nature mathématique. Il reste que le moment essentiel est bien celui où pour la première fois ont été associées solidement les deux idées de mouvement et de conservation indéfinie; la voie était grande ouverte pour une théorie mathématique du mouvement.

§3. Mais ce n'est pas seulement pour la science du mouvement que l'élaboration du principe de conservation revêtait une importance primordiale. En établissant que rien ne devait empêcher le mouvement uniforme de se conserver, dès lors que certaines conditions sont réunies, Galilée se donnait aussi le moyen de consolider, de façon spectaculaire, sa nouvelle cosmologie. En un mot, il lui devenait possible, grâce à son principe de conservation, de neutraliser les objections les plus redoutables qui se dressaient alors contre l'hypothèse d'un mouvement de la Terre, et notamment contre l'hypothèse du mouvement diurne. L'analyse mécanique rejoignait donc l'analyse cosmologique, et rien ne suggère mieux l'ampleur de l'oeuvre galiléenne qu'un rapide aperçu sur cette convergence.

Une distinction doit d'abord être introduite. Contre les Coperniciens, en effet, les tenants du géocentrisme invoquaient, en ce début du XVIIème siècle, deux types d'arguments bien différents. D'abord des arguments optiques. Par exemple, et pour ne citer que celui-là, il semblait évident que les étoiles fixes, étant donné leur diamètre apparent, devraient être perçues en des points différents du Ciel, aux différentes époques de l'année, si vraiment la Terre tournait autour du Soleil, – en termes techniques, les étoiles fixes devraient présenter une parallaxe. De cet argument, qu'invoquait encore Tycho Brahé, la lunette astronomique permettait de faire entièrement justice. En supprimant l'effet d'irradiation, et en faisant des étoiles fixes de simples points lumineux, elle leur conférait un éloignement tel que l'absence de toute variation dans leurs positions apparentes n'avait en fait plus rien de surprenant. Tout différents étaient les arguments du second type. Dirigés surtout contre le mouvement diurne, ils posaient directement le problème des perturbations qu'un tel mouvement ne devrait pas manquer d'engendrer. L'idée est d'ailleurs très simple: si la Terre tournait réellement sur elle-même en vingt-quatre heures, comment croire que nous pourrions ne pas en éprouver les effets? Comment admettre, compte-tenu de la vitesse du mouvement qui nous emporterait sans cesse, que les choses se passeraient à la surface de la Terre comme nous les voyons effectivement se passer?

Dans son *Dialogue sur les deux plus grands systèmes du Monde*, Galilée cite et discute plusieurs de ces arguments, essentiellement trois. Le premier prend appui sur le mouvement naturel le plus simple et le plus familier: la chute verticale des corps pesants vers le bas. Si la Terre, comme le prétend Copernic, tourne sur elle-même en 24 heures, c'est-à-dire si chaque point de sa surface est constamment emporté vers l'est, un corps qu'on lâcherait d'assez haut – par exemple du sommet d'une tour – ne devrait jamais rejoindre le sol en un point situé à la perpendiculaire de son point de départ; celui-ci ayant fui vers l'est, alors qu'il tombait, c'est à plusieurs milles en arrière – du côté de l'ouest – qu'il devrait nécessairement se retrouver. Le deuxième argument, dû au grand astronome Tycho Brahé, illustre la même idée en faisant appel aux tirs d'artillerie. Supposons, raisonne Tycho Brahé, que Copernic ait raison, et que la surface de la Terre se meuve constamment vers l'est; un canon tire alors un boulet vers l'est: durant la projection, le canon sera lui-même emporté par le mouvement de la Terre, si bien qu'il devrait avoir rejoint, ou même dépassé, le boulet lorsque celui-ci retombera; prenons au contraire un tir vers l'ouest: le mouvement qui emporte toujours le canon vers l'est aura cette fois pour effet d'allonger considérablement le tir. En d'autres termes, si la Terre tournait sur elle-même, jamais un tir vers l'est et un tir vers l'ouest, effectués avec le même canon et dans les mêmes conditions, ne devraient avoir la même portée. Faut-il préciser que l'expérience ne révèle rien de tel? Le troisième argument, enfin, nous ramène à un phénomène familier, en l'occurence le vol des oiseaux; durant son vol, en effet, un oiseau perd contact avec la surface du sol; selon Copernic, celui-ci se meut sans cesse vers l'est, à une vitesse prodigieuse: comment comprendre alors que les oiseaux puissent voler dans toutes les directions, puis revenir à leur point de départ, sans le moindre effort apparent? La simple expérience, une fois de plus, paraît incompatible avec l'hypothèse copernicienne. Ne sous-estimons pas ces arguments. Si leur formulation peut paraître naïve à certains égards, ils montraient néanmoins, de la façon la plus claire, que le copernicianisme avait rapport non seulement avec l'observation, mais aussi avec la science du mouvement; autrement dit, que la justification du mouvement de la Terre exigeait encore un progrès radical dans l'analyse du mouvement. Ainsi le comprit Galilée, et son principe de conservation représente bien le premier pas – donc le pas décisif – accompli en ce sens.

§4. Pour plus de clarté, énonçons à nouveau ce principe: sur un plan où toute cause d'accélération et de ralentissement a disparu, c'est-à-dire sur un plan concentrique au centre de la Terre, un mouvement une fois commencé se perpétuera indéfiniment de façon uniforme. Imaginons alors que la Terre soit recouverte dans sa totalité par l'océan. Toute résistance étant abolie par la pensée, notre principe de conservation permet aussitôt d'affirmer qu'un navire mis en mouvement tournerait sans fin autour du globe. Supposons maintenant qu'une pierre ait été placée au sommet du mât: partie intégrante du navire, elle sera mue elle aussi d'un mouvement autour du centre, le même en fait que celui du navire. Si donc elle vient à quitter son appui et à tomber, ce mouvement, *auquel s'applique le principe de conservation*, se comportera comme un mouvement inertial et se composera avec son mouvement de chute proprement dit. Conservant, grâce à son mouvement inertial, même vitesse que le navire, notre pierre demeurera ainsi constamment à la perpendiculaire de son point de départ et touchera le pont exactement au pied du mât le long duquel elle est tombée. Or, cette analyse peut être étendue sans difficulté au cas où notre pierre tomberait du sommet d'une tour, la Terre tournant sur elle-même en 24 heures. La pierre, sur la tour, est en effet dans la même situation qu'au sommet du mât du navire; le mouvement diurne qui l'emporte autour du centre, sans l'en éloigner ni l'en rapprocher, se comportera lui-aussi, dès qu'on l'aura lâchée, comme un mouvement inertial, la maintiendra d'instant en instant à la verticale de son point de départ de telle sorte qu'elle atteindra le sol exactement au pied de la tour. Par son principe de conservation, Galilée était ainsi en mesure de neutraliser d'emblée les objections les plus difficiles contre le mouvement de la Terre.

Car ce qui vaut pour la chute perpendiculaire des graves, vaut de la même façon dans le cas des tirs d'artillerie. Il suffit de comprendre que le boulet, avant d'être tiré, possède, gravé en lui, comme tous les objets terrestres, et en particulier le canon, un mouvement autour du centre doué du privilège de conservation. Soit donc un tir vers l'est: grâce à ce mouvement inertial, qu'il ne perd jamais, le boulet participe comme le canon au mouvement diurne, mais bénéficiant de surcroît du mouvement que lui a communiqué l'explosion de la poudre il tombera à l'est du canon. Passons maintenant au tir vers l'ouest: le mouvement inertial dirigé vers l'est que le boulet porte en lui (donc, cette fois, en sens contraire du tir) annulera à chaque instant l'effet du mouvement diurne sur le canon, et ne laissera subsister à nouveau, comme mouvement relatif du boulet vis à vis du canon, que le mouvement dû à l'explosion de la poudre. Comme ce dernier mouvement est le même dans les deux cas, un tir vers l'est ne différera jamais – *ceteris paribus* – d'un tir vers l'ouest.

Quant à la troisième objection, sa réfutation suit encore plus directement du principe de conservation. Bien qu'ils soient vivants, les oiseaux sont aussi des corps pesants, et à ce titre ils possèdent, comme la pierre ou le boulet dont on vient de parler, le mouvement inertial autour du centre. D'autre part, l'air sur lequel ils prennent appui durant leur vol participe lui aussi, comme corps grave, au mouvement diurne. Les oiseaux n'ont donc en rien à lutter contre le mouvement diurne lorsqu'ils perdent contact avec le sol; pas plus que dans les cas précédents le mouvement diurne ne fait ainsi problème.

Cette analyse, beaucoup trop sommaire, appellerait bien des remarques. Je me bornerai à une seule. Dans les trois réfutations qui viennent d'être examinées, Galilée – et on l'aura peut-être noté – utilise son principe de conservation sous la forme suivante: lorsqu'un corps acquiert un mouvement capable de durer indéfiniment, il le conserve même si les conditions dans lesquelles il l'a acquis sont modifiées, et de plus ce mouvement peut se composer avec n'importe quel autre mouvement auquel le corps vient à être soumis. Que tel soit bien le sens profond du principe de conservation, la chose n'est pas douteuse; mais il n'est pas douteux non plus que par rapport à la forme sous laquelle Galilée l'avait d'abord énoncé, il y a bien développement et enrichissement. Tout en appliquant son principe, Galilée met ainsi à jour, avec toutes ses conséquences, la notion de mouvement inertial qu'il contenait implicitement, et une fois de plus nous fait assister à la naissance d'une idée essentielle.

Résumons-nous. Je notais tout à l'heure que le principe de conservation du mouvement uniforme, en rendant possible une science rationnelle du mouvement, était une pièce indispensable pour la mécanisation du système du Monde. Le fait qu'il l'ait énoncé – même sous une forme encore imparfaite – illustre de façon éclatante le rôle que joua Galilée dans la construction de la mécanique classique. En même temps, l'usage qu'il sut en faire pour éliminer les objections les plus significatives contre le mouvement diurne, inaugurait une nouvelle étape dans la discussion des problèmes cosmologiques. Pour la première fois – et sans mésestimer toute la distance qui restait à parcourir – un problème relatif au système du Monde venait d'être discuté en termes mécaniques.

L'IDEAL EXPLICATIF

Il nous reste à examiner une troisième et dernière initiative galiléenne. Jusqu'ici nous avons vu comment Galilée avait directement préparé, par ses découvertes, le contenu de la mécanique classique. Or celle-ci, autant que par son contenu, repose sur une conception bien précise de l'explication scientifique que, faute de place, je me bornerai à décrire d'une formule: montrer comment la proposition rendant compte d'un phénomène déterminé (ou d'une classe de phénomènes) peut être déduite dans le cadre d'une théorie construite *more mathematico*. Je voudrais donc montrer à présent comment Galilée, créateur de certaines idées capitales de la mécanique classique, est encore l'auteur de la mathématisation dont cette mécanique est indissociable.

§1. Pour bien comprendre ce dernier point, le plus simple est sans doute de revenir une fois encore à la philosophie traditionnelle. Que signifiait en effet pour elle "expliquer"? En un mot, établir les causes. Ou de façon plus développée: expliquer les effets naturels, c'était partout et toujours les rapporter à des essences ou à des principes se comportant comme autant de causes. Sans doute n'était-ce pas une tâche simple, puisque les causes, comme on le sait, se répartissaient en quatre grandes catégories – causes matérielles, causes formelles, causes efficaces, causes finales – et qu'à tout phénomène il fallait assigner quatre causes, une de chaque espèce. L'idée méthodologique n'en était pas moins claire; expliquer un effet naturel, c'était toujours le motiver *physiquement*, mettre à jour les *facteurs* dont dépend sa production. Cette conception avait une conséquence immédiate sur laquelle il faut bien mettre l'accent: elle isolait radicalement l'intelligibilité physique de l'intelligibilité mathématique. Et il est facile de comprendre pourquoi. Si l'explication scientifique a effectivement pour but de relier le phénomène aux causes qui l'ont produit, la démonstration mathématique, dont l'idéal est l'implication logique, n'offre qu'un modèle sans intérêt. Elle peut bien établir le plus rigoureux des liens entre les propositions dont elle traite; elle reste par nature incapable de faire apparaître la cause, donc de guider l'investigation physique. A quoi s'ajoute l'idée qu'on s'est faite si longtemps, à la suite d'Aristote, des notions mathématiques: abstractions issues de l'expérience sensible, portant sur des propriétés "séparées" de leur contexte naturel, soustraites à toute contingence et par là-même susceptibles d'une étude théorique où seule intervient la nécessité rationnelle. Or, cette "séparation" à qui les mathématiques doivent leur force, fait aussi qu'elles sont totalement inadéquates pour l'analyse naturelle: traitant d'objets en qui rien n'appelle le mouvement, quelle aide pourraient-elles bien apporter à l'explication des mouvements réels?.[8] Introduire les mathématiques dans la philosophie naturelle, ce serait en fait vouloir expliquer le mobile par l'immobile, l'irrégulier par le régulier, le contingent par le nécessaire. Tout au plus – comme en astronomie – les mathématiques pourront-elles servir à "sauver les apparences", autrement dit à construire des modèles destinés à restituer la succession des phénomènes; seule la recherche des causes – tâche réservée au physicien-philosophe, et non au physicien-géomètre – apporte une intelligibilité véritable dans les choses de la nature.

Contre cette conception Galilée s'est plu à multiplier les critiques et les sarcasmes. Expliquer en termes de cause, c'est-à-dire invoquer des qualités cachées agissant de l'intérieur des corps – les fameuses qualités occultes – c'est faire de l'imagination la plus naïve la maîtresse de la raison.[9] C'est également – autre façon de caractériser la méthode – demeurer au niveau des mots, et croire que la science n'est qu'une certaine façon d'en opérer la distribution.[10] "Je vois pour ma part", remarquait Galilée

une grande ressemblance entre cette façon de philosopher, et une certaine manière de peindre qu'avait l'un de mes amis; prenant une toile, il écrivait à la craie: "Ici je désire que se trouve la source avec Diane et ses nymphes, là quelques lévriers; de ce côté, je veux un chasseur avec une tête de cerf, et pour le reste une campagne, un bois, des collines" – puis laissait au peintre le soin de mettre les couleurs; ainsi se persuadait-il d'avoir peint lui-même l'histoire d'Actéon, alors qu'il n'avait fourni que les noms.[11]

Mais la vraie faiblesse de la méthode traditionnelle est ailleurs. Parmi les causes devant fournir l'explication, la plus importante, incontestablement, est la cause formelle; or, la cause formelle d'une chose n'est autre que son essence; la méthode exige donc, en dernier recours, que l'esprit humain soit en mesure de quelque façon d'appréhender les essences. Mais n'est-ce pas là pure illusion? Ce qu'on baptise essence n'est-il pas toujours, d'une manière ou d'une autre, simple répétition, en langage plus abstrait, de l'expérience sensible? Cette critique dirimante, Galilée a su l'énoncer avec une grande lucidité:

Car ou nous voulons essayer, par nos spéculations, d'atteindre l'essence vraie et intrinsèque des substances naturelles, ou nous nous contentons de connaître certaines de leurs propriétés. Je tiens la recherche des essences pour non moins impossible, et inutilement fatigante, dans les substances élémentaires les plus proches et dans les substances célestes les plus éloignées. Il me semble ignorer tout autant l'essence de la Terre que celle de la Lune, celle des nuages que celle des taches solaires; et je ne vois pas que dans l'examen des substances les plus proches, nous ayons d'autre avantage qu'une foule de détails faisant eux-mêmes problème d'ailleurs, et entre lesquels nous errons sans beaucoup de profit, passant de l'un à l'autre. Si je demande quelle est la substance des nuages et qu'on me réponde qu'il s'agit d'un peu de vapeur humide, je voudrai savoir en quoi consiste cette vapeur; on me dira, par exemple, que c'est de l'eau allégée par la chaleur, et donc transformée en vapeur; mais désireux de savoir encore ce qu'est l'eau, je trouverai, en cherchant, que c'est ce corps fluide en mouvement dans les rivières, et avec lequel nous sommes constamment en rapport; or, une telle connaissance de l'eau est seulement davantage liée aux sens mais en rien plus directe que ma connaissance antérieure des nuages. De la même façon, je ne comprends pas plus la véritable essence de la terre et du feu que celle de la Lune ou du Soleil; l'état de béatitude et lui seul nous réserve une telle connaissance.[12]

Faute de l'intuition supra-sensible, qui seule lui fournirait les bases nécessaires, l'idéal causal traditionnel apparaît ainsi sans pouvoir explicatif, et donc la science qui l'utilise incapable de *rendre raison*, à partir de principes à portée universelle, de tel ou tel groupe déterminé de phénomènes. Or sur la source de cette impuissance constitutive de la physique traditionnelle, Galilée n'a aucune hésitation: elle ne peut venir que du préjugé selon lequel l'intelligibilité physique serait d'une autre nature que l'intelligibilité mathématique, *forme par excellence de l'intelligibilité.* D'où suit aussitôt que pour le physicien soucieux d'expliquer, la seule solution possible est d'accorder aux mathématiques un rôle dirigeant dans la philosophie naturelle, et par là-même d'abolir la barrière artificiellement dressée depuis toujours entre les normes du discours physique et les normes du discours mathématique.

§2. Parvenus à ce point, il faut toutefois éviter une méprise possible. Affirmer que l'intelligibilité physique n'est en rien différente de l'intelligibilité mathématique, ne signifie nullement que l'on identifie la physique et les mathématiques. Non seulement Galilée n'a jamais rien dit de tel, mais peu d'hommes ont eu au même degré le sens des problèmes physiques et de leur spécificité, y compris vis à vis des problèmes mathématiques. Qu'il n'y ait pas de bonne physique sans mathématique, n'entraîne même pas qu'il suffise d'être un bon mathématicien pour être un bon physicien, ou plutôt, pour qu'il en aille ainsi, il faudrait que la construction de la physique n'exige pas l'élaboration préalable de concepts et de principes, indéductibles des concepts et des principes proprement mathématiques. Mais cette remarque révèle précisément toute la portée de l'identification opérée par Galilée entre l'intelligibilité physique et l'intelligibilité mathématique. Elle signifie en effet que ces principes et ces concepts – fondements irréductibles de l'explication physique – doivent être tels que les faits naturels, une fois analysés et transposés rationnellement grâce à eux, puissent aussitôt être pris en charge par la science mathématique, tels, en d'autres termes, que par leur moyen les problèmes physiques puissent être transformés en problèmes mathématiques, et cela sans altérer en rien la spécificité des phénomènes physiques. Depuis Galilée, le grand physicien est d'abord celui qui sait satisfaire à cette double exigence.

Ce n'est pas tout. Une fois construits les principes et les lois directrices d'une théorie physique, le but de l'explication devient la formulation, sur la base de ces principes et de ces lois, des propositions devant rendre compte de tels ou tels phénomènes (ou classes de phénomènes). Comment concevoir alors la justification de ces propositions à partir de ces principes et de ces lois? La réponse de Galilée tient en un mot: sur le modèle des démonstrations mathématiques. Ainsi ouvrons le *Dialogue* ou les *Discours*; justifier une proposition, c'est toujours établir entre elle et les propositions primitives, ou précédemment justifiées, un rapport de principe à conséquence, en tout point semblable à l'implication mathématique.[13] Aucune considération causale, aucun facteur relatif à la nature des corps ne sont à aucun moment invoqués: tout repose désormais sur la seule force des liaisons rationnelles. Que la mise en évidence de ce rapport d'implication revête finalement, comme dans les *Discours*, une forme mathématique, n'est à coup sûr ni un détail ni le fait du hasard; l'essentiel reste néanmoins que le physicien, quand il construit et justifie ses explications, bannisse toute idée d'un idéal d'intelligibilité radicalement différent de celui du mathématicien. Un texte écrit vers 1611–1612 résume admirablement cette conviction dont nous avons tant de peine aujourd'hui à comprendre l'audace et le rôle décisif. "Je me vois déjà rabroué par mes adversaires", constate Galilée,

> et je les entends me crier dans les oreilles qu'une chose est de traiter de la nature en physicien et autre chose en mathématicien, que les géomètres doivent demeurer dans leurs songeries et ne pas se mêler des sujets philosophiques où la vérité est fort différente de la vérité mathématique. Comme si la vérité n'était pas une. Comme si la géométrie, à notre époque, portait préjudice au développement de la vraie philosophie. Comme s'il était impossible d'être philosophe et géomètre, comme si celui qui sait la géométrie ne pouvait savoir la physique, ni raisonner en physicien des problèmes de la physique.[14]

Ni Newton ni aucun de ceux qui, au XVIIIème siècle, perfectionneront son oeuvre, n'ont exprimé plus nettement l'idéal de la mécanique classique, ni en même temps mieux résumé l'une des raison essentielles de sa réussite.

En terminant, j'aimerais souligner trois points.

Le premier – et je serai bref, tant il est évident – est le rôle de premier plan qui revient à Galilée dans la création de la science classique. Trois quarts de siècle après la constitution de l'histoire des sciences en discipline autonome, Galilée plus que jamais fait figure de grand initiateur. Plus rien après lui n'a été comme auparavant.

Le second point – et il ne contredit le premier qu'en apparence – est que la science de Galilée, tout en préparant directement la science classique, représente aussi un système original. Je veux dire par là que pour passer de Galilée à Newton, un simple développement des concepts galiléens ne pouvait suffire, mais que des concepts nouveaux, inconnus de Galilée, étaient bel et bien nécessaires. Peut-être l'exemple le plus frappant est-il fourni par la *gravité.* En faisant de celle-ci un attribut direct des corps. Galilée ne pouvait ni distinguer clairement la masse et le poids, ni concevoir le poids comme une force; pas plus, et nous l'avons bien vu, qu'il ne pouvait énoncer correctement le principe d'inertie. Précurseur direct de Newton, mais au moyen d'un système conceptuel partiellement pré-newtonien: tel nous apparaît aujourd'hui Galilée, et cette double appréciation qui, seule, rend compte avec exactitude de son apport, donne aussi la mesure de son génie.

Le troisème point est un peu différent. Peu d'auteurs ont été l'objet d'interprétations aussi nombreuses et aussi divergentes que le fut Galilée. Rationalistes et empiristes, positivistes et antipositivistes, sans oublier les marxistes: tous ont cherché (et chercheront encore) à l'enrôler dans leurs rangs. Ces exercices d'école peuvent avoir leur utilité. Il existe toutefois une bien meilleure raison de s'intéresser à l'oeuvre de Galilée. A travers ses recherches, ses brusques illuminations, ses tâtonnements, ses erreurs aussi, c'est l'aventure scientifique que nous voyons prendre corps sous nos yeux, et du même coup retrouver sa véritable dimension avec ses postulats et ses démarches caractéristiques. Préservant à la fois du refus et de la naïveté, la méditation de Galilée aide ainsi à rétablir la juste perspective et à comprendre, selon la parole de Weizsäcker, que si "la nature précède l'homme, l'homme précède les sciences de la nature". Aujourd'hui la science nous parle de partout, et sa voix universelle tend inévitablement à devenir une voix anonyme. Lui rendre un nom, et du même coup la réintroduire parmi les créations humaines, tel devrait être l'intérêt, précieux entre tous, d'un retour à Galilée.

Université de Paris-Sorbonne

NOTES

[1] Laplace, *Exposition du système du Monde*, Imprimerie royale, Paris, 1846, p. 447.
[2] Pour toute cette analyse, et celle des paragraphes suivants, voir le *Traité du Ciel.*
[3] *Sidereus Nuncius, Opere* di Galileo Galilei T. III, (Florence 1890–1906).
[4] *Dialogue sur les deux plus grands systèmes du Monde*, T. VII, pp. 470–71.
[5] *Le Mecaniche*, T. II, p. 180.
[6] *Dialogue sur les deux plus grands systèmes du Monde*, T. VII, p. 174 (on peut consulter aussi l'excellente étude de Stillman Drake, '*The concept of inertia*', in *Saggi su Galileo Galilei*, G. Barbéra ed., Florence 1967).
[7] Cf. T. X, p. 170.
[8] Cf. *Traité du Ciel*, III, 8.
[9] *Dialogue sur les deux plus grands systèmes du Monde*, pp. 470–471.
[10] *Ibid*, pp. 260–261.
[11] *Ibid*, p. 436.
[12] *Troisième lettre sur les taches solaires*, T. V, pp. 187–188.
[13] Pour un exemple très simple, mais très probant, *cf.* plus haut, 2ème partie §4, la réfutation des objections entre le mouvement diurne.
[14] *Opere*, T. IV, p. 49.

B. KUZNETSOV

GALILEO AND THE POST-RENAISSANCE

In books and papers written about Galileo in the last few decades – from Alexandre Koyré to today's historians of science – the general idea becomes clearer if we compare these works to those written prior to this period, in the first quarter of the 20th century. As an example, let us analyse Leonardo Olschki's opinion of Galileo or, to be more exact, the impression he conveys by means of a certain accent rather than in the form of an explicit idea. Olschki places primary emphasis on Galileo's relation to modern thought, to Newton's *Principia*, and in general to the classical science which can be described by means of a stable paradigm. As a result, the ties between Galileo and the Renaissance appear somewhere in the background. The starting point from which Galileo's role in the history of science is to be evaluated is to be seen in the topmost achievements of the classical science. Historians who worked in the early 20th century proceeded from the idea that Galileo *already* knew the idea of inertia, though *not* rectilinear inertia, but had no concept of a gravitational field. Their Galileo is a predecessor of the classical science or, probably, its founder. But there is another aspect to be accounted for, the retrospective relation to the 16th century, to the period at which science cannot be generally characterized by reference to a paradigm and at which its full-fledged evaluation involves contradictory, even mutually exclusive, paradigms which are never the less inseparable from each other.

For historians who worked at the beginning of this century, classical science was something homogeneous, penetrated by a single conception of the world. For a modern historian of science, the starting point of retrospection is the inseparability of classical concepts from non-classical ones: An indivisible alloy of mutually exclusive notions forms the basis for revealing the specific features of some given stage in the history of science.

Relativistic theory covers Newtonian mechanics as a particular approximation; quantum theory cannot be formulated without classical concepts. This starting point of historical retrospection enables us

J. Hintikka, D. Gruender, and E. Agazzi (eds.), Pisa Conference Proceedings, Vol. I, 253–257.

to see the original nonclassicism of the Galilean prologue to classical science, and its closeness to the Renaissance.

Such a non-classical retrospection brings to light actual connections between Galileo's creative style and the Renaissance, and introduces into history of science the notion of the proto-Renaissance, i.e. the transfer of Renaissance traditions into the culture of the 17th century; it discloses the connection between the sources of classical science and the Renaissance, traces the origin of classical science back into the 16th century. Thus Galileo's creative activity appears both as the sunset of the Renaissance and as the dawn of the Modern Era.

However, in the history of science and culture the dawn and the sunset merge together, "giving only half an hour to night", to use Pushkin's reference to the "white nights" to Petersburg. The collision of Renaissance trends with the factors which link Galileo's ideas to the future, to classical science of the second half of the 17th century, comes to the fore in the study of Galileo's creative work. Accordingly, contemporary studies of Galileo demonstrate, more distinctly than those written at the beginning of the century, the difference between the *Dialogo*, where Renaissance trends appear more evident, and the *Discorsi*, where Galileo approaches the style characteristic of the classical science of the subsequent decades.

The third volume of Olschki's *Geschichte der neusprachlichen wissenschaftlichen Literatur* [History of scientific literature in modern languages] deals with Galileo, and the general impression is that Olschki associates Galileo with the Renaissance trends which he had discussed earlier, in the first two volumes. But this is no more than an erroneous first impression, because Olschki sees the evolution of style in Galileo's creative activities as the creative pathos of his personality. Galileo's style gradually evolved between his *Dialogo* (and preceeding works) and the *Discorsi*, in other words, to that associated with Galileo in the new era. The closer to Galileo's time, the more often Olschki rebukes his predecessors for the inadequacy of their concepts, the reference point for his evaluation being the standards of classical science. Olschki even deprives Giordano Bruno of some of the appreciation accorded to him by Hegel in his lectures on the history of philosophy.

Very different is Koyré in his *Études galiléennes*, where he analyses associations between Galileo's and those of Descartes. Koyré's Galileo preserves his autonomous identity in history. Galileo's creative work is no longer treated as a kind of 'dress rehearsal' of classical mechanics. Koyre felt the importance of a new standard or criterion set forth by the Renaissance on scientific theories and hypotheses as an object of historical analysis. Recognizing this criterion is *mutatis mutandis* necessary for satisfactory analyses of the Postrenaissance, i.e., for analyses of the links between the 18th and the 15th–16th centuries. The Renaissance cannot be judged from the vantage point of any rigid and stable *system of historical coordinates*, such as, for instance, the medieval world outlook or the classical conception of the world formed in the 18th century. The significance of the Renaissance for science does not depend on its closeness to or remoteness from the medieval or classical concepts.

In order to be more exact in determining the specific nature of the new criterion and to apprehend fully the evolution of different scholars' interpretations of Galileo's work in our century, let us recall a concept put forward by Reichenbach, that of *the strong irreversibility of time.* It is applicable to real historical time, to the irreversible evolution of cognition and to the irreversible evolution of culture.[1] Reichenbach gives the name of 'weak irreversibility' to the difference between *earlier* and *later*, and the name of 'strong irreversibility' to the irreversibility which we can observe at present, when *the earlier* and *the later* merge into *the now.* The Renaissance as a stage in the evolution of science (like the contemporary situation in science in which the classical 'earlier' and the quasi-relativistic 'later' are inseparable) is a period of strong irreversibility: the medieval 'earlier' has not yet completely faded into the past, whereas the classical 'latter' is not merely a future, but already exists. Their coexistence, their inseparability, and their conflict are specfic features of the Renaissance. This specificity hinders any analysis of Renaissance science in terms of a 'coordinated' correlation with a stable medieval 'earlier' and a stable classical 'later'. But this specificity continues later on, onto the Postrenaissance. For Duhém, Renaissance science is nothing more than diluted medieval scholasticism; for Olschki, it is embryonic classical science; while for Koyré and the historians who worked in the mid- and later 17th century, the Postrenaissance, like the Renaissance, is primarily not an interval between the earlier' and the later'. According to them, the scientists of that time are not successors of the Middle Ages or predecessors of classical science. To these authors, the Renaissance and the Postrenaissance are something specific as an object of a 'non-coordinate' analysis. Let us turn it into a point: 'Renaissance and Postrenaissance'. The new approach, the 'non-coordinate' analysis, unites them and provides supplementary proof of the value of the Postrenaissance as a useful concept. What then is the specific nature of the Postrenaissance? What is the difference between the Renaissance and the Postrenaissance? The Postrenaissance is a transitional period in a more complicated sense, involving a transition from the transition stage proper to a more stable stage, 'the beginning of the end which terminates the beginning'. And, accordingly, in the 'historical aspect' 'it is an object which combines 'non-coordinate' analysis with 'coordinate' analysis, with a correlation between an epoch in the evolution of science and its stable end-point.

We need a truly epistemological definition of what has been referred to by somewhat indefinite expressions such as 'transition period' and a 'fixed, stable period' regarded as the origin of coordinates. An epistemological definition of this sort might rely on concepts introduced by A. Einstein such as that of *internal perfection* of a scientific theory (its logical deducibility from the most general principles) and its *external justification* (empirical confirmation).[2] These critical jointly secure a relatively stable world outlook, and enable this world view to act as the origin of our coordinates in evaluating the historical significance of some period in the history of science. The collision of those criteria and other obstacles to their combination are a specific feature of transitional periods, which are deprived of stable paradigms. The Renaissance is one such period. Galileo's creative work was, in essence, an elaboration of a uniform conception in agreement with astronomical observations and with the experience of applied mechanics. Galileo's conception included the principle of inertia and a differential representation of movement from point to point and from moment to moment. But this conception could not enter classical science: according to Galileo, cosmic inertial movement was circular rather than rectilinear; he had no idea of a gravitational field; and the differential representation had not yet taken the form of an infinitesimal calculus. This is why the evaluation of Galileo's place in the irreversible evolution of cognition associates the *Dialogo* and the *Discorsi* with a specific period, intermediate between the Renaissance philosophy of nature and classical science, that is, with the period of the Postrenaissance.

The 'renaissanciation' of Galileo's personality which has taken place in the middle of the second half of the 20th century does not deny his ties with classical science, but it transforms the earlier approach in two essential aspects. It brings the history of science closer to the history of culture and to epistemology. On Galileo's interpretation, science acquired certain independent criteria of acceptability: experimental proof, which is an antecedent of 'external justification'; and quantitative mathematical deduction, which is an antecedent of 'internal perfection'. Now we see more clearly the relation of these epistemological criteria to the overall process of cultural progress, which in the era of Renaissance did not allow differentiations: in that period. Truth was hardly distinguished from the Good or from Beauty; experimentation was almost exclusively associated with applied mechanics and art (Leonardo); and apology of mathematics was inseparable from philosophy (Nicholas of Cusa).

Modern literature on Galileo, in fact the entire Galilean historiography, has grown very close to the history of culture and the history of art. Meanwhile, research into Galileo's creative work inevitably leads to a search for a solution to the general, long-term problems of cognition. In this respect, such studies resemble epistemological investigations. It is not surprising that the modern historiography which studies Galileo has turned up as an item on the agenda of a joint conference of historians of science and of epistemologists.

NOTES

[1] B. Kuznetsov, 'On the problem of irreversibility of cosmic evolution, cognition, and culture history process', *Philosofskij nauki* (Philosophical sciences, Moscow) 1976, no. 6, pp. 43–55; 1977, no. 1, pp. 43–54 (in Russian); 'Ideas and images of Renaissance' (forthcoming).

[2] A. Einstein, *Collected Scientific Works*, vol. IV, Moscow, 1967, pp. 266–267 (cited from the Russian edition).

DAVID GRUENDER

GALILEO AND THE METHODS OF SCIENCE

There are many reasons why the work of Galileo has been of such great and enduring interest to philosophers and historians. The first, surely, is the drama of the conflict over the religious condemnations of his scientific work. We are often more smug about this issue today than recent events justify, although such conflicts arise most frequently with civil authorities, since religious institutions today have little direct temporal power. And, in all honesty one must admit that, occasionally, conflicts of this character occur even within the academy itself. But that is not the issue I wish to pursue here.

Another reason, which, in part, flows from this first one, is that Galileo's work is seen as marking the very emergence of modern science. As such, it has both symbolic and substantial aspects that are of great appeal. In its symbolic aspect, Galileo and his work may be taken to stand for modern science itself, and hence as an object or token which serves as the occasion for us to unburden ourselves of whatever reactions we have to that complex and, in many ways, overwhelming phenomenon of contemporary life. The appeal of the substantial aspects of Galileo's work is somewhat different from that of the symbolic: given the enormous influence and puzzling historical complexities of science, how did it begin? What were the intellectual, social, and historical elements of its origin, and what were the dynamics of their interactions with one another? By looking closely at Galileo and his work, we hope to find important clues to our understanding of the origin and character of science itself.

When we exercise our interest in Galileo as the very symbol of the origin of science – an origin fraught with adversity – we often make him a hero and a model for aspiring young scientists to emulate. Or sometimes, when moved by aspects of science that we find abhorent, we see in him the very example of just those abhorent qualities. Now, to make a person into a cultural hero requires some oversimplification of the record in order to make one's points with sufficient force and dramatic clarity. And it requires, too, some speculative filling of the gaps in our historical knowledge, for a proper hero never leaves his

J. Hintikka, D. Gruender, and E. Agazzi (eds.), Pisa Conference Proceedings, Vol. I, 259–270.

spectators wondering, either about his noble motives or his brilliant and courageous actions. Of course eventually all this nobility and brilliance and courage seem a bit too much, and the debunkers and demythologizers appear on the scene. For them the flaws of the hero loom larger than life, and anything which can be made to look tawdry or disreputable takes the center of the stage. The debunker is especially critical of the oversimplifications made earlier in describing the life and work of the hero, but is generally carried by the zeal for his cause and the same lamentable gaps in the historical record to making a few of his own.

Galileo has served as the symbol for what is best in the human search for more knowledge, and has had his debunkers too. Now it is easy to say that this interest in Galileo's work for its symbolic value is not very important philosophically or historically. But while that is true, I think it would help us to notice that this symbolic interest does partially parallel our serious interest in the substance of his work. For just as we have had a tendency to glorify or debunk his achievements as a symbolic matter, so our attempts to analyze and better understand the substance of those achievements have been affected by our own attitudes. For, however objective we may try to be, our own studies of the methods used by a pioneer in scientific method cannot help but be made from some point of view. And that point of view cannot help but be one developed in our time, affected both by our knowledge of the work of his predecessors and of that which came after him. As a result, such a perspective could not really fail to reflect the author's view of what constitutes the best in scientific method, against which it would be very natural to judge Galileo's work, overtly or covertly, according praise or blame as may seem appropriate. Of course we recognize that such a judgement would not be sound. Galileo knew only something of the work of his predecessors and nothing at all of his posterity. But it is easier to decry the results of such historical work than to avoid it. Look at the record of Galileo scholarship during the past century. Mach, one of the founders of positivism and empiricism, was troubled by some puzzling aspects of Galileo's work, but saw him as a pioneer of the experimental method and an empiricist. Indeed, Galileo's own students give a similar picture. Koyré, on the other hand, held a philosophy of science that seems closely related to Plato's classic theory of forms. The only empirical option he recognized was the simple version of induction we see in Bacon and his latter-day followers. And Koyré's view of Galileo? It was that Galileo, at heart, was a mathematician and Platonist who recognized that experimentation had little value, although it might sometimes serve to convince a skeptic. Shea agrees with this appraisal and adds, with some scorn, that Galileo was frequently mistaken on matters of elementary physical observation. Drake chides historians for offering Marxist, Thomistic, Platonic, Baconian, or medieval explanations of the influences on Galileo without looking at the events in his life to see whether they were there or not. And Feyerabend sees the best method for science to be none at all: a healthy dose of theoretical anarchy, a little deception and sleight of hand as to how the physical evidence is to be interpreted, and some very impressive rhetoric, preferably in the vernacular – these are the characteristics of the best research. Needless to say, Feyerabend finds himself well pleased with Galileo's work.

In all these enterprises the role of speculative reconstruction looms large. Galileo left us no laboratory notes, and most of what he wrote for publication was persuasive and polemical in purpose and tone. While the rest of his writing that we have has been helpful to us in understanding the development of his thinking, and will prove more helpful as it receives more study, serious gaps in the record remain. I fear that what has happened in many cases is that, out of a desire to fill these gaps so that we may present a coherent and credible picture of the pattern of Galileo's thinking, we have unwittingly projected our own scientific, philosophical, and historical predilections, ideals, hopes, and fears onto Galileo's work.

If he was wrong in an observation, what more can be said for the so-called father of modern empirical science? If his actions upset the ruling classes, who can condone them? Or, if you have a different view of the matter, what could be more valuable? Surely he could not have avoided the influence of the intellectual forces that affected his times! And if Galileo himself could take experimental data lightly, what could be left of the vaunted empiricism of science – or even of its rationality?

But we cannot expect Galileo to be always right. We cannot expect that of any inquirer, nor of anyone else either. But in science, unless history has been unrepresentative, most will turn out to have been wrong, in the strict sense of that term, as their work is superseded by later work. In another sense, which is even more important for our purpose, however, rightness is the successful outcome of a search: a search which is long, arduous, painful and slow, and which may or may not succeed in lighting a small part of the vast unknown and converting it to human knowledge. Error is frequent during the course of this search, its recognition is far from easy, and, since many a search is essentially social in character, engaged in both by contemporaries and by those of other generations (which in Galileo's day might be millenia apart), the problem is compounded by severe difficulties in communication. Indeed, the wonder is that our Galileo's ever succeed. Scientists have made numerous errors of all kinds. I once had a professor of physics who confessed to his class, with some shame, that he had determined beyong any doubt that Newton had been entirely wrong about the refractive index of lead sulfide. He had, he said, kept this a secret for many years, but could do so no longer. If this fact is not already known to you, I report it with equanimity nevertheless, for I hold it can in no way diminish the esteem or respect we ought to have for Newton as a scientist. It is not reasonable to ask a scientist to be right; only that he search with all the diligence, imagination, skill, and care he can muster. And no one can fault Galileo in this respect.

For the rest, to the extent that we apply our theories to Galileo in an a priori fashion, to that extent we tell the world more about ourselves than about his thinking, his work, his successes and failures. Yet how is this temptation to be avoided? And, in particular, how are we to avoid it in seeking an understanding of his view of scientific method? My answer is that there are two actions we can take which are likely to reduce that temptation. The first is to recognize that we ourselves are engaged in a second order search, both historical and philosophical in character, but that must not keep us from recognizing that a search is what it is, and that, like any other, it may not be successful. If it is, we should notice that its outcome is more knowledge, not a platform. The second action is to be as explicit as we can in stating our own views about scientific method, so that we and our audience can be alert to the possibility of our projecting them, however inadvertently, on our hapless subject. Of course, both of these methods may fail, but in what follows I shall apply them in the contrary hope.

Let us begin with my own view of the methods of science, which I shall state but not attempt to defend here. I used the phrase "methods of science" because I think they are rather less unitary and coherent than is suggested by "scientific method". But, what is more important, I do not think they are timeless. To put the matter another way, scientific method did not spring full-grown from the brow of Zeus – nor of Galileo. Rather, it has been won bit by bit as we have been successful in understanding the world around us and, from time to time, have been able to be reflective about how that success was achieved. Nor is scientific method yet complete! There are two important reasons for this. First, in our reflection on the actual work of science, we find we disagree with each other in basic ways about how best to interpret, understand, and characterize scientific method. Second, we must note that science itself is not complete. Even as the twentieth century wanes, it is only fair to say that there is a great deal about the world we do not understand yet. While following in Newton's footsteps, we have picked up only a few more grains of sand. We are, therefore, in no position to claim a knowledge of the whole of scientific method, for unless we find ourselves without a posterity, there is a great deal of science yet to be done. And unless intellectual history so far be a poor guide to the future, our posterity will have a richer methodological armory than we enjoy today.

Another reason for talking about the methods of science, in the plural, is that, even considered abstractly, there is a plurality of them. At the level of elementary observation one must devise appropriate categories and then sort what one has observed among them. There may be regularities in the frequency with which items in the one category appear in relation to items in others. At this point we shall need the methods of mathematics, probability, and statistics, to the extent they may be known. We may also use such inductive methods as Mill's canons, whether in the earlier forms of Bacon or Hume, or the more recent version of Von Wright.

Throughout, some form of theorizing and hypothesizing is occurring, for at the very least one has to choose the categories to observe, and alter or amend them as may be needed. The application of arithmetic and geometry to a problem introduces further theoretical elements. Merely to be counted, for example, objects must be discrete and stable. As the mathematics becomes more complex, additional requirements appear. Probability and statistical theories also have some requirements, although they are less stringent. But, sooner or later, whether in the course of early curiosity and wonder or while formulating categories for inductive use, people begin to develop theories of an abstract and dynamic character. That is to say, with some degree of generality, they begin to devise descriptions of what they suppose to be happening in nature over time. This has the double advantage of explaining their observations in terms of an overall pattern, and then of enabling them either to use this pattern to help them forecast, and, hence, better deal with future events, or, if they have leverage on any elements of the pattern, give them the means to help shape those events. Of course the theory has to issue, eventually, in means to test it. And all along the line, from the first selection of categories to be observed to the testing of an abstract theory, a scientist must be alert to the possibility that he has chosen badly, that his evidence is inadequate, that another theory fits it better, and so forth. That is, he must be rational.

I am sure that in my description you recognize the chief elements of the deductive theory of explanation, or what is sometimes called the hypothetico-deductive method. It remains only to add that the theory be couched in terms abstract enough that one be able to move deductively from the theory to the events in nature it is concerned with, using any techniques of mathematics or logic that may be available and appropriate. This has the special advantage that, since deduction is a process that conserves truth, if one begins a deduction with true premises and makes no logical errors on the way, the conclusions must be true. If it turns out that observation and experiment show them to be false, this characteristic enables one to know it is time to discard the premises of the theory one started with. The technique has proven to be a powerful tool of science.

There are other important scientific methods, but I think only two more need be mentioned before turning to Galileo. First is a theory of error: we need to be able to take account of accidental and systematic mistakes and decide how much leeway to allow in observation. And then we must remind ourselves that, since science is a social process in which inquirers rely on selected predecessors and contemporaries for some observational, theoretical, and background information, we need to be able to appraise the views of others. This appraisal may result in accepting a view without question and on authority, or in attempting to test it on one's own, or in a decision to reject it without further consideration. For this process to be rational, one must be in a position to give reasons for one's choices.

I turn now to Galileo's methods. To begin with, Galileo did not leave us with a systematic treatise on the philosophy of science or scientific method. While he makes comments on this subject, occasionally extended, they are generally remarks which occur in a context of explaining a mechanical theory or arguing that an opponent is mistaken. Even his work *Il Saggiatore* (*The Assayer*), which, as Drake emphasizes, is very helpful in understanding Galileo's philosophy of science, is chiefly polemical in purpose. But there is nothing surprising in this. Most scientists do not write much on the subject of scientific method. When they do, we have the added problem of finding out whether they practiced what they preached, as with Descartes' *Regulae* or Bridgman's *Logic of Modern Physics.* When philosophers write such treatises, we have no such problems: we have only to ask whether it has anything to do with science at all!

But in Galileo's day this neat division of labor between science and philosophy did not yet exist. Galileo thought of himself as a mathematician; philosophers, by some accident of history, were those who confined themselves to expounding and defending the views of Aristotle, or, that portion they were comfortable with, having taken his writings as a canonical text as though they were articles of religion. I imagine Aristotle would have been horrified could he have foreseen their existence. Aristotle, like Galileo, was an advocate and practitioner of observation and experimentation as methods for gaining knowledge, but these men who used his name as an authority were more interested in his results than his methods. Greek systematic inquiry did not end with Aristotle, of course: there was Euclid, Diophantos, Appollonios, Archimedes, Hero, Ptolemy, and many others. But these were taken to be mathematicians, and it was such men that Galileo seems to have taken for his model, with Archimedes clearly standing highest in his esteem.

What sort of mathematics do we have here? Using 'mathematics' in its current sense, we have plain and solid geometry, algebra, conic sections, theory of numbers, the method of exhaustion (a good start on integration and differentiation), and a start, too, on trigonometry, with the method of chords. From the point of view of physics, we need to remind ourselves that the geometry was a geometry of real space, and that Euclid himself extended it to mechanics, with proofs of theorems about levers—a practice Archimedes extended considerably. And from the point of view of logic we need to recognize the form in which the geometry appeared: an axiom system; that is to say, a complete system for proving theorems deductively within the scope of the axioms, postulates, and definitions. This enormous deductive power was very attractive to Galileo, as it was to Descartes and Newton, for with it one had a veritable model of the high standards of rigor and certainty that knowledge could be, but one could apply it not only in the realm of the mind (or wherever Plato's perfect geometrical figures are to be found), but to the space we inhabit, to simple and compound machines, bodies floating in fluids, and who knows what else. One has only to add physical postulates.

Thus, throughout Galileo's writings we read frequent praise of Archimedes, and encounter numerous attempts to use deductive techniques like his to solve new problems. In these attempts, Galileo is ingenious and clear in his use of physical, not merely geometrical, postulates to deduce mechanical conclusions. But these theorems, like those of Archimedes, are not fitted together into an axiom system like Euclid's plane geometry. They are isolated, as geometric theorems must have looked in Plato's day, rather like the proof of the pythagorean theorem in the *Meno*, before Euclid's great work of synthesis.

Similarly, throughout Galileo's writings we find praise of observation and experiment as a sound means of attaining new knowledge of nature. It is better, he sometimes says, than reading books. Yet occasionally he makes remarks about experiments that sound derogatory, as when he suggests that an experiment is useful only for convincing skeptics. Koyré has pointed these out for us. In addition, he occasionally reports observing what we know he could not have, or fails to report what we know he should have. He also categorizes observations in new ways, directing his audience's attention to new aspects of their experiences. And he suggests on several occasions that one should ignore certain elements of experience, for example, the secondary qualities, in order to concentrate on the primary ones. And sometimes he suggests we should not be troubled by the absence of expected observations, as in the case of the circular motion of a body falling on earth, which he explains (although not to Feyerabend's satisfaction) as the result of our sharing it.

These considerations bring us to the dilemma of Galileo the rationalist, on the one hand, as opposed to Galileo the empiricist on the other, to use modern terms. I propose to solve the dilemma by going between the horns.

I think the rationalist-empiricist dichotomy to be a false one: we need both theory and observation to gain more knowledge of this world. And I think it plain that Galileo used both as best he could. To cite errors of observation to count against the view that Galileo found observation and experiment a vital element in our attempts to win more knowledge of the world seems to me futile. There are some logical defects in Euclid's *Elements*, too, but no one argues he was illogical. I have already tried to explain why errors are a persistent part of scientific inquiry. Let us remember that Settle's attempt to duplicate Galileo's inclined plane experiments as he described them were successful. And I might add that my students repeated those experiments with similar results the same year, and within the limits expected by standard error theory.

Let us look at the other charges. That he redescribes experiences? I say it is an essential part of inquiry. There are, in my opinion, no protocol sentences in which we can be sure we have used the right categories; if we are not making progress in understanding some phenomena, it is always appropriate to ask whether we are noticing the important things about them and whether we are describing those adequately. That he suggests certain experiences be ignored or that the absence of others be ignored? Any scientist can recognize, when he first begins his research, that there are thousands of kinds of data that might be relevant to his problem. No one has time to look at thousands of kinds of data: one needs to be able to select out a few kinds to concentrate on. Galileo's theory of primary and secondary qualities attempts to do that for mechanics. He asks us to attend to the size, shape, number, and motion of things. If he had tried to keep track of different colors of balls rolling down an inclined plane he would have had a large experimental task indeed! This is an especially difficult sort of problem early in science or in those particular portions of it in which we still have little knowledge and few clues to go on. Galileo recognized this explicitly, explaining that he was not denying the existence of such secondary qualities as color, but postponing their investigation to a later period in science. And as for his asking us not to worry about our failure to see the curve in a body falling toward the earth although the earth is rotating, he explains this, as I mentioned, on the grounds that we are moving with the earth also. It does not matter whether this explanation satisfies anyone, his offering it is a rational act. Once it is offered it can be examined on its own merits and accepted or rejected as investigation warrants. But his offering it certainly cannot count as evidence that Galileo did not take observation seriously; it is evidence, rather, that he took the complexity of nature and the paucity of our knowledge about it seriously. Finally, it must be added that Galileo carried out his experiments in a time before anyone knew enough to develop techniques to carry out fine physical measurements, and in the absence of any theory of error. Only a strong devotion to experiment as a source of knowledge could enable Galileo to persist in it long enough to reach the success he did.

Thus we have Galileo the experimental scientist: the empiricist. What about Galileo the rationalist? For an insight I suggest we go back to his now famous letter of 1604 to Paolo Sarpi, in which he spells out correctly the principles for falling terrestrial bodies. Sarpi wanted a proof. Galileo replied that he had not been able to find any "truly unquestionable principle" to assume as a postulate for his proofs. He suggests that Sarpi think about this, and especially the common assumption that speed increases with *distance* fallen. This is an assumption whose incompatibility with his own correct principles it took him many years to recognize. I think that Drake's surmise as to the experimental method by which he might have discovered the odd number rule from which he could mathematically have derived the times squared rule – is a likely one. But with Archimedes, or Euclid, as his model, he found himself in the awkward situation of knowing a theorem to be true, yet lacking the postulates to begin its proof.

Herein lies the special problem of the rationalist, who takes a developed deductive system like Euclid's as the proper model for natural knowledge. The point of beginning with "truly unquestionable" principles as one's postulates is that, barring deductive errors, you will end up with equally unquestionable theorems or conclusions. But outside of geometry, where are such principles to be found? We did not discover until the last century, when non-Euclidean geometries began to be explored, that one's postulates do not need to be "truly unquestionable". One can assert the parallel postulate or deny it in various ways, and from each of these, in addition to other postulates, a valid geometrical system can be deduced. The parallel postulate and its various denials can hardly all be true, much less unquestionably so. But the important issue that arose next was this:

granted the lack of need for certified truth in one's postulates, which geometrical system fits the space we are studying? That is to say, as mathematical systems, the empirical or logical truth of the postulates is not important. But as soon as we attempt to interpret them as applying to some aspect of nature, our confidence in the truth of the postulates rests on how well the deductive system fits that portion of nature we are attempting to explore with it. And the only way to find that out is to see whether the theorems of the system can be found experimentally to fit what we can observe. If that fit is satisfactory, we will be well-content with the postulates, and will call them physical laws.

This is, in fact the method of Newton's *Principia*, although many then and since, including the illustrious Kant, have taken Newton's laws to be independently "unquestionably" true, or, as Kant described them, synthetic statements a priori. That brings us to the next point. Newton needed three postulates to work out the deductive system we know as classical mechanics. Galileo had identified two of these, and, if Drake is right, had his finger on the third as well. But his statement of the second, the principle of inertia, was restricted to the earth, and here I think Shapere is right to remind us that nothing is gained by asking whether this restricted version captures the essence of the idea of inertia. Thus, even had Galileo not insisted on having postulates that were unquestionably true in some a priori way, he still had not succeeded in identifying all those that were needed, and stating them with sufficient generality to provide the needed deductive power. This is no criticism of Galileo; that task and the working out of the deductive system of mechanics took much of Newton's working life, and he said he found the first two of his laws in Galileo.

I conclude that Galileo, like Descartes, worked strenuously to follow the model of Archimedes in applying deductive methods to the exploration of nature, and that this work provided vital stepping stones to our better understanding of how such methods can be made to work. The situation is not crystal clear to this day, as witness the heated disagreements that still surround the deductive theory of explanation. But Galileo did attempt to develop such systems and did recognize that they must be compatible with experimental data within limits. It was his confidence in such systems that, I think, led him sometimes to suggest that experiments were for skeptics, for up to that time every mathematician thought the power of the deductive method began with the truth of its postulates. Our knowledge of the delicate interplay between theory and experiment is dependent, in part, on his work. It is unreasonable, I think, to suppose that he ought to have had that knowledge himself. As a mathematician, engineer, physicist, astronomer, philosopher — and as a human being — Galileo made original contributions. Through his efforts, our world and our knowledge of it have been unalterably broadened. For whether successful or unsuccessful, his efforts were rational, and I think that is the secret of the benefits he has passed on to posterity.

Florida State University, Tallahassee

BIBLIOGRAPHY

Burtt, E. A., *The Metaphysical Foundations of Modern Physical Science*, rev ed, Humanities Press, New York, 1932.

Drake, S., *Galileo Studies*, The University of Michigan Press, Ann Arbor, 1970.

Duhém, P., *The Aim of Modern Physics*, tr. Philip Wiener, Princeton University Press, 1953.

Feyerabend, P. K., *Against Method*, NLB, London, 1975.

Galilei, G., *Le Opere*, Edizione Nazionale, ed Antonio Favaro, Firenze, 1890–1909.

Geymonat, L., Galileo Galilei: *A Biography and Inquiry into his Philosophy of Science*, McGraw-Hill, New York, 1965.

Hanson, N. R., *Patterns of Discovery*, Cambridge University Press, 1965.

Hempel, C. G., *Aspects of Scientific Explanation*, The Free Press, New York, 1965.

Kaplon, M. F., ed, *Homage to Galileo*, Massachusetts Institute of Technology Press, Cambridge, 1965.

Koyré, A., *Etudes Galileennes*, J Vrin, Paris, 1939.

McMullin, E., ed, *Galileo: Man of Science*, Basic Books, New York, 1929.

Poincaré, H., *The Foundations of Science*, Science Press, New York, 1965.

Popper, K., *The Logic of Scientific Discovery*, Harper and Row, New York, 1965.

Shapere, D., *Galileo: A Philosophical Study*, University of Chicago Press, 1974.

Shea, W. R., *Galileo's Intellectual Revolution*, Science History Publications, New York, 1972.

A. C. CROMBIE

PHILOSOPHICAL PRESUPPOSITIONS AND SHIFTING INTERPRETATIONS OF GALILEO

It is entirely appropriate that a discussion of shifting interpretations of Galileo should take as its *terminus a quo* the *très cher maître* of so many of us who took up the history of science professionally immediately after the Second World War: the great Alexandre Koyré. With three or four other rare spirits he was one of those who showed by example the enlightenment that can be gained only by looking beneath the surface of immediate scientific results, by seeking to identify the intellectual and technical conditions that made certain discoveries possible and explanations acceptable to a particular generation or group, others not, and the same not to others. By displaying the science in historical relation to the philosophical assumptions, technical equipment, and social context of its designers and discoverers, they showed how the history of science could illuminate the nature both of European culture and of scientific thinking. This contextual approach establishes the identity of the history of science as a field of study. It shows it clearly to belong in its sources, content and methods of scholarship to history and philosophy, and equally clearly to require some special knowledge of the problems whose records are its primary materials. The history of science might be seen as a history of intellectual behaviour, in the sense both of a history recreating the past and of a comparative natural history of examples, a kind of philosophical anthropology. It is enlightening to compare the different attitudes to nature developed through different periods of European history, and to compare European attitudes with others. Such comparisons involve conceptions both of science and of history, a point which Koyré made fundamental in the evaluation of any interpretation of the history of science.

In his own sophisticated reinterpretation of Galileo's sophisticated thinking, Koyré showed that experimental science was an argument, and that experiments contributed to the argument at logically precise points in its development. In doing this it might be said that he saw

J. Hintikka, D. Gruender, and E. Agazzi (eds.), Pisa Conference Proceedings, Vol. I, 271–286.

Galileo too much through his own Platonic vision: as pointed out by Ludovico Geymonat, and indeed also among others by myself.[1] Nevertheless it was from Koyré's *Etudes Galiléennes* (1939) that we all learnt to look at Galileo in a new way: to understand more clearly just how Archimedes was a model for his scientific thinking: that Galileo's view of his procedure in investigating simple but subtle subject-matter like that of motion was first to construct a theoretical world, and then to contrive experiments (real or imaginary) to test whether this was the existing world in which he lived. But we should not forget that in more complex subject-matters, such as those of sunspots, heat, light and floating bodies, experiment had also for Galileo a more directly exploratory role: using then the Aristotelian logical criteria of presence, absence, and agreement in degrees or concomitant variations.[2]

If then the *terminus a quo* of interpretations of Galileo is to be Alexandre Koyré, we might look for likewise sophisticated shifts from that point not in any *terminus ad quem* but in the continuing scholarship relating Galileo and his contemporaries truly to their intellectual and social context: scientific, philosophical, theological, economic, technological, and artistic. And here we might remember those other genial intelligences for whom Galileo was an inspiration: Leonardo Olschki, Edwin Burtt, Giorgio de Santillana, and behind them all the masterly editing and scholarship of Antonio Favaro.

Let me now very briefly indicate a small selection of the characteristics shared by Galileo with his intellectual and cultural context which seem to me to illuminate his intellectual personality and the way he went about his activities, and to illuminate at the same time much other contemporary intellectual behaviour. When we look at Galileo as a man of late Renaissance Italy, we should try to relate the style of his scientific thinking to contemporary styles of thinking in the arts, in philosophy, and in practical affairs. Then through the particular example of Galileo we can offer an analysis of the various elements that make up an intellectual style in the study of nature: conceptions of nature and of science, of scientific inquiry and scientific demonstration and explanation with their diversifications according to subject-matter, of the identity of science within an intellectual culture, and the intellectual commitments and expectations generating attitudes to innovation and change. In exploring these questions we should also keep in view the historical problem of the unique origins of modern science in the society of Western Europe. The example of Galileo gives a central focus to this view in Galileo's style as a Renaissance man of *virtù*. This, in Renaissance Italian, meant a man with active intellectual power (*virtù*) to command any situation, to control what he did and what he made, whether in mind or matter, in the natural sciences or the constructive arts, in private or political life, as distinct from being at the mercy of events, of the accidents of *fortuna*. The conception of the *virtuoso*, the rational artist aiming at reasoned and examined control alike of his own thoughts and intentions and of his surroundings, seems to me of the essence of the European morality, meaning both habits and ethics, out of which European science was generated and engineered.

The scientific movement generated in Western Europe, above all, a capacity to act with rational intent in the control at once both of argument and calculation, and of a variety of materials and practical activities. It generated, thereby, an effective context for seeing and solving the exemplary technical problems shared by the mathematical sciences with the visual, musical, plastic, and mechanical arts. All exemplified a common mastery of nature by the rational anticipation of effects, whether by means of quantified theory alone, or by modelling a theory with an artifact analytically imitating and extending the natural original. The rational artist and the rational experimenter and observer thus acted alike in conceiving alike an artistic construction and a scientific inquiry first in the mind before executing it with the hands. But, beyond that, there was in the conception of a man of *virtù* a programme for relating man to the world as perceiver and knower and agent in the context of his integral moral and social and cosmological existence. The programme entailed a commitment to reasoned consistency in all things. Let me simply illustrate with some quotations:

Domingo Gundisalvo (late 12th century):

Natural things are those which nature produces by motion visibly operating from potency to actuality.... But artificial things are those made by the art and will of man.... The artist is the natural philosopher who, proceeding rationally from the causes of things to the effects and from effects to causes, searches for principles. By optics ... what appears in vision otherwise than it is is distinguished from what appears as it is. For this science assigns the causes by which these things are brought about, and this by necessary demonstrations. ... The science of engines is the science for contriving how one can make all those things agree, of which the measures are expressed and demonstrated in mathematical theory, agree I say in natural bodies.... The sciences of engines therefore teach the ways of contriving and finding out how natural bodies may be fitted together by some artifice according to number, so that the use we are looking for may come from them.[3]

Leon Battista Alberti (1435):

Man is to render praise to God to satisfy him with good works for those gifts of excelling *virtù* that God gave to the soul of man, greatest and preeminent above all other earthly animals.[4]

In writing about painting... we will, to make our discourse clearer, first take from mathematicians those things which seem relevant to the subject. When we have learned these, we will go on, to the best of our ability, to explain the art of painting from the basic principles of nature.... We will now go on to instruct the painter how he can represent with his hand what he has conceived with his mind.[5]

Giorgio Valla (1501):

Nature also follows a path, that is she follows her own order because she carries out everything of that sort with reason as leader. But not with imagination like art, for she does not prepare anything inwardly which she wishes to produce. But the artist reasons when he wants something for himself, fashions and forms it inwardly, and accordingly makes an image for himself of everything that is to be portrayed.[6]

Archytas of Tarentum... says at the very beginning of his *On Mathematics*: They seem to me to be correct in their estimation and apprehension of mathematical sciences, who consider these to weigh particulars exactly because mathematicians are very well versed in the nature of the whole.[7]

Marsilio Ficino (1469–74):

What is a work of art? The mind of the artist in matter separate from it. What is a work of nature? The mind of nature in matter united with it.[8]

Other living things live either without art or with one single art, to the practice of which they do not apply themselves of their own accord but are drawn to it by a law of fate. Proof of this is that they make no progress with time in the business of constructing things. Men on the other hand are the inventors of countless arts which they pursue of their own will. This is proved by the fact that individuals practise many arts, change, and through long practice become more skilful. And what is remarkable, human arts construct by themselves whatever nature herself constructs, as if we were not slaves of nature but rivals.[9]

Leonardo da Vinci (lived 1452–1519):

Astronomy and the other sciences proceed by means of manual operations, but first they are mental as is painting, which is first in the mind of him who theorizes on it, but painting cannot achieve its perfection without manual operation.[10]

In fact whatever exists in the universe through essence, presence or imagination, the painter has first in the mind and then in the hands.[11]

But first I make some experiment before going any further, because my intention is to cite experience first and then to demonstrate with reason why such an experience is constrained to operate thus in this way; that is the true rule according to which theorizers about natural effects have to proceed. And although nature starts from the reason and finishes at experience, for us it is necessary to proceed the other way round, that is starting ... from experience and with that to investigate the reason.[12]

There is no effect in nature without reason; understand the reason and you do not need experiment.[13]

Oh speculator on things, I do not praise you for knowing the things that nature through her order naturally brings about ordinarily by herself; but, I say, rejoice in knowing the end of those things which are designed by your own mind.[14]

Vitruvius on the architect:

His works are born from both construction and reasoning (*De architectura*, i.1.1).

The philological commentary on the earliest Italian translation (1521): *Machinatio* ... may be derived from I cunningly contrive, ... I deliberate, I think out, ... stratagem, ... whence undertaking, thinking, machine and ... mechanic or mechanical operator.[15]

Machina .... Mechanics ... is commendable whether for its basic imitative resemblance to the divine work of the construction of the world, or for the great and memorable usefulness reached by the human beings enumerated .... And that furthermore ... has been put into practice through a burning desire to produce in sensible works with their own hands that which they have thought out with the mind.[16]

Alessandro Piccolomini (1542):

On the cause on which human happiness depends. Given that the happiness of man consists in acting according to *virtù* in a perfect life, it is reasonable to ask on whom it depends, that is in whose power is this happiness. So it must be noted that according to Aristotle it must derive from one of three causes: divine or human or fortuitous. And if it comes from a human cause, this is either through reasoning or through practice. That it cannot depend on fortune can be seen from the fact that so noble an effect as our own happiness cannot be produced by so vile a cause as fortune, since fortune is not an essential cause but accidental and consequently vile and ignoble. An essential cause surely is what produces the effect according to the intention of that cause, just as an architect produces a house according to his own intention.[17]

Niccolò Tartaglia (1554):

... things constructed or manufactured in matter can never be made so precisely as they can be imagined by the mind outside that matter, by which effects may be caused in them quite contrary to reason. And from this and other similar considerations, the mathematician does not accept or consent to demonstrations or proofs made on the strength and authority of the senses in matter, but only on those made by demonstration, and arguments abstracted from all sensible matter .... Similarly, as to those questions that have already been demonstrated with mathematical arguments which are more certain, one should not attempt or believe it possible to certify them better with physical arguments, which are less certain.

Nevertheless

all those things which are known in the mind to be true, and especially by demonstrations abstracted from all matter, should reasonably also be verified in matter by the sense of sight: otherwise mathematics would be wholly useless and of no help or profit to man.[18]

Daniele Barbaro (1556): on Michelangelo:

For the artist works first in the intellect and conceives in the mind, and then signs the external matter with the internal habit.[19]

Art can so far imitate nature, and this comes about because the principle of art which is the human intellect has a great resemblance to the principle that moves nature, which is a divine intelligence; from the resemblance of power and of principles is born the resemblance of operating, which for the present we will call imitation.

But

the intellect of man is imperfect and not equal to the divine intellect, and matter so to speak is deaf, and the hand does not respond to the intention of art;

hence

the architect must think very well and, in order to make more certain of the success of the works, will proceed first with the design and the model, . . . and . . . he will imitate nature, which does not do anything against its makers. Yet he will not search for impossible things, either as to the matter or as to the form, which neither he nor others can accomplish . . . . Whence art, observer of nature, wanting also to make something, takes the matter of nature put into existence with sensible and natural form, as is wood, iron, stone, and forms that matter with that idea and with that sign which is reposing in the mind of the artist.[20]

Peter Ramus (1569):

For Archytas and Eudoxus, says Plutarch in his *Marcellus*, transferred mathematical contemplations from the mind, and from things falling within the understanding of thought alone, to examples of sensible and corporal things, enriching geometry with a variety of demonstrations, not only logical but also practical, and they taught that geometry was of the very greatest use in life . . . this faculty of geometry is called mechanical and instrumental. But Plato, indignant that they revealed and published the noblest of philosophers to the vulgar and as it were betrayed the secret mysteries of philosophy, deterred both from their undertaking . . . . How much better and more correctly you judged, when you said that philosophers should be urged from the school and leisure to the government of state, from the contemplation of arts to practising their use.[21]

Guidobaldo del Monte (1588):

But if art overcomes nature by imitating her so that those things which are done by art happen contrary to nature, the genius of art will appear by this much more excellent; if indeed in imitating nature it may be said to act contrary to the order of nature (for that will seem perhaps paradoxical, although it is very true). For art with wonderful skill overcomes nature through nature herself, by so arranging things as nature herself would do if she decided that such effects should be produced by herself.[22]

Galileo Galilei (1593):

I have seene all engineers deceiv'd, while they would apply their engines to works of their own nature impossible; in the success of which both they themselves have been deceiv'd, and others also defrauded of the hopes they had conceiv'd upon their promeses . . . ; as if, with their engines they could cosen nature, whose inviolable laws it is, that no resistance can be overcome by force which is not stronger than it. Which belief how false it is, I hope by true and necessary demonstration to make most manifest . . . ; and this is according to the necessary constitution of nature . . . . Nay if it were otherwise, it were not only absurd, but impossible . . . . And . . . all wonder ceases in us of that effect, which goes not a poynt out of the bounds of nature's constitution.[23]

Vincenzo Viviani (1654) on Galileo's decision to live outside the city on his return to Florence in 1610:

There he lived with the more satisfaction because it seemed to him that the city was a kind of prison for speculative minds, and that the freedom of the country was the book of nature, always open to those who enjoyed reading and studying it with the eyes of the intellect. He said that the letters in which it was written were the propositions, figures and conclusions of geometry, by means of which alone was it possible to penetrate any of the infinite mysteries of nature . . . . He praised indeed all that had been written in philosophy and geometry that was well designed to enlighten and awaken the mind towards similar and higher speculations, but he very aptly used to say that the main doors into the richest treasury of natural philosophy were observations and experiments, which could be opened by the noblest and most inquisitive intellects by means of the keys of the senses . . . . At all times he took great delight in agriculture, which served him at once as a pastime and as an occasion to philosophize about the way plants feed and grow, the prolific power of seeds, and the other wonderful works of the Divine Architect . . . . He had a marvellous understanding of the theory of music, and of this he gave a clear example in the First Day of his last Dialogues [i.e. *Discorsi*, 1638]. Besides the delight he took in painting, he enjoyed with exquisite taste works of sculpture and architecture and all the arts subordinate to design.[24]

Living from Michelangelo's death to Newton's birth, Galileo marks the transition between two great European intellectual movements each in its own way dominated by mathematical rationality: the transition from the world of the rational constructive artist to that of the rational experimental scientist. A product of the integrated intellectual culture of the 16th-century Italian cities, he enjoyed both its interest and its expertise in a rich variety of activities ranging from natural history and the mathematical arts and sciences of music, perspective painting, cartography, architecture, engineering, gunnery, mechanics, and astronomy to philosophy, cosmology, theology and the Italian language as a vehicle of scientific and philosophical expression. When Galileo was negotiating his return to Tuscany in 1610 in the service of the Grand Duke he wrote famously that "as to the title and function of my service, I desire that in addition to the name of mathematician His Highness will add that of philosopher; for I claim to have studied more years in philosophy than months in pure mathematics."[25] It is above all as the author of a philosophical strategy for the sciences of nature, of the design of a scientific style, that Galileo seems to me to illuminate the identity of science within the intellectual culture of early modern Europe.

A pugnacious man like his musical father, Galileo placed himself within what might be called the battle conception of history. He came to see himself battling for two essential elements of an effective strategy for scientific thinking: for principles of explanation applicable to the whole of nature, and for methods of inquiry by which to find accurate and convincing solutions to specific limited problems. One essential criterion for accepting the first was their strict and technical application to the second. These were two aspects of the same search for dependable knowledge. Whether we see Galileo as a Platonist for whom the book of nature was written in mathematical language, or as a Renaissance artist-engineer who sought to control his materials by taking nature to pieces in a workship in order to reassemble it from then known principles, he acted also as a humanist scholar debating the best ancient models for true scientific thinking. He could then claim the title of philosopher as an heir to the ancients who transformed their general questions into questions capable of technical scientific answers. This was, and remained long after Galileo, one fruitful approach to the continuing problem of establishing the identity of natural philosophy within the whole varied context of intellectual culture, and with that an agreed conception both of nature and of scientific explanation.

In seeing himself as a pioneer of a new natural philosophy that excluded all others because it uniquely could both define and solve particular and quantitative physical problems and in doing so relate them to a general system of explanation, Galileo was at one with conservatives and reformers alike in their paradoxical claim *stare super vias antiquas*. But his ancient guides for selecting the answerable questions defining the true new science came to be less Plato or Democritus than the models for problem-solving set out by Archimedes and the atomic or corpuscular theory of matter offered within a scientific treatise by Hero of Alexandria. His strategy was to establish effective criteria for selecting answerable questions, as well as acceptable answers, to be admitted into an inquiry. The admissible questions came to be, in principle, those answerable by means of mathematical and experimental analysis and an exclusive conception of physical causation. The last criterion came from Aristotle, but Galileo came to make it explicitly mechanistic. True natural science was then distinct from scholastic disputations conducted simply as logical exercises, or building general systems that solved no particular physical problems, or philosophical erudition and theological exegesis like those aiming at concordance in one divine and natural truth, or the uncritical curiosity of natural magic, or the arbitrary transactions of human commerce or law, or constructions for artistic or engineering effect rather than scientific explanation. Galileo's rational experimental science was defined by its integrated search at one and the same time both for reproducible practical results and for corresponding principles of theoretical explanation.

To unravel and date the relative influences on Galileo's scientific style of Plato and Aristotle, and Archimedes and Hero of Alexandria, we must more than ever look below the immediately obvious historical surface of his investigations and disputes and try to grasp the intentions of his arguments. We should distinguish first between the mathematical techniques and conceptions used in exploring the relations within phenomena, for which his essential guide was Archimedes, and the logic of demonstration supposed. The latter was provided by Aristotle's conception of apodeictic demonstration, with its epistemological demand that this should contain a complete explanation in physics through the efficient and material as well as formal causes. By discovering and investigating the date of Galileo's use of textbooks by three Jesuit professors at the Collegio Romano for his essays on Aristotelian natural philosophy misnamed by Favaro *Juvenilia*, and by relating these essays to his other writings, Adriano Carugo and I have established the longevity and depth of the Aristotelian knowledge from which Galileo approached the search for the true cosmology which became the overriding intellectual preoccupation of his life. Work on another unpublished Aristotelian essay, the *Disputationes de praecognitionibus et de demonstratione*, establishes his more fundamentally enduring acceptance of apodeictic search for "necessary demonstrations" of "the true constitution of the universe. For such a constitution exists, and exists in only one, true, real way, that could not possibly be otherwise".[26] From the epistemological demands of such demonstrations came Galileo's consistent search for physical causation, filling the gap left by the Renaissance Platonic acceptance of mathematical harmony and proportion as a sufficient basis for scientific explanation. Before him, his father Vincenzo Galilei had rejected such Platonic conceptions in music and had based his own investigations of musical sensations insistently on searching for physical causes. Galileo in his footsteps based his defence of the telescope as a valid observing instrument on the grounds of there being a valid and stable causal relation between sight and the seen.

Yet in his analysis of the relation of perceiver and knower to the perceived and known Galileo's Platonism also becomes evident. No other inquiries illustrate more clearly both the unity of his intellectual style and the triumphs that led him into matching contradictions. In a contribution to a fashionable debate whether painting or sculpture was the superior, Galileo wrote: "The farther removed the means by which one imitates are from the thing to be imitated, the more worthy of admiration the imitation will be".[27] Hence the representation in painting of a three-dimensional scene on a two-dimensional surface was superior to a merely sculptured relief, just as the representation of emotion by instrumental music was superior to that by song. Likewise, in natural philosophy, Galileo's consistent rating of reason above immediate experience marked a conscious preference for Plato and Democritus, as well as for Archimedes, over Aristotle as a guide to the true nature of things which was to be found in abstract theory. Hence his conclusion that the irreducible minimum existential "conditions" which he found that he must necessarily attribute as he came to "conceive of a piece of matter or of a corporal substance" were not its appearances of colours and tastes and smells and sounds, but the "primary and real properties" of "sizes, shapes, numbers and slow or swift motions".[28] Hence again his judgement on the motion of the Earth: "I can finde no end of my admiration how reason could so much withstand sense in Aristarchus and Copernicus, that, that notwithstanding, this is become the mistresse of their credulitie".[29]

This was a triumph. But Galileo's acceptance of mathematics as in some sense nature's true if hidden language led him, when accompanied by his search for apodeictic demonstration, into making an ideal of the impossible. Mathematicians in the 16th century had pointed out the worthlessness of contemporary attempts to put Euclid's geometry into syllogistic form. Galileo's contemporaries likewise rejected his belief that an apodeictic cosmology was even a distant possibility, while he at the same time pursued his ideal by means of the quite different criterion of making the new cosmology ever more probable by its ever increasing range of confirmation.

Finally, in this cosmological debate that was to make Galileo so much an historical symbol, there appeared a further aspect of natural science as an exemplary exercise of *virtù* in all its contexts. An obvious diagnostic characteristic of Western science is that it has been throughout its history as much a moral enterprise as a means of solving physical problems. One form of this has been the view established in different ways by philosophers of many different persuasions since Plato that nature was not just a deductive system, but also a moral order, with accompanying them others insisting like the atomists that, on the contrary, nature was morally neutral. The debate has profoundly affected both the specific intellectual character and the political role of science in Western culture. It has led to tensions repeated formally again and again between doctrines and loyalties derived from radically different sources, tensions often cruel though sometimes fruitful when generated by the insistence of the heirs at once of Aristotle and Moses that the truth must be either one thing or another, universally and exclusively, over the whole range of principle and practice for all that exists.

It may be argued that it was, above all, Galileo who showed how to disembarrass nature of its moral charge, and who through his public controversies and their consequences focussed the moral enterprise of science instead as one of the inalienable freedom of responsible inquiring minds to search for the objective truth. For "Nature, deaf and inexorable to our entreaties, will not alter or change the course of her effects".[30] Nature could not be cheated, and he

> being used to study in the book of nature, where things are written in only one way, would not be able to dispute any problem *ad utranque partem* or to maintain any conclusion not first believed or known to be true.[31]

Moreover:

> We must not ask nature to accommodate herself to what might seem to us the best disposition and order, but we must adapt our intellect to what she has made, certain that such is the best and not something else.[32]

Theologians (and politicians) should then:

> consider with all care the difference that there is between opinable and demonstrative doctrines; so that, having clearly in front of their minds with what force necessary inferences bind, they might the better ascertain themselves that it is not in the power of professors of demonstrative sciences to change opinions at their wish, applying themselves now on one side and now on the other; and that there is a great difference between commanding a mathematician or a philosopher and directing a merchant or a lawyer, and that the demonstrated conclusions about the things of nature and of the heavens cannot be changed with the same ease as opinions about what is lawful or not in a contract, rent or exchange.[33]

Hence it was not the scientifically demonstrated conclusions of the new cosmology and their like that threatened humane and responsible intelligence, but rather:

> Who doubts that the novelty just introduced, of wanting minds created free by God to become slaves to the will of others, is going to give birth to very great scandals? And that to want other people to deny their own senses and to prefer to them the judgement of others, and to allow people utterly ignorant of a science or an art to become judges over intelligent men and to have power to turn them round at their will by virtue of the authority granted to them: these are the novelties with power to ruin republics and overthrow states .... Be careful, theologians ...":

for if one doctrine were thus ignorantly condemned, "in the long run, when it has been demonstrated by the senses and by necessity",[34] its opposite might have to be declared heretical instead.

Galileo as a public figure dramatically brought out into the open a subtle shift in the commitments and expectations of disagreement as well as agreement in the intellectual style of early modern Western science. A man of aggressive creative energy who could never be neutral in his own society, contact with him now through the living page of his marvellous baroque Italian, as then in the flesh, may be compared to his own description of the musical interval of the fifth "which, tempering the sweetness by a drop of tartness, seems at the same time to kiss and to bite".[35] In the page, as in the flesh, his account of scientific objectivity is a brisk antidote to the naiver sociological relativism promoted by people evidently ignorant of the distinction between the history of science and the history of ideology. As according to his last disciple and first biographer, Galileo "renovated mathematics and true philosophy in his own country" because he "was endowed by nature with such a marvellous ability to communicate learning",[36] so his arguments communicate the identity of the scientific movement. He communicates this as an intellectual enterprise integrated by its explicit historic criteria for choosing between theories and investigations at different levels of the true, the probable, the possible, the fruitful, the sterile, the impossible and the false. By means of these criteria scientists have exercised a kind of natural selection of theories and investigations which has directed scientific thinking as the history of at once solving problems and embodying them in ever more general explanations.

All this raises fundamental questions of historical interpretation which I have discussed at length in my forthcoming book, *Styles of Scientific Thinking in the European Tradition.*[37] The historiography of science is there introduced in the context of intellectual culture, conceptions of nature and of science, technical possibilities, social habits and dispositions especially those favouring or opposing change, and physical and biological ecology. The subject is treated as a kind of comparative intellectual anthropology, the study of human behaviour in situations of habit and opportunity and decision. This provides the historiographical context for the history of styles of scientific thinking as an integral part of cultural identity. Essentially I offer, then, a study of the variety and historical commitments of methods of scientific inquiry and explanation in the situations of intellectual orientation leading to their development. Styles of scientific thinking in Western intellectual history have been dominated and progressively diversified by the interaction of philosophical and practical programmes, embodying antecedent conceptions of nature and of science, with the success or failure of their scientific realization in widening varieties of subject-matter. Scientific experience made explicit the organization of scientific inquiry historically round a series of overlapping types of scientific method and explanation, with characteristic modes of self-correction and criteria of acceptability. These types of science have been differentiated, out of the rational programme initiated by the Greeks, by the demands imposed by diverse subject-matters; the conceptions of nature presupposing what was there to be discovered and so guiding inquiry and supplying the ultimate irreducible explanatory principles; the consequential procedures of research, including the crucial point at which experiment came into a scientific argument; and the theories of scientific demonstration distinguishing kinds of causal and non-causal relations and governing what to accept as having been discovered.

The active promotion and diversification of the scientific methods of late medieval and early modern Europe reflected the general growth of a research mentality in European society, a mentality conditioned and increasingly committed by its circumstances to expect and to look actively for problems to formulate and solve, rather than for an accepted consensus without argument. The varieties of scientific methods so brought into play may be distinguished as the simple postulation established in the mathematical sciences, the experimental exploration and measurement of more complex observable relations, the hypothetical construction of analogical models, the ordering of variety by comparison and taxonomy, the statistical analysis of the regularities of populations and the calculus of probabilities, and the historical derivation of genetic development. The first three of these methods concern essentially the science of individual regularities, and the second three the science of the regularities of populations ordered in space and in time.

Trinity College, Oxford

NOTES

[1] A. C. Crombie, *Galilée devant les critiques de la posterité* (Les Conferences du Palais de la Découverte, Paris, 1957), cf. on the Galileo Prize, *Physis*, xii (1970) 107; L. Geymonat, *Galileo Galilei* (Milano, 1957; New York, 1965); and below n.2.

[2] See for full documentation of this paper with bibliographies Crombie, *Galileo and Mersenne: Science, nature and the senses in the sixteenth and early seventeenth centuries*, and *Styles of Scientific Thinking in the European Tradition* (both forthcoming: Clarendon Press, Oxford); cf. above n.1, below nn.4, 26, 28, 30.

[3] Dominicus Gundissalinus, *De divisione philosophiae*, hgr. L. Baur (*Beiträge zur Geschichte der Philosophie des Mittelalters*, iv.2–3; Münster, 1903) 10, 27, 112, 122.

[4] Leon Battista Alberti, *I libri della famiglia*, ed. C. Grayson (*Opere volgari*, i, Bari, 1960) 153; cf. Crombie, 'The search for certainty and truth, old and new' in *Science and the Arts in the Renaissance: Folger Institute Symposium 1978* (forthcoming).

[5] Alberti, *De pictura*, i.1 and 24, ed. C. Grayson in *On Painting and On Sculpture* (London, 1972) 58.

[6] Giorgius Valla, *De expetendis et fugiendis rebus opus*, i. 3 (Venetiis, 1501).

[7] *Ibid.* vi. 4.

[8] Marsilius Ficinus, *Theologia Platonica*, iv. 1 (*Opera*, Basileae, 1576) 123.

[9] *Ibid.* xii. 3, pp. 295–7.

[10] Leonardo da Vinci, *Treatise on Painting, Codex Urbinas Latinus* 1270, i. 19, transl. A. P. McMahon (Princeton, 1956): a posthumous compilation.

[11] *Ibid.* i. 35.

[12] Leonardo da Vinci, *Les manuscrits... de la Bibliothéque de l'Institut*, Codex E., f. 55^r, publiés ... par ... C. Ravaisson-Mollien (Paris, 1888).

[13] Leonardo da Vinci, *Il Codico Atlantico nella Biblioteca Ambrosiana di Milano*,... transcrizione ... di G. Piumati, f. 147^v (Milano, 1894–1904).

[14] Leonardo da Vinci, *Les manuscrits*... Cod. G., f. 47^r, par ... Ravaisson-Mollien (1890).

[15] Marcus Lucius Vitruvius Pollio, *De architectura libri dece, traducti de latino in vulgare, affigurati, commentati*, i. 3 (Como, 1521) f. 18: begun by Cesare Cesariano and completed by Benedetto Giovio and Bono Mauro; see P. Galluzzi, 'A proposito di un errore dei traduttori di Vitruvio nel '500", *Annali dell' Istituto e Museo di Storia della Scienza di Firenze*, i (1976) 78–80.

[16] Vitruvius, *ibid.* x. i. f. 162^v.

[17] Alessandro Piccolomini, *De la institutione di tutta la vita de l'homo nato nobile e in un città libera*... i. 3 (Venetiis, 1545) f. 16^r (first ed. 1542).

[18] Niccolò Tartaglia, *Quesiti et inventioni diverse*, vii. 1 (Venetiis, 1546) ff. 78^r–9^v.

[19] Daniele Barbaro, *I dieci libri dell'Architettura di M. Viruvio, tradutti e commentati*... Proemio (Vinegia, 1556) 9.

[20] *Ibid.* i. 3, p. 26.

[21] Petrus Ramus, *Scholarum mathematicarum libri*, i (Basileae, 1569) 18.

[22] Guidobaldus e Marchio Montis, *In duos Archimidis Aequeponderantium libros paraphrasis scholiis illustrata*, Praefatio (Pesauri, 1588) 2.

[23] Galileo Galilei, *Le mecaniche* (1593), ed. naz., direttore A. Favaro (*Le opere*, ii, Firenze, 1968) 155, transl. Robert Payne (1636): transcribed from British Museum MS Harley 6796, f. 317^r, by A. Carugo: see Crombie, *Galileo and Mersenne*, ch. 2. iii.

[24] Vicenzo Viviani, 'Racconto istorico della vita di Galileo' (1654; *Le opere*, xix) 625, 627.

[25] Galileo to Belisario Vinta, 7 May 1610, *Le opere*, x, 353.

[26] Galileo, 'Prima Lettera circa le macchie solari' (1612; *Le opere*, v) 102, *Lettera a Madama Cristina di Lorena* (1615; *ibid*) 316, 330; *cf.* Crombie, 'The primary properties and secondary qualities in Galileo Galilei's natural philosophy' in *Saggi su Galileo Galilei* (Firenze, preprint, 1969), 'Sources of Galileo's early natural philosophy' in *Reason, Experiment and Mysticism in the Scientific Revolution*, ed. M. L. Righini Bonelli and W. R. Shea (New York, 1975) 157–75, and above n. 2. Adriano Carugo established during 1968–69 that two of the sources of the *Juvenilia* were Benito Pereira and Francisco de Toledo, and I discovered in June 1971 that a third was Christopher Clavius. William A. Wallace began later to look in the right direction for the first two sources but failed to identify them, writing in 'Galileo and the Thomists', in *St. Thomas Aquinas 1274–1974 Commemorative Studies* (Pontifical Institute of Medieval Studies, Toronto, 1974) 327: "... there is no evidence of direct copying from any of the Thomistic authors mentioned in this study". He did not mention Clavius. In response to a letter of 16 July 1971

with a typed copy of this article I sent him our information on 31 March 1972: "So far as the sources of the *Juvenilia* are concerned, we have shown that three main sources, sometimes copied word for word, are Clavius's commentary on Sacrobosco's *Sphaera*, Pereira's *De communibus omnium rerum naturalium* and Toletus's commentaries on the *Physics* and on *De generatione et corruptione*. Certainly there is no evidence for, and there is negative evidence against, his using Bonamico" (as Favaro had supposed). Later in 1972 after he had visited both myself and Carugo I sent Wallace at his request the relevant sections of our book. A note on p. 330 added subsequently to the published version of his article gives a misleading account of our discoveries, which directed attention to the Collegio Romano; cf. also his *Galileo's Early Notebooks*: *The Physical Questions* (Notre Dame, Indiana, 1977).

[27] Galileo to Cardo dà Cigoli, 26 June 1612, *Le opere*, xi, 341, transl. E. Panofsky, *Galileo as a Critic of the Arts* (The Hague, 1954) 36, *cf.* 9.

[28] Galileo, *Il Saggiatore*, q. 48 (1623; *Le Opere*, vi) 347–8, 350; *cf.* Crombie, 'The primary properties . . .' (1969).

[29] Galileo, *Dialago sopra i due massimi sistemi del mondo, Tolemaico e Copernicano*, iii (1632; *Le opere*, vii) 355, transl. Joseph Webbe (c. 1634) in British Museum MS Harley 6320, f. 259^r: see Crombie, *Galileo and Mersenne*, ch. 6. i.

[30] Galileo, 'Terza Lettera delle macchie solare' (1612; *Le opere*, v) 218; *cf.* for these questions Crombie, 'The relevance of the middle ages to the scientific movement' in *Perspectives in Medieval History*, ed. K. F. Drew and F. S. Lear (Chicago, 1963) 35–57, 'Some attitudes to scientific progress: ancient, medieval and early modern', *History of Science*, xiii (1957) 213–30, 'The Western experience of scientific objectivity' in *Proceedings of the Third International Humanistic Symposium* 1975 (Athens, 1977) 428–45: all with further references.

[31] Galileo in 1612 (*Le opere*, iv) 248.

[32] Galileo in 1612 (*ibid.* xi) 344.

[33] Galileo, *Lettera a Madama Cristina di Lorena* (1615; *ibid.* v) 326.

[34] Galileo, notes related to the *Dialago* (*ibid.* vii) 540, 541.

[35] Galileo, *Discorsi e dimostrazioni matematiche intorno a due nuove scienze*, i (1638; *ibid.* viii) 149.

[36] Viviani, 'Racconto . . . ' (*ibid.* xix) 627–8; *cf.* above nn. 24, 35.

[37] Above n. 2; *cf.* for the concept of scientific styles, and for the historical method of looking for the questions asked or implied by the answers given in any historical situation and for their changes: R. G. Collingwood, *An Autobiography* (London, 1939), *The Idea of Nature* (Oxford, 1945) and *The Idea of History* (Oxford, 1946); also Crombie, *Augustine to Galileo*, introduction (1st. ed. London, 1952, new revised ed. 1979); and the discussion of Galileo's scientific style by Winifred Lovell Wisan, 'Galileo and the Emergence of a New Scientific Style' (in this volume) and 'Mathematics and the study of motion: emergence of a new scientific style in the 17th century' (forthcoming).

V. S. KIRSANOV AND L. A. MARKOVA

CREATIVE WORK AS AN OBJECT OF THEORETICAL UNDERSTANDING

I

If the history of science is understood as that of scientific ideas, the researcher who sets about studying it is faced with the duality of this history in the following sense: on the one hand, the existence of scientific ideas is independent of each individual human being; they are characterized by a chronological sequence; follow from and justify each other forming a single system of knowledge. On the other hand, the historian cannot fail to take into account the fact that scientific ideas arise in the brain of a scientist and that various events and circumstances that, at first sight, bear no relation whatsoever to the strictly logical structure of scientific knowledge can either promote or impede their appearance. These circumstances may pertain to the sphere of social, cultural, and political relations, and may express the peculiarities of a scientist's individual biography. The history of science splits into two histories: the history of ideas which is objectivized and independent of the subject, and the history of the production of knowledge which is related to the activity of a scientist as a personality.

It is exactly this element of historical and scientific studies which has drawn the attention of S. Drake when he discussed various possible ways of analyzing Galileo as one of the founders of modern science.

Without denying the importance and necessity of the history of science as that of scientific ideas by themselves, Drake nevertheless is inclined to attach a greater significance to the biographical history of science. In this sense Drake sees his position in the history of science as opposite to the position of A. Koyré.

Thus, in Drake's opinion, Koyré formulates his studies in such a way that his attention is concentrated on the aspects that are external with respect to the personality of a scientist: first of all, on the logic of the development of scientific ideas. Drake, on the other hand, is

J. Hintikka, D. Gruender, and E. Agazzi (eds.), Pisa Conference Proceedings, Vol. I, 287–310.

interested mainly in the circumstances that are internal to the scientist's personality, in the psychological aspects of his scientific activity. When Drake sets these two methods of research as opposite to each other, he does not believe they are mutually exclusive; in his understanding one type of study complements the other.

The division of the history of science into the history of scientific ideas and the history of the personal activity of a scientist, as fixed by Drake, appears to us quite essential for describing the modern situation in the historiography of science. We shall make an attempt to offer our own version of the interpretation of the scientist's personality in the history of science, basing our analysis on an assumption of the possibility of a theoretical, logical interpretation not only of the history of the scientific ideas as such, in their objective existence outside the subject of theorization, but also of the creative processes in the mind of the scientist, of his scientific activity as a personality.

In the second part of our article we shall analyse how Galileo's personality formation in the context of the *Weltanschauung*, cultural, and logical ideas of the Renaissance led to his creation of the foundations of new mechanics. Thereby we shall show the compatibility of the point of view that stresses the importance of philosophy in the history of science (A. Koyré) with the opinion about the significance of the investigation of a scientist's personality against the background of an epoch (S. Drake).

The contrast between the process of a scientist who formulates new scientific theories in his creative work, which does not permit of a logical interpretation, as opposed to the available theoretical knowledge, which is constructed purely logically, became nearly a commonplace of the great majority of historical-scientific, philosophical and logical analyses of the present time. Inasmuch as the social aspect of the history of science is the social institutionalization of science, the possibility of its theoretical, logical analysis is not subject to doubt. In the same way as the sociological theories of society on the whole are created, in the frame of the sociology of science the theories of science as a social institution are formulated. The personal activity of the scientist as a theoretician is opposed both to the social organization of science and to the logical ordering of theoretical knowledge, which are in the same position from the point of view of the possibilities of theoretical comprehension.

In connection with the personal activity of a scientist and the social organisation of science as a social institution let us turn to Marx's opposing universal to cooperative labor. In Marx's words,

> ... one should distinguish between universal and cooperative labor. Both of them play their own role in the process of production, each of them charges into the other, but there is also a difference between them. Universal labor is every scientific labor, every discovery, every invention. It is conditioned, in part, by the cooperation of contemporaries, and in part by the predecessors' labor. Cooperative labor implies direct cooperation between individuals (K. Marx and F. Engels, *Collected Works*, vol. 25, part I, p. 116).

Both kinds of labor are characterized by their own role, and each of them changes into the other not only in the process of production on the whole but in science in particular.[1] Universal labor determines the specificity of science, its essence: inasmuch as this labor is devoted to the production of new knowledge, it is an inventive activity. But to the extent to which science is not only an invention, discovery, obtaining of the new knowledge, it includes cooperative labor as well. Marx subjects cooperative labor to a detailed analysis in connection with the study of the capitalist method of production, and it is exactly here that it is embodied in its purest form. The meaning of cooperative labor is the exchange of activity between the individuals in the form of the exchange of the products of this activity. Cooperative labor as a result of the division of labor in manufacturing or in more developed forms of capitalist production implies that all the participants of the labor process are under one roof, and each of them uses the product of the labor of another, being interested neither in how this product has been obtained, nor in who will make use of the product of his own labor nor in how they will do so. The process of creating a half-finished product and its "history" are of no significance.

This type of relation need not exist "under one roof" in the literal sense of the word, nor among contemporaries. The heart of the matter is that it is *as if* the exchange of activity – even between people who live on different continents or belong to different generations – occurs under one roof and outside of time. As far as science of the modern period is concerned, the relations of use, as well as cooperative labor play an important role. It is of great interest to consider modern science, from this point of view, as a process of the production of knowledge, the course of which is similar to that of the process of the production of material values. In society, the science of the modern period as a social institution functions and exists according to the laws of cooperative labor, and this is especially true of the great science of the second half of the twentieth century. In modern research institutes the cooperation of scientists is nearly always governed by the principles of cooperative labor: some scientists use the results of computations, experiments, and analyses of others without going into the details of how those were obtained. The results of the scientific activity of the past are also treated on a timeless basis: the past achievements of science are presented in the textbooks in the form in which they enter the structure of the modern knowledge, history being projected onto the timeless space of contemporaneity. T. Kuhn aptly described such activity of scientists under the name "normal science" in his book *The Structure of Scientific Revolutions* (enlarged second edition, University of Chicago, 1970).

The same type of activity covers the social, or what Holton calls the "objective" aspect of science, and also Merton's characterization of science as a social report in the form of a scientific text. According to both Merton and Holton, the opposite type of activity that is associated directly with the scientific invention, discovery, is related to the personality of the scientist, his individual behavior; Merton considers only the historic description of such behavior as possible at all, but by no means its theoretical reconstruction, while Holton, although formulating the task of the rational, logical analysis of the invention, stresses constantly nevertheless its specifically individual, subjective, non-social nature.

We would like first of all to emphasize that the activity directed towards creation of new knowledge – the creative activity – in spite of its apparent subjective nature, is by no means less social than the activity devoted to the spreading of the knowledge already available – to its assimilation through the printed text, to its application in other branches of science or in industry, and so on. The point is that the inventive activity in science belongs to the sphere of universal labor which, although opposite in many respects to cooperative labor, is not at all less social and which, as cooperative labor, is impossible without the cooperation of people, although this cooperation is different.

The modern study of science literature has already recorded certain forms of cooperation of scientists that have been produced by tasks of a creative nature. Those are the invisible colleges, the problem groups, where the hierarchy of relations is governed not by the administrative principle (the director – the subordinate; the head of the division – the senior, junior scientist) but by the requirements of the creative solution of the problems. One can identify quite distinctly, however, the social (and hence the objectively necessary, rational, and logical) character of labor directed at obtaining new knowledge without a direct study of the external manifestations of its social nature: without an analysis of cooperation in the form of invisible colleges or problem groups. The labor of a scientist who sits alone at his desk and invents something new, being universal labor, is social and objectively necessary to the same extent as any cooperative labor.

At present, with the institutionalization of science at a very high level, which is exactly what makes it possible to speak of the "big" science of our time as compared to the "little" science of the past, the cooperation of scientists according to the principles of cooperative labor becomes more and more obvious and appears to be the only form of the social nature of science. The more evident the social character of science as a social institution which can be studied theoretically in the frame of sociology becomes, the more actively the personal activity of the scientist directed towards obtaining new knowledge is rejected by sociological theories as intuitive and asocial. It turns out to be quite out of place, as such, in logical theories of scientific knowledge also. The better and more consistently the analysis of scientific knowledge is conducted in the frame of classical logic, the apex of which is mathematical logic, the more distinct is the opposition between the structure of available knowledge and the process of obtaining this knowledge, between the context of justification and the context of discovery, between the personal subjective creative activity of the scientist and the objective world of the present scientific knowledge.

Such clear demarcation of two spheres of activity, however, results not only in the expansion of the front of research in the field of sociology of science and in the field of logic of scientific knowledge, but also in the realisation of the fact that the form of activity of the scientist directed at the production of new knowledge, which is absolutely necessary and vitally important for science, cannot be theoretically comprehended in the frame of sociological and logical theories. And the more persistently the personal activity of the scientist is rejected by modern theoretical constructions, the more clearly it is realised, exactly through this rejection, what it is that we cannot comprehend theoretically. Hence a great number of phenomenological descriptions of the creative activity of scientists, and recommendations to look for diaries, notebooks, or autobiographies. The creative laboratory of a scientist is something imperceptible, intangible, unobjectivizable, something which has no material embodiment whatsoever, which can only be watched through a key-hole.

Is it still possible to objectivize the subjective creative processes in the mind of the scientists, and if so, then how?

In order to become a scientist, to secure certain success in the scientific field, the human being should assimilate a certain sum of knowledge to make it his own. He should master in the course of education that very knowledge which is presented in textbooks and scientific publications in impersonal, impassive, rigorously logical form.

The peculiarities, the characteristic definitions of one or another animal species which have been developed and accumulated in the course of its history are inherited by the animal as behavioral instincts and are from the very beginning inseparable from it – make an integral whole with it. And the human being still has to assimilate his social-human definitions that exist outside of him and independently of him. We are interested here in those definitions of the human being which make him a theoretician. There is an objective opposition between the history of mankind as that of the theorizing subject and the human being; this history has yet to be assimilated in one or another way. The human being himself, as the future theoretician, becomes a problem for himself. This is due to that objective reality of the scientific knowledge accumulated by the preceding generations which he has to reproduce in himself as his personality characteristics.

As is well known, for one or another phenomenon to be understood as social, it is necessary to consider the relations between objects or the relations between the human being and the objects as the relations between people. In our case we should try to see behind the relation between the future scientist or simply the scientist and the scientific text of a certain type, the relations between people, between the scientists. Our discussion will be, again, based on Marx' division of cooperation of people into two types: the cooperation which corresponds to the norms of cooperative labor and the cooperation which corresponds to the norms of the universal labor.

In the same way as the human being, according to Marx, doubles himself in reality through active interaction with nature in the course of material production, he doubles himself in scientific consciousness as well when he creates an objective, real being in the form of the objectivized scientific knowledge recorded in a text. Although the knowledge recorded in a scientific text, as the objectivized result of scientific labor, exists objectively, independently of each individual human being, it nevertheless has a meaning only in conjugation with a human being who can read and assimilate it in one or another way. It is only on the condition of this assumption that a scientific text can bear a relation to science, in the same way a marble block bears a relation to sculpture only if one assumes that a human being can carve a sculpture out of it. The marble block should be understood as the *inorganic body* of a human being: as an element of nature included in human history in the course of which the skills, the know-how of treating marble are accumulated, the artistic style is developed, etc.

Thus, scientific knowledge is the objectivized subjective activity of the scientist. As was made clear in the above, however, there can be two types of the scientist's activity, namely, activity in the frame of universal labor, and activity in the frame of the cooperative labor. When the process of formation of a scientist is considered, it is very important what kind of subjective characteristics are observed in a theory. If the theory, for us, represents only the sum of ready-made answers, it also certainly possesses subjective characteristics, since it is a result of the activity of its creator and can be used as a result for most diversified purposes, increasing the possibilities of our thought in the same way as the instruments of material production increase our muscular power. When such an approach to scientific knowledge is used, the education process prepares the specialists-executors, but not the creators. And in this case in the theory itself we see the subject, but not a creative subject; an analogy to this is that the machine is, undoubtedly, another form of being with respect to man, but nobody can see the creator in it.

Scientific knowledge has to be subjectivized in such a way that it preserves the *subject as the creator*, and not merely the subject, since the latter can be observed even in the simple hammer. In order to achieve this end, one has to see in the theory itself the various types of activity. The theory is not just a sum of answers, but a sum of questions as well, and as such it represents something unbalanced. The theory appears as a problem, not just as understanding, but also as the lack of understanding, as a system of disparity between the presence and the lack of understanding. The theory is assimilated not as a certain level of knowledge but as a certain level of theorizing. For all its logical completeness, every given theory is pre-theoretical in nature since the new theory emerges from it. Such an approach to theories and to scientific knowledge in general contains a possibility of forming creative scientists.

In accordance with this, our attempts to objectivize in a certain way subjective and creative processes should be reduced to the study of interiorization of the types of social cooperation existing in a scientific community and in the mind of the scientist. The scientist's theorizing is determined by the objectively existing scientific knowledge, which can be regarded as objectively necessary. It represents the subjective aspect of the same scientific knowledge in the form of scientific publications, but it is very important to remember that the methods of determination can vary: they can be realized both in the context of cooperative and in that of universal labor.

In the first case, the scientist is formed through the assimilation of the objectivized history of scientific knowledge as a certain body of information, of the ready answers no matter by whom and when they were obtained. The scientists of the past who had obtained answers to the questions that were tormenting them are present in modern textbooks only as the original cause: the source of knowledge set forth in the textbook. The scientist assimilating knowledge presented in this way in the scientific texts becomes a person with scientific education who has turned the past objectivized history of the scientific ideas into his own. In correspondence with the available knowledge, he shapes himself as the carrier of information obtained in the past: as a person who knows the ready answers but who cannot doubt and formulate questions, and, therefore, has no need for a collocutor to conduct a dialogue with the creators of theories produced in the past. The scientist-executor is formed who can use the knowledge assimilated by him more or less successfully, the knowledge having been obtained by others for the solution of problems that have been formulated also by others.

When the interiorization of social knowledge of the second type, cooperation in the frame of universal labor, occurs, scientific knowledge is assimilated in the context of culture, in its subjective-creative aspect. Not only the answers, but the questions, doubts, and perplexities that are contained in a scientific theory become interiorized. In the mind of the theoretician, a "microsocium" of the subjects of various types of theorizing that join in a dialogue with each other appears, in the frame of various theoretical systems. And it is only in this microsocium that the objectively existing theory as a sum of answers to the questions that were formulated at one time becomes real and thus capable of further development. A scientist is being formed for whom creative activity is possible.

It goes without saying that in actual life the above two ways of forming a scientist are never realized so distinctly. Even an education which is most apparently formal and openly gives first place to the assimilation of already available results, inevitably produces in the minds of the students a greater or lesser need to doubt the apparently unconditionally true answers (depending on many circumstances, first of all, on natural abilities).

In order to discuss the fact that creative thinking is directed not only outside, towards the creation of a new theory, with new knowledge of the objective, material world, but also towards the subject, in the sense of the development of his capabilities, namely the theorizing skills, consider the analogy with labor in material production.

To become an expert worker, the human being has to master a certain set of labor instruments that have been created by past generations of people and are intended to strengthen his qualities through becoming his artificial organs. The instruments of labor will never fuse with him tightly, for there will be always a possibility for them to be objectively opposed to the worker. They will always be his and not his. During the process of mastering these instruments and during the process of labor, however, the worker also develops those natural abilities which are inseparable from him and belong to him as an individual: his muscular power, sleight of hand, knack. Finally, a time may arrive when the specifically personal, physical qualities of the worker do not correspond to those instruments of labor which he uses. It would then become necessary to change the instruments.

The human being also makes use of his natural qualities – memory, imagination, thinking, and will power – during education when he becomes a mature theoretician and masters scientific knowledge, and this use is diversified since the human being faces frequently the problem of choice and is "perplexed" by some situations or other that he comes across. As the laborer, the scientist possesses characteristics which exist objectively as the results of the social history of mankind and which have yet to become assimilated in the course of becoming a theoretician; there are also characteristics which are specifically biological or psychological and cannot be separated from him under any circumstances. And the latter, as well as the muscular power and the skill of the worker in the process of labor are subject to development and improvement, and the time also comes when the discrepancy between the characteristics of the human being as a scientist, inseparable from him, and those which are at the same time his and not his, which possess apparently a double being when they belong to a given scientist and oppose him as the objectivized being of the scientific knowledge, becomes obvious.

The gap – the space between these two groups of characteristics – the discrepancy between them – exists always, in the same way as there is always a gap between the being of a natural object and its ideal image. The formulation in philosophy of the last problem, the problem of the correspondence between the natural being and its theoretical image, is very diversified: the thing in itself of Kant, the impossibility of its total comprehension by theoretical reason, or the infinite approximation to the absolute truth which can never be attained, the approximation to more and more perfect knowledge of natural being to the knowledge which, nevertheless, can at no time exhaust this being or the problem of the reflection of natural reality in theoretical images (concepts), or the same problem of the relation between experience or practice, on the one hand, and theoretical activity, on the other, and in the more narrow sense, the relation between scientific experiment and scientific theorizing.

It is important for us now to emphasize that in all the above formulations of the problem of the relation between the natural reality and its theoretical reproduction (and their number could be increased), it is always recorded that nature in its material being is always "greater" than its theoretical image: that it can never be exhausted in theory. Hence there arises inevitably the problem of the relation between theoretical knowledge and its object in material reality, as well as the object which has not yet been theoretically understood. The theory itself already contains knowledge of what cannot be comprehended theoretically by means of a given theory, and the task consists in theoretical understanding of the transition from the lack of knowledge to its presence, or, to put it differently, from the being of natural phenomena which is given to us in our practical activity to their theoretical comprehension in the theoretical reproduction of the discrepancy recorded in the available theory.

The situation is the same with the two types of characteristics of the theoretician: of his physical capabilities that are fused with him absolutely and cannot be separated from him under any circumstances (memory, reason, imagination, etc.), and his specifically theoretical capabilities that have a history of their own, independent from a given individual, and should yet have been assimilated and always subject to the process of objectivization. The set of characteristics of the first type makes it possible for us to speak of the being of the subject of theorizing in the same way as we spoke of the being of natural objects. This being is extralogical and extratheoretical, but in the final analysis it is exactly what serves (in a mediated way through the creation of the ideal) as the object of cognition. The constantly present gap between the theoretical construction and the material being (both natural and human) is the source and the stimulus of creative thinking: thinking directed at the creation of new knowledge. The theoretical understanding of the transition from the extratheoretical (that existing outside of the *given* theory) to the theoretical (comprehended in the frame of the new theory) is realized in one form or another.

The activity of the human being devoted to shaping himself as a theoretician (during training in school, in college, and as a result of reading the textbooks and scientific papers) is determined by the scientific knowledge which exists objectively at a given historical moment and which is ideally available for everybody as the object of assimilation. In real life, however, the theoretician is never shaped under such pure "experimental" conditions for, on the one hand, there is a human being who is oriented from early childhood towards theoretical activity and who excludes from his life all other circumstances not related to this activity; and, on the other, there is his objectivized other form of being as a theoretician which he should first assimilate and then transform to a certain degree in the course of work as a theoretician. The process never occurs in such pure form, in the same way as inertial motion can never be realised in reality.

Actually, a very small percentage of people formulate for themselves the goal of taking the ideal opportunity to become a scientist-theoretician. Most people simply have no access to the scientific texts and to educational institutions because of their social, material conditions. And those who live under such conditions by no means always strive to become theoreticians since there is a great number of other ways of existence that appear to many much more attractive than the career of a scientist. Finally, those who do adopt such a goal realize it very non-uniformly, by different methods and with different results. In others words, there is obviously no equality between theoreticians, although ideally it seems that everybody has equal opportunities.

When the problem is formulated to understand theoretically the theorizing scientist, it does not mean at all that the theory should include the whole social, psychological, biological and other history of the theorizing person. Being is always greater than any theoretical construction, and this holds also for the existence of the human being. In the same way that the mechanical theory of motion does not lay claim to the explanation of all individual peculiarities of every special motion, the theory of a theorizing scientist cannot be based on the personal fate and the circumstances of the life of an individual scientist, even if the latter performed a revolution in science. A certain theoretical construction of the being of a theorizing scientist is necessary, which would serve directly as the object of study. Any person who is becoming a theoretician and is engaged in theorizing follows a certain theoretical ideal in a way similar to that of a physical body, which, for all individual deviations from the "norm", moves in accordance with the mechanical laws of motion. At a definite time, the theoretical image of theorizing ceases to be perceived as the ideal, the discrepancy between the real capabilities of the theoretician to think, to indulge in fantasies, to invent and the existing theorizing norms begins to be realized (in analogy with developed muscular power and skill), not being in correspondence with the available instruments of labor. The gap between the theoretical way of theorizing and being-oriented theorizing, which always exists but is not always realized as such, begins to be perceived as discomfort; not only the being which is understood but also the being reproduced in the theoretical construction as incomprehensible, are observed in the latter. The appearance of the new type of theoretician occurs through a transition into the category of scientist's characteristics that are theoretical: of those that are subject to objectivization, can belong to all, represent certain being, and, up to a certain time, be inseparable from the personality of the scientist. In the same way, if one considers scientific knowledge from the point of view of its being oriented towards nature, the appearance of new knowledge would mean the raising to the rank of theoretical of certain being-oriented characteristics of material reality which up until that time had been present in the old theory only as the incomprehensible, alogical element which had been pushed out of it.

Galileo, being one of the founders of modern science, was frequently studied by historians of science from most diversified points of view. We lay no claims whatsoever for a thorough analysis of formation of new principles of theorizing in the creative laboratory of Galileo as a historic personality. We shall consider the question how the philosophical, cultural and logical conceptions of an epoch, being refracted through the personality of a scientist and participating most fundamentally in the formation of a certain type of personality, promote the appearance of the corresponding scientific ideas. In this connection we shall try to determine the significance of studying philosophy for the historian of science, and thus in this manner join in the argument between S. Drake and A. Koyré.

II

1. One of the main topics which we are going to discuss is the question to what extent the philosophy of science can turn out to be helpful for the studies of the historian of science, in particular for the study of Galileo's creative work. That shift in the methodology of research which is best expressed by the slogan "from Koyré to Drake" corresponds to the ever growing deviation from the logical schemes offered by philosophy and the going deeper and deeper into biographical details in order to achieve what Drake called "psychological plausibility"[2]. He regards the philosophy of science rather as a dogma, an obstacle in the path of a historian's search for truth that frequently leads to erroneous constructions.

There are, apparently, weak points in such an approach, and its vulnerability is determined, first of all, by the uncertainty of the concept of historical truth. In this connection, one could even make up one's mind to produce an extreme statement to the effect that no objective historical truth exists in reality since the initial materials that form the foundation of the historian's work (letters, diaries, contemporaries' statements, etc.) are always essentially subjective. The concept of historical accuracy considered in this context is meaningless, and it is only the completeness of the historical description that can be implied. On the other hand, the purpose of study in the history of science, as in any other historical discipline, is not the set of facts as such, but the reflection directed to this set of facts, i.e., the construction of a certain logical conception. To put it briefly, the historian cannot help being a philosopher, otherwise his work will be of no interest whatsoever, except for the interest in the primary source.

It seems to us that the reason for such misunderstanding is mainly terminological ambiguities. For instance, according to Drake, the creative path of Galileo reflected in his works, as well as in those of his contemporaries, was distorted by the conception of the continuous development of science that was expressed in Koyré's works: Drake then arrives at the conclusion that philosophy of science is useless in historical research. Apparently, we encounter here an illegitimate interpretation of science on the whole as identical to a conception existing in the frame of this science. As far as the parallel with natural science is concerned, the question of the extent to which these attempts with inappropriate means are fruitful, and of the scientific value of historical studies conducted in the frame of these wrong conceptions is also interesting. Because of the analogy with natural science, we are prompted to answer this question in the positive: in the same way as many striking discoveries were made in astronomy, chemistry or geography on the basis of wrong assumptions, (one would recall here Kepler, Carnot, and Coulomb), a lot of valuable information was obtained by historians whose logical and philosophical position was rather biassed. Further, it is possible to say that the publications of that very same Koyré, for example his *Etudes Galiléennes*, also possess today a substantial scientific value which is by no means totally determined by the accuracy of their correspondence to biographical data.

There is also another aspect of the problem. It is impossible to make any great discovery that would lead to a revision of habitual conceptions concerning the Universe and the way of interaction of material bodies in it within the frame of previous conceptions. And since in a given science there are no conceptions whatsoever except the previous ones by the time of revolution, these revolutionary stimuli should originate from spheres that go beyond the boundaries of a given science: they should be determined by the global change of the whole style of thinking and be localized at the intersection of linguistic and conceptual structures that are inherent for various spheres of human activity and creative work. Only a discipline that is concerned with the study of universal rather than special regularities, i.e., the philosophy of science, can cover the broad circle of diversified problems in culture in general and in various sciences in particular.

It seems to us that these two extreme points of view (those of Koyré and Drake) could be reconciled if one uses again the analogy with natural science. Einstein once remarked that there is a double criterion: of the "external justification" (the correspondence with the observed experimental facts), and of the "internal perfection" (the logical consistency of a theory) for the evaluation of the correctness of one or another physical theory. The approach of Drake, that of a scientist-biographer who strives for the maximal accuracy of presenting facts and psychological plausibility in the description of Galileo's life and creative work corresponds to the criterion of "external justification". The approach of Koyré, who tried to force the creative activity of Galileo into the boundaries of a conception of the continuous development of science free of logical gaps, fits the second criterion, that of "internal perfection", more closely. Although each of these approaches is necessary, neither of them is sufficient, and their synthesis can be regarded as a way out of the crisis.

2. Now, on the basis of these considerations we shall make an attempt at reconstructing the course of the thought of Galileo, who was the founder of the Modern Science. This reconstruction will, in a way, answer the question how and why it was exactly Galileo who was the forefather of the Modern Science.

First of all, we would like to note that the transformation of science into a new quality that occurred at the boundary between the XVIth and the XVIIth centuries is associated with the realization of the fact that the actually existing world of things and objects can be set in correspondence with the ideal physical universe (i.e., the world of ideal objects that are under ideal conditions) which, on the one hand, reflects the actually existing objects and processes, and on the other, permits an adequate mathematical description.

Now, in order to answer the above question, one has, first of all, to consider two other questions: namely, a) what did the change of style of thinking in post-Renaissance Italy consist in, and b) how was this change refracted in the thinking of Galileo himself?

3. Static properties and hierarchicality were typical features of medieval thought. The hierarchicality and the static properties of the universe were reflected in many manifestations of human activity, i.e., in culture, art, and science. They were revealed especially clearly in painting, where the gulf between the divine ideal and the actually existing world was presented in a visually obvious manner. The Renaissance philosophy and culture, on the contrary, proclaimed everywhere the analogy between the ideal and nature, the divine and the human. An apt statement about this can be found in the study of a historian of the Renaissance A. Kh. Gorfunkel':

> .... if there had been no other proofs of the radical change that had occurred in the Italian culture of the XIVth–XVth centuries, just the comparison of the "Triumph of Thomas" by Andrea da Firenze with the "Athenian School" by Raffael would have been sufficient... The contrast between the hierarchical stability of the Spanish Chapel frescos and the dynamic harmony of the world that appears before us in the Chamber of Signatures in the Vatican Palace is striking. Here we have a different measure and a different order: the hierarchy was replaced by harmony, the static property by motion, the closeness by the presence of perspective[3].

The main obstacle against the appearance of the new science concerning nature was the scholastic model of the world. And it was attacked not by gradual modification of the medieval conceptions but from the outside, from those spheres of culture which, from the very time of their appearance, turned out to be outside the structure of the medieval scholastic knowledge. Scholasticism itself, both the "late" one (of XIVth century), and the "second" one (that emerged in the latter half of the XVIth century), proved incapable of radical change.

Nevertheless, in many works on the history of science that became classical (P. Duhèm, L. Thorndike, A. Crombie, A. Maier), the significance of the late scholastics is highly exaggerated. The point is not even that there is no evidence of the real influence of the late scholastics' discoveries on Galileo (all attempts to find in his works and conceptions traces of his being directly acquainted with the texts that had been forgotten by that time ended in failure). Another thing is important: even when certain formulations and methods of proof in the works of Oresme and Galileo, apparently coincide, for instance, the differences that are discovered between them are so profound and essential that the fundamentally different approach to the interpretation of physical phenomena of thinkers of different epochs becomes quite obvious. The very structure of scholastic knowledge differs from that of the new European science, and one should not be misled either by the similarity or genetic kinship.

Because of its traditional Aristotelian tendencies, the university professional science in its strict forms turned out to be incapable not only of preparing the transition to the new science, but even of establishing contact with it. The total lack of understanding made even polemics meaningless; the fact that after Galileo's discoveries Césare Cremonini unperturbedly publishes his peripatetic treatise *About the Sky* is a sufficiently eloquent piece of evidence, and his refusal to look through the telescope acquires profound symbolic significance: i.e., the representatives of scholastic wisdom and the new science spoke different languages.

The very thought of turning the telescope to the starry sky required a different, new approach to knowledge, a different attitude to experience: the new view of oneself as the investigator of nature. The boldness of rejecting tradition was necessary. And Cremonini was one of the best educated peripatetics. He was brave enough to speak against the dominance of Jesuits in Padua. He himself was taken to court for his Averroist liberties in his lectures and books. Nevertheless, it was just he who signed a judgement on the basis of which the book of B. Telesio *On the Nature of Things, in Accordance with its Own Principles*, a manifesto of the Italian natural philosophy opposed to Aristotelian thinking, was banned. As far as Cremonini was concerned, the scientific truth was already contained in the works of Aristotle and his adherents. He did not need a telescope, since he did not expect to see in the starry sky anything which would disprove his conceptions; neither did he want to see any such thing. Galileo's telescope produced only sneers on his part. In order to look through a telescope, one had to be not only a great scientist, but a scientist of the new type.

In connection with this episode, we would like to digress somewhat from the main topic in order to demonstrate that from the point of view of everyday life, and, moreover, from the point of view of logic and, maybe, ethics also, the behavior of Galileo in similar situations resembled that of Cremonini. Let us recall that place from the *Dialogs* where Galileo, in his argument about the relativity of motion, states that a stone dropped from the top of a moving ship's mast will hit the place near the base of, and not at a distance from, the mast. The question of Simplicio whether such an experiment has been conducted in reality is answered by Galileo through Salviati's words: "I am sure even without the experiment that the result will be as I am telling you, since it is necessary that it should be so"[4]. Thus, the conviction of Galileo is equivalent to that of Cremonini, although they are rooted in different grounds, i.e., in different cultural and ideological traditions.

Thus, it was not late scholastics that created the spiritual prerequisites for the revolution in the subject-matter and method of natural sciences represented by the origin of classical mechanics. Its causes lie in that "greatest progressive revolution" exemplified by the Renaissance epoch and, above all, by the culture of humanism.

The contribution of humanistic culture in the creation of new science has not yet been estimated and recognized properly. Most historians of science, especially the adherents of Duhèm and Thorndike, not only ignore the merits of humanists, but are inclined to adopt a directly negative estimate. Such an attitude is incorrect, however.

Certainly the humanists were what is now called researchers in humanities, i.e., they were interested in philology, history, oratory, morals, political science, and theology. But it was exactly the general culture of the epoch being renewed that led, in the final analysis, to the revision of the traditional natural scientific conceptions. The merits of the humanists consisted in the reconstruction of classical antiquity, the restoration and discovery of texts (e.g., translation into Latin of the works of Archimedes, who had been unknown to scholastics and to whom Galileo addressed himself, bypassing his imaginary and real contemporaries). It is impossible to explain, outside the tradition of humanist literature, the development of the new literary form of a scientific text, and its *genre* diversity (for example, Galileo's open dialog and his letters written, as "A message to Ingoli", not for an addressee given in the headline but for social opinion, and the scientific and parascientific intellectuals among his contemporaries). The typical features of this tradition are the struggle against the "barbarian" scholastics' Latin, and, at the same time, the working out of scientific texts in the national language. In this case also Galileo appears as the heir of Italian humanist culture.

The culture of the Renaissance produced a new ideology which determined new relations between the divine and the human: between the ideal world and earthly reality. It is important to see two aspects in this new relationship, which were equally essential for the development of the new science, in spite of the fact that their meaning might seem to be opposite.

On the one hand, the hierarchical cosmos in which God symbolized an unattainable ideal that was separated from all being, was replaced by the humanists and artists of the Renaissance with a universe in which the divine and the human turned out to be similar, if not identical. Apparently, hence the tradition originated that produced such impressive results in Newton's creative work: namely, a habit of considering Nature as a building constructed according to the law of the universal analogy. This identity of the divine and the human, the ideal and the real, can be regarded as, in B.G. Kuznetsov's words, the "invariant of cognition" which was typical of the whole epoch.

On the other hand, Nature was liberated from God's autocracy, and the cosmos was desacralized. Such a conception made it possible to consider the laws of nature independently of religious dogmas. And in this consideration itself the role of the authority was reduced sharply (be it not even God but just Aristotle).

As far as the system of moral conceptions of the Renaissance epoch is concerned, knowledge becomes the personal property of a scientist, and the cognition of truth his personal merit, in the same way as, instead of being noble, personal merit becomes the measure of value of the human personality, independently of the tradition represented by the scientist. The new attitude to the human being is related to the new attitude to Nature. The "rehabilitation" of nature, the justification of both the world and the human being in the struggle with ascetic traditions, meant also a new understanding of nature as the object of scientific studies. Nature with its own regularities being autonomous promoted the overcoming of theological conceptions.

The problem of the relation between the world and God is formulated in a new way by humanism, which finds in nature not only a perfect creation of God, but, most important, the set of regularities inherent in it and free from direct divine interference and arbitrariness.

Certainly, the elimination of the structure and method of scholastic knowledge, the new realization of the role and place of tradition, and the formation of the new type of researcher of nature, constitute the merits of humanism. It was the natural philosophy of the XVIth century, however, that created the direct prerequisites for the appearance of the new experimental mathematical science. And it was this natural philosophy that destroyed the old traditional picture of the world fixed in scholasticism and Thomism, and replaced it by the new conception about nature.

The term 'natural philosophy' has not been brought into the history of late Renaissance philosophical thought from other historical epochs. The thinkers themselves chose the words 'natural philosophers' or the 'philosophers of nature' as their names, having in mind not only the subject of their study but the "natural" approach to the cognition of the laws of the world structure, which was opposed both to the book knowledge of scholastics and to theological constructions as well.

Natural philosophy includes also the activity of pure metaphysicians, such as Telesio and Patrizi, Bruno and Campanella, and the professional scientists, physicians and mathematicians, e.g., Cardano, Fracastoro, Cesalpino, and the amateurs who delved deeply into the cognition of mysterious forces of nature. Their methods of cognition are most closely related to the development of the problems of the philosophy of nature, and their philosophical views to the specific character of Renaissance science.

Italian natural philosophy of the XVIth century completes the process of freeing philosophical and scientific thought from the rigid boundaries of scholastic peripateticism, and ensures freedom from the traditions codified in the medieval universities; in this struggle it rests on the antique scientific and philosophical heritage restored by humanists.

4. Consider now the way in which the cultural traditions of the Renaissance were reflected in the education and the world view of Galileo. The answer to this question is entirely unambiguous.

First of all, it should be noted that Galileo, having been born into a family of a musician and theoretician of music, was brought up from early childhood in the highly intellectual atmosphere of Florence, where the interests of music, art, and literature were predominant. In Galileo's youth, painting, which marked in such a typical way the turn to the new ideology (it was said, not without reason, that the hips of Titian's Venus delivered a more damaging blow against catholicism than Luther's 95 theses), represented, apparently, a serious interest and occupation. As is witnessed by Viviani, Galileo told his friends that if he had had control over his life, he would definitely have become a painter after his youth. Only when he was 20 years old did he become engaged in mathematics, about which he had had, up to that time, only a very vague idea. In his classical work L. Ol'shki stresses frequently that it was exactly the humanistic education of Galileo that "served as the basis and the sphere of his literary and critical activity"[5]. Of interest is also another remark of Ol'shki, namely, that the works of the great ancient poets, Virgil, Ovid, Horace, and Seneca, that were known by Galileo by heart, were appreciated by him for their purely poetical values, and not for their philosophical content; in contrast, for instance, to such scientists as Bruno and Campanella. For the same reason, he never mentions Lucretius. It is widely known that Galileo was one of the founders of classical Italian artistic prose (it did not matter that his writer's talents found application in his scientific treatises); this characterized him as a humanist and a direct follower of humanists. The fact that Galileo rewrote his Latin treatise *On Motion* three times without introducing any essential changes shows to what extent the care for style became his flesh and blood. Apparently, two causes that could be also explained by the needs produced by the heritage of Renaissance humanists initiated his transition to the use of the Italian language at a later time. First, the new contents of his works required a new form which would be adequate to it. Thus he arrived at the necessity to present his studies in the form of dialogs, and, certainly, it was not for nothing that he made three attempts at the improvement of the Latin text of *De Motu*; because of his further search for an adequate form of presentation, Galileo had to begin to use his native language in his works. On the other hand, Latin remained, to a great extent, the symbol and the means of obdurate scholastic exercises, the symbol of the adherence to authorities typical of universities. It was not without reason that during Galileo's times the centers of scientific thought moved from the Catholic universities into private secular societies, academies, and associations. Scientific societies in town republics and at the courts of educated sovereigns acquired the role of traditional centers of intellectual life. Because of this, Galileo also is understood first in extrauniversity circles. It is natural that, by making the transition to the Italian language, he tries to bring the essence of his works closer to these new listeners and readers of his. Thus, as far as Galileo's education, style of thought, and whole intellectual culture are concerned, he is an heir of Renaissance humanistic ideas.

As a scientist, Galileo stands at the origin of classical science, that new physics-mechanics which was organized into a system only in the next century through the works of Newton, Descartes and Leibniz. He is still incapable of taking it in at a glance, and therefore he still cannot apply to it the "criterion of simplicity" and thus prove its truth. And neither is he an extremely skilled mathematician. At any rate, it is not mathematics which is the source of his ideas, his generative grammar. It is not for nothing that he leaves for years the letters of his student Cavalieri, who was indeed a brilliant mathematician, without reply: the letters in which Cavalieri presented his "geometry of indivisibles". What was it, then, that enabled Galileo to defend his position with such conviction? Why is his Salviati saying that he does no need to conduct any experiment, that he knows the result and knows it beforehand? Obviously, Galileo made such statements on the basis of a certain general methodological principle or procedure. For Galileo, the role of this principle or procedure belonged to the ideal experiment, which possessed a paradoxical property: namely, that, while being impossible in principle under real conditions, it, nevertheless, reflected correctly the regularities of the real world.

Thereby Galileo introduced into research practice a special kind of experiment that was specifically conducted for the verification of a scientific theory and had unambiguous results that could be reproduced in the future. His discovery was valuable, since everybody could check the correspondence between the ideal experiment and reality, and with sufficient care, nobody could doubt the correspondence.

This ideal experiment was by no means identical to direct experience and to traditional common sense. For instance, when falling bodies are considered, it follows from direct experience that the fall of light bodies is slower than that of heavy ones. Instead of this, Galileo assumes that in a vacuum all bodies fall with the same acceleration, and their motion is described by a special law[6].

It was thus that in Galileo's mind an idea of the existence of an ideal *physical* world appeared, the world which is adequate to reality and permits mathematical description. Galileo's Platonism expressed itself, not through the worship of the world of numbers as the primary basis of the Universe, but through his constructing of ideal images. But while even Aristotle reproached the Pythagoreans for their formulation of the theory and search for the explanation of phenomena without paying attention to facts, and selecting them in accordance with one or another of their favorite conceptions, Galileo's ideal experiment was constructed in such a way as to reflect reality also. Certainly, he first arrived at this method by way of intuition, since the restructuring of the perception of the world by humanists and natural philosophers of the Renaissance provided for him this thought of the identity between the divine and the secular, the ideal and the real. The idea of the correspondence between the ideal experiment and the real experience is a projection of the world view and culturological ideas of the Renaissance on the sphere of science in Galileo's mind.

Galileo understood the physical, not the mathematical aspect of mechanics, and when only five years after he had started the study of mathematics Galileo undertook to write a mechanical treatise about motion, the words of Ol'shki that "the scientific instinct replaced the technical skill in his mind"[7]. Further, possessing this scientific instinct or instinctive experience, and being in complete agreement with the new current of opinion, Galileo started his investigation of special problems, in the course of the solution of which he became more and more convinced of the correctness of his method.

Institute for the History of Science and Technology,
U.S.S.R. *Academy of Sciences,*
Moscow

NOTES

[1] In the Soviet literature see, on the relation between universal and cooperative labor, V. S. Bibler, *Thinking as Creative Work*, Moscow, 1975, pp. 234–261.
[2] Drake S., *Galileo Studies*, Ann Arbor, 1970. p. 11.
[3] Gorfunkel', A. Kh. *Humanism and Natural Philosophy of Italian Renaissance*, Nauka Publishing House, Moscow, 1977, p. 117.
[4] Galileo G. *Selected Works* (in Russian), vol. II, Moscow, 1964, p. 10.
[5] Ol'shki L. *The History of Scientific Literature in New Languages*, vol. III, Moscow–Leningrad, 1933, p. 117.
[6] The direct experience is replaced here by its idealization which, as with every reasonable idealization, enables one to understand the mathematical structure of a phenomenon. There is no doubt that, at this early stage of modern science, the mathematical regularities discovered for the first time served as the true basis of its cogency.
[7] Ol'shki L. Op. cit., p. 111. What follows is concerned with the instinctive experience: "His instinctive experience obtained by virtue of special upbringing is purified by Galileo through artificial reconstruction of a certain phenomenon, which exposes the old errors and confirms new hypotheses at the same time".

W. L. WISAN

GALILEO AND THE EMERGENCE OF A NEW SCIENTIFIC STYLE

The scientific revolution that took place in Europe during the seventeenth century, like the renaissance in art preceding it, was a period of striking innovations built in large part of a revival of interest in and deeper knowledge of the classical works of ancient Greece and Rome. Moreover, just as Renaissance art was long regarded as exemplifying a complete break with the medieval past, so science in the seventeenth century was also long believed to be entirely independent of medieval ideas and achievements. Galileo, in particular, was celebrated as the "reformer" of natural philosophy who, more or less single-handedly, reestablished science on a new basis of observation, experimentation, and "the method of induction."[1] Today, however, just as most texts in art history point out subtle continuities between medieval art and that of later periods, so many histories of science now take cognizance of important medieval developments behind the emergence of modern science. Thus, Galileo is now usually found to have medieval predecessors, especially in mechanics and the study of motion.[2]

Difficulties arise, however, as soon as we try to locate specific links between medieval writings on motion and mechanics and those of Galileo. Like others of his time, Galileo greatly admired the ancient mathematicians, especially Archimedes, and upon occasion he indicated indebtedness to Aristotle, but he seldom mentioned medieval authors and almost never attributed ideas to them.[3] In fact, he did not formulate his own concepts in quite the same way as did his medieval predecessors, and, if he did know and draw upon medieval mechanics, he failed to adopt some key concepts that one might expect.[4] Moreover, similar difficulties arise in examination of Galileo's sucessors, notably Newton, in whom the evidence of direct influence by Galileo turns out to be unexpectedly scant.[5] Although not conclusive, lack of firm evidence suggests caution in assuming a continuous development of science from ancient through medieval into the modern period. Thus some scholars have come to interpret

J. Hintikka, D. Gruender, and E. Agazzi (eds.), Pisa Conference Proceedings, Vol. I, 311–339.

Galileo as a singular figure with no immediate predecessors and with minimal impact on generations immediately following him.[6]

Nonetheless, for most historians today it seems almost absurd to suppose, as was taught until late in the nineteenth century, that Galileo did in fact entirely pass over medieval (and even most of Renaissance) thought, reaching back to the Greeks alone for inspiration and guidance in developing his new science of motion. Similarly, to suppose Galileo had no important influence on later science seems a distorted result of myopic historical methods. Yet, this conclusion is reinforced by recent philosophical studies of scientific change in which Galileo no longer occupies a crucial position. Galileo's *De motu locali*, his major work on motion,[7] did not become a Kuhnian paradigm for future research, nor did it lay out a new research programme as defined by Lakatos, and it does not exemplify an important new research tradition in Laudan's study of the way in which science progresses.[8] Similarly, it does not appear as a development from, or a continuation of, earlier paradigms, research programmes, or traditions. Nor does it seem part of the continuing process of conjecture and refutation, the pattern suggested by Popper, whose writings stimulated much recent analysis of scientific change.[9]

An underlying difficulty in these analyses is the tendency to focus almost exclusively on the process of *theory* change as the key event through which science unfolds historically. At the least, one is expected to find the development of new and fundamental principles in any work of real importance. Galileo, however, in his mathematical sciences, was not primarily engaged in formulating a new theory, or in finding new principles. The mathematical sciences were conceived of as following from principles or axioms that ideally would be as evident as the axioms of Euclidean geometry. Thus, whenever possible, Galileo drew on "known and evident" principles to derive new consequences, and it was an embarrassment to him when his mathematical propositions could not be derived from suitably familiar and evident truths.[10]

Only as he struggled with problems for which older assumptions clearly would not work, did Galileo adopt new principles. Moreover, he was long undecided about these principles, and, in fact, he worked for many years with contradictory propositions.[11] In spite of various experiments Galileo is known to have performed, he did not generally pursue experimental verification or falsification in developing his mathematical science. The function of experiment in the science of motion was limited to rendering as evident as possible his most fundamental propositions. These, however, did not form a "hard core"; by Lakatos' criteria, what Galileo had was an "immature" science, scarcely worth analyzing, much less worthy of a place in the pantheon of great scientific achievements.[12] Similar difficulties are found in the Kuhnian formula for "paradigm shifts."[13] Galileo's work on motion is hard to characterize as either a normal continuation of or a revolutionary departure from earlier traditions. As will be shown below, it simply takes off in a new direction and gradually unfolds without becoming a fully developed theory. Thus, current analyses of the way theories change cannot be fruitfully applied to Galileo's work on motion.

There are, nonetheless, important clues in some of these studies of scientific change. Kuhn's notion that something like a *Gestalt* shift occurs at particularly creative moments in science is not quite applicable to the general course of development in Galileo's work on motion, which proceeds primarily in a cumulative manner. In fact, it seems more likely that the major *Gestalt* shift occurred in a generation gap between Galileo and his students, rather than in Galileo's own "way of seeing." Kuhn's analysis, however, underlines similarities between scientist and artist, and this similitude emerges even more clearly in Margaret Masterman's review of Kuhn's different meanings of the term, "paradigm." Masterman emphasizes the "concrete achievement" that, in the absence of a theory, may function as a paradigm, and which consists in a new "trick, or embryonic technique, or picture, and an insight that this is applicable in this field . . . it is this trick, plus the insight, which together constitute the paradigm."[14] In such a sense, the paradigm is "a construct, an artefact, a system, a tool; together with the manual of instructions for using it successfully and a method of interpretation of what it does."[15]

Galileo's *De motu locali* may have been just such a paradigm, in which the scientific text is the concrete artefact embodying a scientific achievement that goes beyond the text itself. We might very well ask, then, whether the development of scientific thought, exemplified in successive treatises, may not proceed much as does the evolution of those works created by artists. If so, could one of Galileo's major contributions have been, in addition to important discoveries and clarifications, a new *style* in the treatment of motion?[16] In that case, study of changing styles in the history of art may illuminate some important aspects of scientific development and enable us better to locate the continuous threads that run through disparate yet related scientific works.

This paper is the culmination of a preliminary exploration of such a possibility. It is, admittedly, a rash undertaking for a historian of science without training either in the arts or in art history. The results, however, turned out to be so striking that it seemed worthwhile to set them forth in order to stimulate discussion and further inquiry into this area. These results are admittedly crude and can bear much refining. The main thesis, however, should stand, at least in some form: in the case of Galileo, an important but hitherto unnoticed aspect of his contribution to modern science is the creation of a new style in the study of motion, a style that caught the imagination of those who came after him and thus inspired much of the great advance that has shaped the course of modern mathematical physics.

There are difficulties with this thesis, not least of which is the traditional distinction between science and the arts which places them at opposite poles. Art, for example, is said to be primarily concerned with the expression of values, science with the establishment of objective truth; art deals with the particular, science with the universal and general; art is subjective, whereas science must meet the test of public verification.[17] For the most part this is, of course, true; but the sharp edge of these distinctions is already beginning to be blunted. They stem mainly from a nineteenth century positivist conception of science that is no longer generally accepted. Moreover, this characterization is particularly inapplicable to the earlier mathematical sciences, such as rational mechanics, where verification of mathematical principles through experimental establishment of derived "facts" is relatively rare until quite recent times.[18] Also, verification is now generally considered rather less critical in the process of theory evaluation than was long taught by scientists of the positivist persuasion. Philosophical presuppositions or general conceptual schemes are found to play a crucial role in the development and acceptance of scientific theories, and it is no longer so clear that value judgments are entirely absent in this procedure.[19] Even so, there still remain important differences between art and science, but there may well be sufficient similarity to make an investigation of changing styles in science as profitable as in the arts, where it has functioned well as a tool for sorting out innovations and continuities.

Let us see, then, how this tool may be used in the history of science. First, since the success of this venture will depend on the adequacy of the categories employed to characterize different styles in science, and since this may well be the major difficulty in a first attempt to interpret scientific change in terms of changing styles, examination of the evidence which lays the foundation for my analysis may be the best way to start. Therefore, I will first set forth the problem by examining the transition in Galileo's own writings on motion from his *De motu antiquiora* of around 1590 to the *De motu locali*, which first began to take shape from about 1604–1609, but was not completed and published until 1638.[20] For here, in works by a single man, we find a change that appears on the surface to be more radical and discontinuous than that between his own writings and those of others before or after.

These two treatises by Galileo, then, will provide an instance in which there must be some degree of continuity between two very different works, even though employing different assumptions and principles, using different methods, and written in what will later be identified as two distinguishable styles. Moreover, Galileo nowhere refers to the earlier essay, and he is by no means an obvious candidate as its author. Nonetheless, external evidence and handwriting are such that there has long been unanimous agreement among scholars that the *De motu antiquiora* was, in fact, composed by the famous author of the later work on motion.[21] This conclusion gains further strength from my own study of manuscript notes relating to the later treatise, which show far more overlap between the two works than previously thought. Location of the continuous thread in this case, then, should help clarify the general problem of detecting continuity between different works, whether by the same or different writers, and provide a basis for characterizing some less obvious aspects of the way in which one kind of research grows into another.

The *De motu antiquiora* is written in essay form and treats conventional topics in a manner typical of sixteenth century studies of similar materials.[22] The first few chapters establish the ratios of speeds for bodies descending through media of varying densities. As in similar material written earlier by Benedetti, Galileo employs Archimedean methods and assumptions and criticizes the views of

Aristotle. The remaining chapters deal with a number of problems which, with one notable exception, are also discussed in similar ways by Benedetti or others. The main exception is a chapter on motion along inclined planes,[23] a topic not dealt with elsewhere, so far as we know, except for a hint in Blasius of Parma and in the notebooks of Leonardo da Vinci.

Blasius argued that a heavy body would descend more rapidly along the chord of a circle than along the arc.[24] This problem, later transformed into the problem of the brachistochrone, and eventually solved by Jean Bernoulli in the eighteenth century, was generated by examination of weights on the arm of a balance, which, descending, move along a circular path. The chord considered by Blasius would be the inclined plane spanning the lower quadrant of a circle. Leonardo, who paraphrased this passage in Blasius, sought to prove that the arc would be traversed in less time than the chord, and I will refer to this as Leonardo's theorem.[25] Leonardo was unsuccessful, but one of the theorems he seems to be using, and which is found elsewhere in his notes, compares speeds along inclined planes of the same height, and in one version Leonardo concludes, as does Galileo later, that the speeds would be inversely proportional to the lengths of the planes.[26]

There are no clear indications that Galileo knew the writings of either Blasius of Parma or Leonardo. Galileo himself remarks that the problem he is exploring – that is, why a heavy body moves more rapidly down more steeply inclined planes and what precisely are the *proportio talium motuum* – is not one previously discussed by any "philosopher."[27] Nonetheless, some of the basic ideas with which the *De motu antiquiora* began can be found in medieval statics. The key assumption, that a heavy body moves downward with as much force as is required to lift it, is implicit in the first postulate of the *Ratione Ponderibus*, attributed to Jordanus de Nemore.[28] The question that generates the discussion turns the third postulate of the *Ratione* (that a body is "heavier in descending" as it descends more directly towards the center of the earth) into a problem to be solved. Why, Galileo asks, does a heavy body move more rapidly along planes with greater inclination to the plane of the horizon?[29] The solution follows from the principle of the inclined plane and from the Aristotelian 'dynamic' principle that the speed of a moving body is proportional to the force acting upon it.

The first correct statement of the inclined plane principle is found in the writings attributed to Jordanus.[30] The medieval authors established the conditions for equilibrium in the case of two attached weights on planes of equal height but different inclination. Galileo carried this further by deriving the ratio of the "speeds" of the same weight along the inclined plane and its vertical. Assuming the Aristotelian 'dynamic' principle, it followed that the speeds would be inversely proportional to the lengths of the plane and its vertical. Since Galileo assumed "natural" motion to be uniform rather than continuously accelerated, the theorem as intended is false.

Furthermore, some of the first conclusions drawn by Galileo from this theorem were both false and banal. One is as follows:

> . . . given two bodies of different material but equal size, to construct a plane so inclined that the body which, in a vertical fall, moves more swiftly than the other will descend on this plane with the same speed with which the other would fall vertically.[31]

Galileo assumes here that, *in vacuo*, bodies of different densities would fall with speeds proportional to their weights. Thus, one need only adjust the length and height of the plane so that the weights moving along them are in the same ratio as the lengths. Similar problems, Galileo tells us, can be easily solved, but he does not give details

Despite his erroneous results, Galileo's treatment of the inclined plane principle was a significant advance over medieval attempts. He further improved upon it in a somewhat later essay on the mechanics of simple machines. *Le mechaniche*, the final version of this essay, written about 1600, but never published, has been noticed mainly because of its near approach to the principles of inertia and virtual velocities, and the conservation of work.[32] Perhaps equally important, it served as the starting point for Galileo's use of more sophisticated geometrical methods in the study of motion along inclined planes. As his manuscript notes show, he first proved one of his most fertile theorems, the law of chords, by methods employed in *Le mechaniche*, together with a traditional medieval rule relating times, speeds, and distances.[33] Soon after, he derived several more elegant (but still crude) results on paths of "quicker" motions. Then, again drawing on the medieval rule and using more sophisticated geometrical tools, he proved increasingly complex propositions on motion in a more rigorous manner.[34]

These early theorems bridge the gap between Galileo's *De motu antiquiora* and the *De motu locali*. They follow entirely from principles already generally known and accepted. Although there is no clear link with the earlier efforts of Leonardo da Vinci or treatises on medieval statics, the general aim of establishing theorems on the "proportions of motions" and paths of quicker and slower motions, suggests the continuation of a very old tradition that began with Aristotle's discussion of motion.

Aristotle's concept of motion is very broad, embracing that of change in general, and his treatment is primarily a philosophical analysis of different kinds of motion, of which "local motion" is one of four general categories. Although not interested in quantitative relations, Aristotle does include, in his brief comments on local motion, a few remarks on the relations between the speed, time, and distance travelled by a moving body, the force causing it to move, and the resistance of the medium.[35] Aristotle's followers and commentators questioned some of his conclusions, particularly that heavy bodies fall with speeds proportional to their weights, and that motion could not take place in a medium of no resistance in a time greater than zero. These criticisms were raised again in the middle ages, and the literature on local motion greatly increased, until there were whole books on the "proportions" of motions.[36] By the later middle ages, these works had become quite elaborate, employing proportion theory and often giving rigorous arguments from explicit definitions and axioms.[37] Still, the subject matter was quite limited, continuing as variations on Aristotle's laws governing the "proportions" of motions and presented in a generally discursive form.

With the Renaissance revival of interest in Archimedes' mathematical treatises, together with the Pseudo-Aristotelian mechanics, new problems on motion were generated.[38] In particular, Leonardo's speculations about the ratios of motions along inclined planes were probably stimulated by the mathematical sciences of Archimedes, as well as by what he knew of Aristotelian and medieval mechanics. Similarly, G. B. Benedetti used Archimedean methods to criticize topics treated in the Aristotelian *Mechanics*, and he developed an entirely new treatment of local motion that employed Archimedean hydrostatic principles to establish ratios of speeds of heavy bodies descending through different mediums, such as air and water. Although not entirely unanticipated,[39] Benedetti's treatment of motion was highly original, but limited in its possibilities at that time. Both these lines of development reappear in Galileo's *De motu antiquiora*, which begins with a treatment of motion similar to Benedetti's, but includes the chapter on motion along inclined planes, in which he derives propositions similar to, but going beyond, the efforts of Da Vinci.

Galileo's first radical innovations in the treatment of motion come with his law of chords and the first theorems he derives from it: theorems about motions that take place in equal or lesser times. This development constitutes a conceptual jump. Nonetheless, it is still a small step, and its roots may lie in medieval materials, such as those exploited by Leonardo, and therefore perhaps by others. It is, moreover, a simple extension, not a *Gestalt* shift, or a new "way of seeing." There is no rabbit here that begins to look like a duck.[40] Further innovations come as Galileo tries to use his new tool, the law of chords, to find a proof for Leonardo's theorem. This generates discovery of the times-squared theorem and other basic propositions of the later *De motu locali*.[41] More results then quickly follow as Galileo uses these new tools to develop enough mathematical theorems on motions along inclined planes to make up a whole treatise. A number of these are by-products of the search for the brachistochrone, which may well be the initial problem that triggered the entire development and, even more likely, may have led to Galileo's first realization that free fall is necessarily accelerated.[42]

Galileo now has a genuinely new "science" of motion. However, it has neither an experimental nor a theoretical basis. The times-squared law was most likely confirmed by an early experiment, but he has no proof from established principles, and there remain several fundamental puzzles concerning accelerated motion. From this time on, Galileo is more and more preoccupied with the problem of clarifying and explaining accelerated motion and finding properly evident principles on which to base the growing body of mathematical theorems. It is clearly Archimedes' mathematical treatises that furnish the model for the new science. The role of experience is limited to rendering immediately evident the fundamental propositions, and the emerging treatise, conceived as mathematical, includes a number of theorems impossible in physical reality. Most, for example, assume no loss of speed by a body turning corners, and some mathematical speculations include investigation of the motion of a body dropped from a point at an infinite distance.[43]

Galileo continues to discover new theorems and solve new problems (constructing, among others, propositions on projectile motion), but he cannot explain the cause of acceleration, he does not take into account change in direction, and many of his results cannot be experimentally verified. Moreover, for some years he uses contradictory assumptions to ground his theorems.[44] Without a deeper theory, Galileo can investigate only special cases of motions along differently inclined planes, and most of these theorems were, in fact, soon forgotten. Nonetheless, the published treatise may have been an important prelude to development of a more general theory. Galileo studied accelerated motion by finding the ratio of times in which a given body would descend along given paths (usually an inclined plane, or portion of that plane, and its vertical) under varying initial conditions. These comparisons eventually made up a body of mathematical propositions on the "proportions of motions" entirely different from those of the medieval authors.[45] The Florentine scientist overreached himself in insisting that his mathematical work was comparable to that of Archimedes. Yet Galileo's work may have been a necessary step towards further exploitation of the power of mathematics as a tool for exploring nature. His mathematical investigations, unhampered by physical considerations, opened the door to fuller exercise of the creative power inherent in free mathematical speculation. By the time he was distracted by his telescopic discoveries in 1609, Galileo had an extraordinarily original treatise, unlike anything done in the past. Divorced from its earlier foundation and motivation, and only loosely tied to empirical reality, the study of "proportions" of motions along inclined planes had developed into a wholly new and highly promising treatment of motion.

The finished work, not published until 1638, differs even more from the old *De motu* than the latter did from earlier works on local motion, or isolated results found in Leonardo, Cardano, and others. Yet, we can scarcely deny continuity in Galileo's own development, especially in light of manuscript notes which link the earlier and later writings on motion and mechanics, and which show how slowly he moved toward new concepts and assumptions found in the later work. Thus, we see that the problem of detecting continuity in scientific development is rather more complex than is generally supposed. In particular, it appears that continuity in science does not depend upon identity in content, structure, or method. In the transition between the old *De motu* and the later one, there are radical changes in the propositions established, the most general assumptions on which the treatise is based, the mathematical tools employed, and in the way these tools are used. So wherein lies the continuous thread that connects a scientific development with the earlier stage out of which it grows? Surely, it has to lie, not only in theory, propositions, principles, or methods, but also in execution of the main purpose and the manner in which this is carried out.

The primary purpose of *De motu locali* is creation of an Archimedean science dealing with motion. Like Archimedes, Galileo's aim was not new principles, but solutions to previously unsolved problems and the extension of geometry into a new area by employing it to discover new "properties" of motion.[46] Just as Archimedes' fame lay more in his elegant mathematical treatises than in his few principles,[47] however important these may be, so Galileo's chief goal was the derivation of *bellissimi* propositions on motion. His purpose, then, was not to develop a new theory of motion in the modern sense, but to create a new science of motion in the sense of a new mathematical structure in which a body of geometrical propositions are rigorously derived from suitable principles, again, preferably already known and widely accepted. Always fierce in defense of his priorities, Galileo was quite reticent about the principles on which the new science is based. In the introduction to *De motu locali,* he boasted not of his definition of uniformly accelerated motion, or his postulate that equal speeds are acquired in descent through equal vertical heights. Rather, it was the times-squared theorem and the parabolic path of projectiles that he set forth as his chief discoveries, adding that, even more importantly, he opened up "a vast and most excellent science, in which my work shall furnish the elements by which other more acute minds shall penetrate to yet deeper recesses."[48]

In the first day of the *Discorsi,* Galileo praises himself (through the mouth of Sagredo) for solving problems by means of "reasons, observations, and experiments [*esperienze*] which are common [*tritissime*] and familiar to everyone."[49] From his earliest to his last writings, this is a constant *leitmotif*: for a *mathematical* science the fundamental principles must be immediately evident, and preferably already known and accepted. It is, in fact, an embarrassment that the principles required by the new science of accelerated motion are not already known, and making them properly evident was a problem to the end.[50]

Galileo, then, was guided primarily by the purpose of creating a certain type of treatise in a specific form; *i.e.*, a treatise on motion using Greek geometry in a manner similar to Archimedes' mathematical works, and he expected his work to be the starting point for "infinitely" more such results.[51] Indeed, the *De motu locali* is filled with complex geometrical propositions which, although less elegant than Galileo supposed, reflect considerable ingenuity, if for no other reason than that in applying geometry in this way to the study of motion, there were no prototypes. His was the pioneering venture in this direction, and if most of Galileo's explorations seem tedious and uninteresting today, they were not so in his own time.

What, exactly, was the impact of this work? Descartes, of course, dismissed it as trivial,[52] criticizing Galileo for his failure to consider causes of motion. Indeed, the main line of research in the generation after Galileo focused on the causes of both earthly and planetary motions. The main impetus to further progress arose out of work on the problem of impact and analysis of circular motion, rather than attack on the specific problems explored in the *De motu locali.* But Galileo's problems and methods can be traced through much of late seventeenth century mechanics and even into Euler's early *Mechanica* of 1736.[53] The development of techniques for using the new "analysis" in the study of motion came slowly, and even those using the newer methods devote themselves to exploration of a general kind of problem set, not by Descartes but by Galileo: that is, strictly mathematical investigation of limited, technical questions about the motions of heavy bodies, beginning from given initial conditions. The study of motion no longer consisted in philosophical analysis, simple geometrical or arithmetic investigations of "proportions of motions," or isolated topics in mechanics, but it had become a mathematically organized and conducted exploration of real, moving bodies; and in this general form it has remained a fundamental part of modern physics.[54] Galileo's content, techniques, and general structuring of the new science bridge the gap between ancient or medieval and modern treatment of motion.

These characteristics are not, of course, the substantive elements normally looked for in a scientific work, but are primarily matters of style. Style, however, may well turn out to be more important than generally supposed, particularly in the transmission and development of scientific thought. If so, this suggests that we might profitably employ concepts and methods of art historians as tools for analyzing and clarifying elements of continuity and innovation in history of science. A chief difficulty, however, is that whereas techniques and subject matter, as well as purely formal elements, appear as essential aspects of a scientific work, technique and subject are less important to the art historian, while form and expressiveness are usually considered the dominant aspects of an artistic work. Indeed, art historians have customarily distinguished between art and science, in which the former seems primarily concerned with values, and the latter with establishment of descriptive "facts," on the one hand, and universal theories, on the other.

As pointed out above, however, this distinction between art and science stems mainly from a nineteenth century positivist notion of what science is, and today it is no longer generally accepted as an adequate characterization of science. In particular, aesthetic judgments may be relevant in choice of fundamental presuppositions and thus can affect the course of scientific change. It is no great novelty, then, to suggest that a scientific treatise might have an "expressive value," and it may be possible to locate this element in Galileo's *De motu locali.*

It has long been argued, for example, that Galileo's new mathematical science stems from belief that nature is fundamentally mathematical, a belief that might constitute the expressive value of the treatise. Now, Galileo never clearly states his beliefs about the nature of ultimate reality, and there are many contradictory clues to his thought on this subject. He seems to be guided mainly by the general purpose of creating a science of motion resembling the mathematical mechanics of Archimedes, and the extent to which nature is or should be mathematizable may never have been completely thought out. Indeed, Galileo's conception of nature and how it can be known seems to change as he progresses in development of an Archimedean science.[55] His innovations do not appear to start from understanding the most general and profound implications of these innovations, but rather grow out of insightful moves with traditional elements, each move suggesting further steps. At a crucial point Galileo is able to apply theorems of Euclidean geometry to motion along the chords of a circle, and this move generates more elegant propositions which fundamentally alter the style of the growing work. The style, but not the specific content, becomes more and more Archimedean. Galileo, however, insists, as Archimedes does not, that he is developing a mathematical science dealing with real, natural phenomena. Although from time to time he retreats to the position that his book on motion is purely mathematical, and it therefore does not matter whether its principles can be established as true of natural motions,[56] and although, in fact, many of his propositions are too abstract to be applied in the concrete, nevertheless Galileo is convinced that he has arrived at true fundamental principles governing natural motion, and that conclusions drawn from these principles by rigorous mathematical reasoning must be physically true in some sense.

Galileo characterizes these conclusions as the *accidents* of motion. He sometimes claims that knowledge of accidents leads to knowledge of substances, but, other times, to discovery of causes.[57] In these passages, accidents are mathematical properties, but it is not altogether clear just what is the substance that is to be known through the accidents. In some cases, it appears to be the "true cause."[58] In the science of motion, however, he was unable to find such a cause, a serious lack in his own eyes;[59] nonetheless, he seems confident of the value and magnitude of his achievement. There is, perhaps, genuine ambiguity in the thought of this man whose early training included a thorough grounding, not only in Euclid and Archimedes, but also in Aristotelian physics and logic. But if he was less clear than he might have been (and is often said to have been) in his conception of nature as mathematical, his most astute readers may well have had little difficulty seeing the larger implications of the *De motu locali*. Hence, the actual *Gestalt* shift may lie in a generation gap. Despite Galileo's ambiguity elsewhere, his powerful mathematical treatment of motion mutely testifies that nature is to be understood in just this way.

This, however, is to say that Galileo's followers not only learned from him a new content and techniques for a science of motion, but by the sheer force, the expressive power, one might say, of his mathematical treatise, they learned also that nature is fundamentally mathematical. Such a conclusion is neither a scientific fact nor a purely abstract metaphysical notion. It is a conviction that rests ultimately on a *feeling* about the world. As the empiricist has a special feeling for the meaning of the concrete, the rationalist has a feeling for mathematical form. Moreover, one can respond differently in different situations, depending on the context and one's aesthetic sense of that context. Galileo himself also had a feeling for the factual and particular, as can be seen in much of what he wrote.[60] But the composition of *De motu locali*, written in the manner of Archimedes, is governed throughout by the feeling of the mathematician, not the empiricist.

Thus, in developing an Archimedean science dealing with motions in the empirical world, Galileo, like an artist, seems to have initiated a new style. He began with small elaborations on earlier treatments of motion which are primarily topical and discursive. His first attempts at mathematical treatment suggested further steps in this direction, and he began to draw on Euclid's geometry. The process continued until the initial point of departure was no longer visible, and something altogether new seems to have emerged. In this sense, Galileo surely produced a treatise on motion in a new style. Moreover, its development followed a pattern quite similar to that of an emerging new style in art. Almost all traces of the *De motu antiquiora* disappear into the mature treatise. Similarly, I suggest, his *De motu locali* may have played a like role in subsequent texts by others which reveal little of the works from which they took their start. Yet, again, as happens in the arts, what appears as an entirely new style may owe a very considerable debt to the past.

Let me now recapitulate the entire development from Aristotle to Galileo and briefly sketch subsequent developments to Lagrange. Using categories borrowed form art history, the overall picture is briefly outlined in Table I. This attempt is admittedly crude; it is set forth here, however, as a first approximation, and it is hoped that it will be found sufficiently useful to warrant further efforts in this direction.

The most promising categories for analysis of scientific styles seem to be structure, content, techniques, and expressive quality. The first three, however, are sometimes difficult to distinguish and the last is hard to articulate satisfactorily. Transitional figures, particularly difficult to capture in a first analysis, will be passed by for the present.

Structure, the broadest category, will be considered in terms of the distinction between discursive and axiomatic, with the former subdivided into classificatory or topical, the latter into propositional, geometrical, or algebraic. Although not always mutually exclusive categories, these will help characterize treatments of motion which dominate in those treatises under consideration here. Other works,

TABLE I: EVOLUTION OF SUCCESSIVE STYLES IN

	ARISTOTLE	**MEDIEVAL**	**BENEDETTI**
STRUCTURE	Discursive and classificatory.	Similar to Aristotle with elaborations, some in axiomatic form; more commonly discursive and classificatory.	Discursive and topical.
CONTENT	Definition of fundamental concepts: time, space, motion, infinity, continuity. Some brief treatment of "proportions" of local motions.	Similar to Aristotle with more detailed treatment of local motion, until entire books on "proportions of motions." Some criticisms of Aristotle.	Local motion studied in context of Archimedes' hydrostatics; new theory of "proportions of motions" through different media; far more criticism of Aristotle.
TECHNIQUES	Philosophical analysis and explication of terms.	Similar to Aristotle but more use of geometry, arithmetic, proportion theory.	Archimedean methods in hydrostatics. Elementary geometry and arithmetic.
EXPRESSIVE QUALITY	Feeling for a world of substances and essences.	Similar to Aristotle.	Transitional.

TREATMENTS OF MOTION FROM ARISTOTLE TO LAGRANGE

GALILEO I (*De motu antiquiora*)	**GALILEO II** (*De motu locali*)	**NEWTON**	**LAGRANGE**
Similar to Benedetti.	Euclidean: axiomatic and geometrical.	Similar to Galileo.	Axiomatic, but form algebraic rather than geometrical (not even diagrams in Lagrange).
Similar to Benedetti; adds examination of "proportions" of motion in context of simple machines, especially the inclined plane.	Unified treatment of local motions: three books on uniform, accelerated, and "forced" motion. "Proportions" of motions main topic in Books I and II; in Book III finds "properties" of projectile motion (altitudes, amplitudes, impetus, etc.)	More general than Galileo. Combines earthly and heavenly motions derived from same universal laws. Still finds proportionalities, paths, etc., but purpose now seen as derivation of further "laws" governing all motions (e.g., *lex vis centripetal tendentis*).	Unified treatment of earthly and heavenly motions as in Newton, but still more general. Contains equations describing motions of points under given conditions.
Similar to Benedetti, but applies simple geometry to motions along inclined planes.	Advanced geometry to derive mathematical consequences from new foundation, omitting consideration of forces that cause motion.	Combination of Greek geometry and early stage of the "new" analysis to derive consequences from foundation defining forces acting to produce motion.	Much more advanced analytic methods to derive consequences from essentially Newtonian principles.
Transitional.	New feeling for the power of mathematics.	Similar to Galileo.	Similar to Galileo.

particularly those of a more transitional nature, tend to defy such classification.

Content will be classified in terms of subject matter: analysis of specific concepts; exploration of consequences flowing from fundamental principles or assumptions concerning "local motions," heavenly motion, simple machines, or hydrostatics. Again, these categories do not exhaust all relevant materials, but may suffice for the present purposes.

Techniques, although sometimes difficult to distinguish from either content or structure, will be classified in terms of processes employed in developing subject matter: philosophical analysis, geometrical methods, and mathematical analysis as it existed in the eighteenth century. Geometrical methods will, in turn, be further subdivided according to its complexity.

Expressive quality will be crudely divided between the two extremes and characterized in between by resorting to the vague term, "transitional." At one extreme is the expressive aspect of Aristotelian science, characterized here as a "feeling for the world of substances and essences." This phrase attempts to capture the feel of an Aristotelian world in which substances have an inner, essential nature, and science is expected to somehow grasp this nature. At the other extreme is a "new feeling for the power of mathematics," an attempt to express the sense of mathematics as a new and powerful tool for discovering how the world is made and how it works.

Using these categories, we begin with Aristotle's treatment of motion in the *Physics*, III, 1, which is a philosophical analysis of concepts and terms needed to understand the nature of motion. He examines just what factors are involved and how these are to be understood. Aristotle distinguishes between four kinds of motion, one being "local motion." A few brief chapters include discussion of the "proportions" of motions, in which local motion is analyzed with respect to ratios of times, distances, and moving forces.[61] Most of his discussion, however, is more general. The overall structure, then, may be characterized as classificatory and discursive, technique as philosophical analysis, and content, analysis of concepts. The whole expresses Aristotle's assumption that the physical world is to be understood through philosophical analysis of the essential nature of things.

One can immediately recognize Aristotle's way of treating motion in medieval writings on that subject, which are governed by much the same underlying assumptions about the nature of things. In the medieval writings, however, there is further elaboration of classifications and specific cases, marking the emergence of what might be called a "scholastic" style. Although still primarily discursive, there is considerably more detailed treatment of local motion, often through elaborate application of proportion theory, until entire books appear on the "proportions of motion," sometimes in axiomatic form. By the end of the fifteenth century, the subject matter has been developed almost beyond recognition in some cases. Local motion, an incidental topic treated, as it were, in the background of Aristotle's writings on motion, has come to the foreground in a manner analogous to one of the processes by which stylistic changes come about in art.

The scholastic style continued well into the sixteenth century. But during that century the revival and translation of Archimedes' mechanics together with the Pseudo-Aristotelian *Mechanics* introduced fresh ideas and triggered a radical change in style. The new style seems to begin abruptly with a very significant change in the treatment of motion by Giovanni Benedetti, who used Archimedes' hydrostatics as a base from which to criticize Aristotle and develop a new approach to the study of local motion. Using Archimedean methods and concepts, Benedetti calculated the ratios of motions of two bodies falling through a resistant medium from the ratio of differences between the weights of unit volume of the bodies and of the medium displaced by the bodies. Like so much else in medieval and early modern science, the basic idea goes back to Arabic sources. However, Benedetti seems to be the first to develop it in a treatise on local motion.

Galileo's own earliest work on motion is in the new style of Benedetti. But Galileo himself begins moving in another direction almost immediately, as shown above (see pp. 319–20). For he shifts from the problem of ratios of fall of the same or different bodies through a different or the same medium to exploration of the 'ratios' of motions along inclined planes. These are gradually elaborated into a superstructure of problems and theorems that are completely unlike anything done before. Again, elements previously occurring only incidentally and in undeveloped form have moved into the foreground, and a new style emerges.

With Galileo, the transformation of the subject of motion from a philosophical topic in an Aristotelian style to a mathematical science in the sense of Archimedes is complete. The organization of Galileo's treatise is that of formal geometry, and not only are the basic theorems largely quite different from any previously published, but the structure raised upon these propositions is entirely new. We are still a long way from modern mathematical physics, however. Galileo did not succeed in establishing a satisfactory foundation for his new science of motion, and, in fact, this was not his original purpose. For Galileo set out, not to find new principles of mechanics, but new results of old principles drawn from ancient and medieval sources. His intent was elaboration upon the past, not a radical departure from it. To solve the problem he set for himself, however, particularly to prove Leonardo's theorem, he had to find new propositions that did not follow from the old foundation. The new principles he was then forced to articulate and establish were forged only very slowly, and with what might appear today as excessive caution. The foundation of his new treatise was not fully formulated until some forty years after the beginning of the new science.

In fact, Galileo never completely solved the problem of foundations for the new science, but his treatise demonstrated the potential inherent in his new mathematical approach to motion. His mathematics was soon made obsolete by the new "analysis" and his "dynamics" was replaced by that of Newtonian science, but meanwhile, his work drove out the older scholastic way of treating motion. Future studies of motion take shape in the form of mathematical treatises which attempt to develop consequences from fundamental principles which for the most part stem from different sources and use different concepts and techniques. Yet Newton's treatise, although markedly different from that of Galileo, is much closer to him than to the scholastics. Newton draws on Descartes, Kepler, the atomists, and other sources, and unifies mechanics and astronomy on a general foundation for both. But the *form* of his treatise is much like Galileo's. Beginning with "laws" or "axioms" of motion grounded in the mechanics of simple machines, Newton studies the motions of heavy bodies along paths represented by lines, and explores the results that follow from different initial conditions.

Meanwhile, on the continent, Galileo's impact is particularly noticeable in the writings of Mersenne and in the work of Huygens and others, both in the matter of some fundamental assumptions and in the kind of problems attacked (especially concerning paths of least time). New mathematical methods supersede those of Euclidean geometry rather more quickly than in England, but despite these differences, eighteenth century rational mechanics, like Newton's, continued in a style close to that of Galileo. Mechanics was assumed to be a mathematical science which proceeded from true and evident principles, mostly confirmed through reason or immediate experience, with little indirect experimentation undertaken for this purpose. As in the case of Newton, the continentals draw on other sources for problems and fundamental principles, as well as methods, but the Galilean style is still visible. On the continent, however, there followed such an explosion of results from the new mathematical approach to motion aided there by more rapid development of analytic methods as to add up eventually to a fundamental change in style, as seen in the *Analytic Mechanics*[62] of Lagrange, where diagrams altogether disappear. Geometrical treatment of lines which represent paths along which heavy bodies move and times along these paths under given conditions – Galileo's chief innovation – vanished into the symbolism of a new algebraic calculus.

The *Analytic Mechanics* as an artefact bears little resemblance to the *De motu locali* of Galileo. But Lagrange pays tribute to Galileo as source and founding father, much as Galileo himself defers to Archimedes. Moreover, the *Analytic Mechanics*, as an almost purely mathematical treatment of motion and mechanics, is, despite different methods, principles, and problems, closer in spirit to Galileo than to Aristotle or the later scholastics. It is the "geometric spirit" of Galileo's treatise on motion that pervades the rational mechanics of the eighteenth century. Like Galileo's own work, this now "classical" mechanics assumed the form of a geometry in which the fundamental principles had to be "evident," in the manner of a mathematical treatise, and ambiguity about the status of mechanics lingers into the nineteenth century, where one still finds mechanics as a branch of mathematics.

Galileo's place in this tradition has been somewhat obscured by the tendency of historians to focus primarily on Galileo's experimental verification of his fundamental principle that velocity increases in proportion to time of fall through the times-squared theorem derived from it. As shown elsewhere, however, Galileo's method in his mechanics is not that of the modern hypothetic-deductive-experimental method.[63] He sought to the very end to render his fundamental principles immediately evident as in a mathematical treatise, and it was only from desperation that he gave a single indirect argument from experiment. This, of course, did, indeed, involve some critical experimentation, but it was primarily a kind of mathematical play with new methods and concepts that led to entirely new developments in the science of motion. And it is this exploitation of the creative power of mathematics that lies at the heart of the phenomenal acceleration of the mathematical sciences beginning with Galileo's discovery of a way to solve problems involving terrestial motions taking place in given ratios of time. For it was precisely this move that most facilitated the application of mathematics to a broader spectrum of physical problems and powered much subsequent development, even into the nineteenth century.

There is an intriguing analogy here with Wolfflin's analysis of the way in which the banal picturesque was gradually exploited by artists until a new sense of beauty emerged, giving rise to Baroque art.[64] I suggest a parallel in which the mathematical sciences developed in the seventeenth and eighteenth centuries through small technical changes in which the very facility of mathematics in further exploration generated a new feeling for the power of mathematics and its potential for exploring and understanding nature. This is not to deny that there were other major factors involved in the rise of the mathematical sciences, but rather to suggest that a significant part of the impetus towards the phenomenal advance in new directions lies in Galileo's unique creation of a strikingly new kind of treatise on motion which, like a great work of art, succeeded in catching the imagination of others.

GENERAL ACKNOWLEDGMENT

The writer would like to express her gratitude to the National Science Foundation for grants (SOC 7708068 and SOC 7800237) in support of research during the period in which this paper was initially prepared. The paper subsequently took its present form while the author was an Andrew Mellon Postdoctoral Fellow in the Department of History and Philosophy of Science at the University of Pittsburgh, and she is especially grateful to Larry Laudan and Peter K. Machamer for comments and criticisms. Last, but not least, she wishes to thank art historian Arthur M. Lawrence, of New York City, for much helpful discussion throughout the period of preparation.

NOTES

[1] Libri, G., *Histoire des sciences mathématiques en Italie*, 4 vols., J. Renouard, Paris, 1867, 2nd edition, Vol. 4, p. 159.

[2] Typical views are found in E. J. Dijksterhuis, *The Mechanization of the World Picture*, trans. by C. Dikshoorn, Clarendon Press, Oxford, 1961, pp. 265–66, 335–36; I. B. Cohen, *The Birth of a New Physics*, Doubleday, New York, 1960, pp. 107–113. These interpretations are the fruit of much research which began primarily with P. Duhem, *Études sur Léonard de Vinci*, 3 vol., A. Hermann and fils, Paris, 1906–1913.

[3] A forthcoming book by W. A. Wallace on the sources of Galileo's *De motu antiquiora* (*Prelude to Galileo: Essays on Medieval and Sixteenth Century Sources of Modern Science*) promises to establish sources for most of Galileo's earliest writings on motion.

[4] In the earliest stages of Galileo's work on motion, for example, he failed to adopt the assumption that in free fall, instantaneous velocity increases in proportion to time, an assumption underlying medieval developments of the "mean speed" theorem, and considered by many historians to have been one of Galileo's most important borrowings from medieval writings (see n. 2, above).

[5] See, for example, I. B. Cohen, 'Newton's second law and the concept of force in the *Principia*', the *The Annus Mirabilis of Sir Isaac Newton*, ed. by R. Palter, MIT Press, Cambridge, 1967, p. 176, n. 31, where Cohen takes as "strong evidence" that Newton was not familiar with Galileo's work on motion before he wrote the *Principia*, Newton's statement that Galileo used the first two laws to "prove" the times-squared law. What Newton says, however, is that Galileo *discovered* his law in this way, and this is a common reconstruction of Galileo's thought by those who most surely were familiar with what Galileo actually wrote in the *Discorsi*. See, for example, Guido Grandi's remarks in his edition of the *Opere di Galileo Galilei*, 4 vols., Padua, 1744, Vol. 3, p. 308. Cohen is quite right, however, that there is no direct evidence of Newton's having read the *Discorsi* before he wrote the *Principia* (*op. cit.*, p. 173, n. 10).

[6] See, esp., Stillman Drake on Galileo in the *Dictionary of Scientific Biography*, edited by C. C. Gillispie, Charles Scribner's Sons, New York, 1972, Vol. 5, pp. 237–249; also, by the same author, *Galileo at Work: His Scientific Biography*, University of Chicago Press, 1978.

[7] *Le Opere di Galileo Galilei* (hereafter, *Opere*), 20 vols., ed. A. Favaro, Barbèra, Florence, 1890–1909, Vol. VIII, pp. 190–355. The treatise on local motion was originally published as the third and fourth "days" of the *Discorsi e dimostrazioni matematiche intorno a due nouve scienze.* It is, however, a separate, formal work, independent of the other two informal discourses.

[8] Thomas S. Kuhn, *The Structure of Scientific Revolutions*, University of Chicago Press, 1962; Imre Lakatos, 'Falsification and methodology of scientific research programs', in *Criticism and the Growth of Knowledge*, ed. Lakatos and Musgrave, Cam-

bridge University Press, 1970; Larry Lauden, *Progress and its Problems: Towards a Theory of Scientific Growth*, University of California Press, 1970.

[9] K. Popper, *Conjectures and Refutations: The Growth of Scientific Knowledge*, Basic Books, New York, 1962.

[10] See W. L. Wisan, 'The new science of motion: A study of Galileo's *De motu locali*', *Archive for History of Exact Sciences* **13**: 103–306, esp. pp. 122–125; also, 'Galileo's scientific method: A reexamination', in *New Perspectives on Galileo*, ed. by R. E. Butts and J. C. Pitt, D. Reidel, Dordrecht, 1978, esp. pp. 7–8, 11, 13–14, 37–45. The first of these papers will be referred to hereafter as NSM.

[11] See Wisan, NSM, Section 5, esp. pp. 219, 220, 226–229.

[12] See Lakatos, *op. cit.*, n. 8, above.

[13] See Kuhn, *op. cit.*, n. 8, above.

[14] 'The Nature of a Paradigm', in *Criticism and the Growth of Knowledge* (cit. in n. 8, above), p. 69.

[15] *Ibid.*, p. 70. I concur with Masterman's assessment of the importance of Kuhn's analysis (see pp. 59-61), which stems from his grasp of science as a scientist rather than as philosophical onlooker.

[16] My suggestion that an important aspect of Galileo's contribution to science may be found in his *style* is in no way intended as agreement that he contributed little else. I believe his substantive contributions were of great importance despite the difficulty in establishing precise lines of "influence." In fact, study of the stylistic aspects of scientific progress may well shed fresh light on the role of specific discoveries, whether or not they were, in fact, chronologically the first actual discoveries or formulations of the scientific data or principles in question. For example, it is well known that Galileo did not articulate the first completely adequate statement of the modern inertial principle. Nonetheless, the way in which he actually *used* it may turn out to be of greater importance for the subsequent advance of scientific theory than earlier and more precise statements by others (Beeckman and Descartes, in particular). (See n. 53, below.)

[17] See, for example, *A Modern Book of Esthetics*, ed. with introduction and notes by Melvin Rader, Holt, Rinehard and Winston, 1973, 4th ed., p. 6.

[18] A century ago, Sir W. Thomson and P. G. Tait wrote that many fundamental principles, derived "directly from experiment, lead by mathematical processes to ... results, for the full testing of which our ... methods are as yet totally insufficient", which still follows Newton's dictum that principles are "deduced" from the phenomena. See their *Treatise on Natural Philosophy*, Vol. I, Part I, Cambridge University Press, 1879, 2nd ed., p. v. See, also, C. Truesdell on the methods of rational mechanics in the eighteenth century in his *Essays in the History of Mechanics*, Springer–Verlag, New York, 1968.

[19] See, especially, works by Kuhn and Lauden cited in n. 8, above.

[20] See Wisan's NSM for more complete details.

[21] See esp., R. Fredette's 'Les *De motu* "plus anciens" de Galileo Galilei: prolégomènes.' (Ph.D. dissertation, University of Montreal, 1969).

[22] *Opere* I, pp. 251–419, includes the entire essay together with related materials. Most of these materials are translated by I. E. Drabkin in *Galileo Galilei: On Motion and On Mechanics*, University of Wisconsin Press, Madison, 1960. On studies of Galileo's sources, see pp. 10–11. See also Drabkin's selections from both Galileo and G. B.

Benedetti in *Mechanics in Sixteenth-Century Italy*, trans. and annotated by S. Drake and I. E. Drabkin, University of Wisconsin Press, Madison, 1969. Several topics in Benedetti's writings recur in Galileo's *De motu antiquiora*, often with quite similar treatment. Particularly striking is Benedetti's *Demonstratio proportionum localium contra Aristotelem et omnes philosophos*, of 1554. This is not to suggest that Galileo borrowed directly from Benedetti or others, but to indicate the extent to which his work may have evolved from ideas current in his time. Further light will soon be shed on Galileo's sources by W. A. Wallace's forthcoming book, cited in n. 3, above.

23 See Chapter 14 of the *De motu*, in *op. cit.*, n. 22.

24 In *Tractatus de Ponderibus Magistri Blasii de Parma*, trans. in *The Medieval Science of Weights*, ed. with Int., Eng. trans. and notes by E. A. Moody and M. Clagett, University of Wisconsin Press, Madison, 1960, pp. 239.

25 The paraphrase from Blasius is in the *Codice Atlantico* (Atl. f 335v) and Leonardo's theorem is in the Sabachnikoff manuscript (Sab. 1v).

28 "The movement of every weight is toward the center (of the world), and its force is a

27 *Op. cit.*, n. 22, p. 63.

28 "The movement of every weight is toward the center [of the world], and its force is a power of tending downward and of resisting movement in the contrary direction." In *Medieval Science of Weights*, (*op. cit.*, n. 24, above), p. 175.

29 *Op. cit.*, n. 22, p. 63.

30 *Op. cit.*, n. 24, pp. 190–191.

31 *Op. cit.*, n. 22, p. 69.

32 Trans. as *On Mechanics*, by S. Drake in *op. cit.*, n. 22. There is much important analysis of motion implicit in this essay. We see Galileo developing his concept of *momento* and moving towards the doctrine that free fall must be naturally accelerated rather than inherently uniform, and he elaborates on an earlier principle that bodies resist being moved only insofar as they are moved away from the center of the earth. This principle underlies the development of his inertial concept and his postulate of equal speeds. See NSM, pp. 153–56, 160, 261–63.

33 That is, if two distances are traversed in the same time by two moving bodies, the distances are said to be proportional to the speeds. Versions of this rule can be found in early writings attributed to Euclid and in various medieval texts. Galileo's law of chords (Thm. VI on accelerated motion) states that the times of descent are equal along all chords drawn from the highest or lowest points of a vertical circle. See NSM, pp. 133–37, 163–65.

34 See NSM, pp. 165–171 on the relatively crude proofs for theorems on "quicker" motions. A new technique is developed in the "right-angle" theorem (pp. 171–72) and is then used (pp. 172–75) in the very elegantly and rigorously proved Theorem IX (of the later *De motu locali*). Drake presents a different interpretation in his *Galileo at Work*; he must, however, ignore a good deal of evidence from linguistic clues, technical developments, and formal considerations. It is most improbable, for example, that a crude argument on f. 147t is a *revised* version of that on f. 180r, and that the latter preceded the contents of ff. 151r and 160r.

35 These relations can be summarized as (1) $V_1/V_2 = D_1/D_2 \times T_2/T_1$ and (2) $V_1/V_2 = F_1/F_2 \times R_2/R_1$ (provided there is sufficient force to cause motion), or in more modern terms $V \propto D/T \propto F/R$. The proportionalities translate into the equalities if expressed as $V = kD/T$ and $V = qF/R$, from which $V_1 = kD_1/T_1$, $V_2 = kD_2/T_2$, $V_1 = qF_1/R_1$, $V_2 =$

qF_2/R_2. Now, (1) and (2), above, follow from these equalities if we divide the first by the second and the third by the fourth. See *Physics*, IV, 8; VII, 5.

[36] See, for example, Thomas Bradwardine's *Tractatus de proportionibus*, edited by H. Lamar Crosby, Jr., University of Wisconsin, Madison, 1955; a brief section is translated into English as *Treatise on the Proportions of Velocities in Movements*, in Marshall Clagett, *Science of Mechanics in the Middle Ages*, University of Wisconsin Press, Madison, 1961, pp. 220–21.

[37] See, for example, the passage from Franciscus de Ferraria's *Questio de proportionibus motum* in Clagett, *op. cit.* (n. 36), pp. 499–503. See, also, Clagett's *Giovanni Marliani and Late Medieval Physics*, Columbia University Press, New York, 1941.

[38] The story of the revival of ancient mathematics is well told in Paul L. Rose, *The Italian Renaissance of Mathematics*, Librarie Droz, Geneva, 1975. On the revival of Aristotelian mechanics, see P. L. Rose and S. Drake, 'The pseudo-Aristotelian questions in mechanics in Renaissance culture,' *Studies in the Renaissance*, XVIII (1971), pp. 65–104.

[39] For selections from Benedetti, see *op. cit.*, n. 22, above.

[40] A favorite example of art historians which has crept into the philosophy of science (see Kuhn's *Structure of Scientific Revolutions*, p. 110).

[41] NSM, section 4.1. Again, Drake gives a different reconstruction in *Galileo at Work*, and again he must disregard evidence from language and techniques. He assumes, for example, that f. 186t is later than f. 189t. On the latter, however, one finds facile use of a technique not yet worked out in the former.

[42] See NSM, pp. 176–77, for notes relating Galileo's search for his brachistochrone to early work on the inclined plane. See also pp. 160–62 for evidence linking his doctrine of naturally accelerated motion to further consideration of a body descending along the lower quadrant of a circle.

[43] NSM, section 6; on fall from infinity, pp. 247–48.

[44] NSM, section 5, esp. pp. 199–200, 208–09, 219, 220, 227–29. Particularly striking is Galileo's proof on f. 91 that velocity of fall increases in proportion to time; the proof depends on the double-distance rule, which still depends on the assumption that velocity of fall increases with distance. Moreover, the latter assumption continues to be stated in drafts copied after Galileo moved to Florence (see p. 218n; also p. 129n).

[45] Galileo refers to his treatise as establishing the "proportions" (sometimes the "ratios") of acceleration (*Opere* VIII, pp. 190, 243–44, 266–67). Thus, he describes his work in more specific terms than the medievals, but the underlying concept --that of an investigation of the "proportions of motions"-- is much the same.

[46] In Book II on accelerated motion, these properties (*symptomata, passiones, accidentia*) are just those "proportions" or "ratios" mentioned in the previous note (*Opere* VIII, pp. 190, 197–98). Book III on projectile motion builds on the results from Books I and II on uniform and accelerated motion. Again, the results are described in terms of lines in some given ratio to (or identical with) given lines. The language of "proportion," however, all but disappears, as Galileo describes his results. What we have are paths resulting from composition of given motions under given conditions, calculation of amplitude and altitude for given speeds at different angles of elevation, and the like. Galileo even suggests, but does not actually use, a standard measure for speeds acquired in free fall. (See NSM, section 7.) In this book, finding "proportions" of motions is no longer a dominant feature.

[47] Archimedes defines new terms and sets forth what are, in fact, new *principles* as though he hardly notices his own innovations, while his covering letters are full of excitement about *problems* he has solved. This attitude, of course, is typical of the mathematician, who does not normally discover new principles in the sense that one can do this in the physical sciences. Archimedes, however, treats his physical assumptions, such as his hydrostatic principle and the foundations of his theorems on the equilibrium of planes, in the same way as he does his mathematical assumptions. Galileo adopts much the same attitude in his mathematical science of motion; unlike Archimedes, however, Galileo insists on the physical truth of his principles, at least most of the time. For Galileo's ambiguity about this, see Wisan, 'Galileo's scientific method', section V.

[48] *Opere* VIII, p. 190 (my rendering).

[49] *Ibid*, p. 131, following translation by Crew and DeSalvio, Dover, New York, unaltered republication of first ed., 1914, p. 87; my interpolations.

[50] A year after publication of the *De motu locali*, Galileo was still searching for a more evident demonstration of his postulate. In a letter to Baliani, he conceded that this proposition lacked the *evidenza* required in principles assumed known, that is, *principii da supporsi come noti* (*Opere* XVIII, 78).

[51] *Opere* VIII, p. 267.

[52] Descartes to Mersenne, Oct. 11, 1638, in *Correspondance* III, pp. 94–97.

[53] As is well known, Marin Mersenne publicized Galileo's work in France, both in his correspondence and his publications. Mersenne published a version of Galileo's early mechanics and a paraphrase of the later work on motion in *Le méchaniques de Galilée* . . ., Paris, 1634, and *Les nouvelles pensées de Galilée*, Paris, 1639. Huygens and Isaac Barrow both explicitly acknowledge Galileo's influence, and, according to Montucla's *Histoire des mathématiques*, Paris, 1799, 2nd ed., II, Galileo's problems on the paths of "quicker" motions provided a point of departure for *les jeunes analysts* (p. 187). In fact, Galileo's propositions on motion were soon being translated into an elementary form of "analysis" even in England, as see Paynes' annotated copy of the *Discorsi* now in the Bodleian Library (Savile Bb 13). Almost a century later, in Leonard Euler's more developed (but not yet fully matured) analysis of 1736, one finds a number of Galileo's fundamental propositions proven by more sophisticated analysis. See, for example, *Mechanica Sive Motus Scientia Analytice Exposita Auctore Leonardo Eulero*, II, 1736. Prop. 13, Corollaries 5–7, where Euler derives Galileo's times-squared theorem, the law of chords, and a simple corollary (called the "right-angle theorem" in NSM, section 3.2). Euler, like Galileo, calls his mechanics a "science of motion"; unlike Galileo and the medievals, however, he nowhere speaks of finding the ratios, or "proportions," of motion. Rather, he writes about "De motu in genere" (Ch. I, Vol. I), on the motion of a point that is free or constrained, and its path along given lines, in *vacuo* or in a resistant medium. Here we see an essential change in the study of motion. It is no longer described as an examination of "ratios" of motions, and the distinction between "local" and other kinds of motion is no longer relevant. In fact, of course, Galileo's work on projectile motion also moved away from describing results in terms of "proportions of motions." It is only after Galileo that the distinction between natural and forced disappears; however, Galileo's book on "forced" motions already portends the end of this distinction.

[54] See Stephen G. Brush, *The Kind of Motion We Call Heat: A History of the Kinetic*

Theory of Gases in the Nineteenth Century, North Holland, Amsterdam, 1976. Brush weaves together the many conceptual, mathematical, and experimental threads that lead to modern kinetic theory and statistical mechanics. His analysis of the role of mathematics in the development of physical theory is particularly relevant to the thesis of this paper: "Often the turning point in a revolution is not a new physical idea but the unexpected result of a mathematical calculation. Just as a political revolution, once it is underway, seems to acquire an irresistible momentum which pushes along its nominal leaders perhaps faster than they wanted to go, so his mathematical derivations may force a physicist to accept conclusions for which his physical intuition has not yet prepared him. Despite the frequently repeated assurance that mathematics, if done correctly, is pure tautology – it cannot give you back any more information about the physical world than you put into it – it is precisely what Eugene Wigner has called 'the unreasonable effectiveness of mathematics in the natural sciences' that has provided a significant driving force in scientific revolutions" (p. 42). My suggestion is that, even without calculations, the simple power of mathematics in derivation of new propositions may have a similar effect, particularly in the very early stages of investigation in a new field, and that this kind of effectiveness in mathematics may operate in a manner similar to the way in which new styles evolve in art. This is not a denial of the importance of calculations and verifications, nor of fundamental principles. The exploration of stylistic aspects of science is for the purpose of developing tools for tracing continuities in development and is not intended as an alternate interpretation of what science is about or where its instrinsic importance lies.

[55] See Wisan, 'Galileo's scientific method,' cited in n. 10, above.

[56] *Ibid.*, esp. section V.

[57] In Galileo's early *Trattato della sphera, ovvero cosmografia* (*Opere* II, p. 212), he remarks that "the intellect is guided to knowledge [*cognizione*] of substances by means of accidents." The language used is almost exactly that found in earlier notes on logic (MS 27, cc. 16v, 17r). In the *Discourse on Bodies in Water*, the discovery of accidents leads to cognition of causes (*Opere* IV, p. 109; English trans. by Salusbury, University of Illinois Press, Urbana, 1960, p. 45). In his *Letters on Sunspots*, Galileo expresses skepticism about our ability to know the "true and intrinsic essences" of natural substances (*Opere* V, p. 187; trans. by S. Drake in *Discoveries and Opinions of Galileo*, Doubleday, New York, 1957, pp. 123–24). In the *Dialogue Concerning the Two Chief World Systems*, it is by cognition of effects that we investigate and discover causes, a new and somewhat different emphasis. (*Opere* VII, p. 443; Eng. trans. by S. Drake, Univ. of California Press, 2nd ed., 1967, p. 417). The search for "true causes" replaces the quest for knowledge of substances, and cognition refers more often to the means than the end.

[58] *The Discourse on Bodies in Water* (cited in n. 57).

[59] This may have been part of the motive behind the third of the three "fantasies" in Galileo's *Dialogo*, according to which the "true and real" motion of fall is not accelerated at all, but uniform. The acceleration, then, is not fundamental, and we are thus freed of the necessity for finding its cause. (*Dialogue*, *Opere* VII, p. 192; Drake's trans., p. 166).

[60] Galileo's language speaks both to the theorist and the experimentalist, but in a manner that varies over time, and, to a considerable extent, depends on the subject at hand. He regarded mechanics and motion as subjects for a mathematical science in the

Euclidean tradition of rigorous mathematical derivations from true, evident principles. Astronomy, on the other hand, could not be based on immediately evident principles, but required empirical confirmation of hypotheses. In the science of motion, then, there is less emphasis on experience; principles must be rendered immediately evident in some sense, and to this end Galileo appeals to experience and experiment, but he is generally indifferent towards verification of *remote* consequences. Hence, my claim that Galileo's adherence to the Greek mathematical model kept him from making experimental confirmation central to the development of his new science of motion (in 'Mathematics and experiment in Galileo's science of motion,' *Annali dell 'Istituto e Museo di Storia della Scienza di Firenze, Anno* II, *Fas.* 2, 1977, pp. 149–60, and 'Galileo's scientific method,' cited in n. 10, above). R. H. Naylor altogether fails to take my point in his 'Mathematics and experiment in Galileo's new sciences,' *Annali*, IV, 1, 1979, pp. 53–63.

Despite Galileo's early and persistent mathematical orientation, however, his writings reveal a general drift towards greater empiricism, which is no doubt greatly affected by the success of his telescopic observations, and later writings reflect greater sensitivity to observation and experiment. In the old *De motu*, for example, we find Galileo remarking that "when I could not discover a reason for such an effect, I finally had recourse to experiment . ." (Drabkin, *op. cit.*, n. 22, above, p. 83). In his last work, on the other hand, we find him saying that "where the senses fail us reason must step in" (*Discorsi*, First Day, trans. by Crew and De Salvio, cit. in n. 49, above, p. 60). The works on astronomy, of course, are rich in description of close observations, and Galileo's language and attitude are sometimes suggestive of the so-called "nature philosophers," especially in his emphasis on the sense of sight, but sometimes also that of touch (see Wisan, 'Galileo's scientific method,' sections III and IV).

[61] See, for example, Physics, IV, 8; VII, 5.

[62] *Mécanique Analytique* (Paris, 1788).

[63] See Wisan, 'Galileo's scientific method,' cited in n. 10, above.

[64] *Principles of Art History*: *The Problem of Development of Style in Later Art*, trans. from the German by M. D. Hottinger, G. Bell and Sons, Ltd., London, 1932; from the 7th ed., 1929; 1st ed., 1915. One need not subscribe to the precise analysis offered by Wolfflin to see the force of such analogies between the way in which styles develop in the arts and in science.

Sources consulted on style in art or on similarities between art and science include E. H. Gombrich, *Art and Illusion*: *A Study in the Psychology of Pictorial Representation*, Bollingen Series XXXV, 5, Pantheon Books, New York, 1960; Meyer Shapiro, 'Style,' in *A Modern Book of Esthetics*, cited in n. 17, above, pp. 270–80; *On Aesthetics in Science*, ed. by Judith Wechsler, MIT Press, Cambridge, 1978; Nelson Goodman, *Ways of Worldmaking*, Hackett, Indianapolis, 1978. The last became available only a few days before sending this paper to press; however, the chapters on 'The status of style,' pp. 23–40, and 'When is art?,' pp. 57–70, look particularly promising. So, also, do advance excerpts from A. C. Crombie's forthcoming 'styles of scientific thinking.'

N. JARDINE

PHILOSOPHY OF SCIENCE AND THE ART OF HISTORICAL INTERPRETATION

I

Why do historians and philosophers of science expect to profit from each others' enquiries: and why are their expectations so apt to end in disillusionment of the kind voiced by Wisan in the context of Galilean studies?[1]

On the assumption that the primary task of the historian of science is to describe and explain the growth of scientific knowledge, that of the philosopher of science to elucidate the nature and grounds of that knowledge and to explicate the strategies of enquiry apt to engender it, prospects for collaboration between the two disciplines look good indeed. Historians of science can reasonably appeal to philosophers' accounts of the nature of science and of scientific rationality, in order to show how scientific progress has been fostered by adherence to sound views about the goals and methods of scientific inquiry. Conversely, philosophers of science can reasonably appeal to the chronicles of historians both to suggest and to test accounts of the nature of scientific method. This is the model for the relation between the two disciplines that underlies Lakatos' oft-quoted aphorism 'Philosophy of science without history of science is empty; history of science without philosophy of science is blind';[2] and it is the model that informed the Heroic Age of the history and philosophy of science, the age of William Whewell, Ernst Mach and Pierre Duhem – I shall call it 'the Classical HPS Model'.

The shortcomings of history of science written in accordance with the Classical Model are notorious. Take Mach's elegant and concise account of Galileo's achievements in *The Principles of Mechanics*.[3] There is the anachronistic presentation of Galileo's theorems and proofs in modern algebraic notation. There is the hindsight which leads him to credit Galileo with a special case of the Law of Inertia. There is the textually unwarranted claim that Galileo adhered to Machian canons of scientific method and espoused a Machian view of

J. Hintikka, D. Gruender, and E. Agazzi (eds.), Pisa Conference Proceedings, Vol. I, 341–348.

the mathematical sciences as eschewing the search for hidden causes. And finally there is the ruthless selection from the corpus of Galilean beliefs of those that can be forced into the modern category of Mechanics and made to appear, by our standards, true or approximately so, a principle of selection which makes Galileo appear as a miraculous anticipator of fragments of later coherent systems of belief, not as the holder of a coherent system of beliefs of his own. Subsequent Galileo studies further attest the distorting effects of the Classical HPS Model; indeed it is arguable that Mach's distortions are modest compared with those induced by imposition of the 'Experimental Method' at the hands of Drake or of 'Methodological Anarchy' at the hands of Feyerabend.

What, apart from its unfortunate historiographical effects, is wrong with the Classical HPS Model? By way of commentary on Wisan's frustration with the philosophy of science I shall sketch one particular line of objection to it and reflect on the extent to which it undermines philosophers' expectations of the history of science and historians' expectations of the philosophy of science.

II

The Classical HPS Model is doubly open to sceptical attack. Presupposing, as it does, that historians can offer objectively valid chronicles of the growth of scientific knowledge, it is vulnerable to the various theses of indeterminacy and relativity of translation and interpretation which threaten the possibility of objective sifting of the true from the false in past bodies of scientific belief. And presupposing, as it does, that philosophers can hope to articulate and justify the canons of scientific method, it is vulnerable to arguments for the impossibility of such articulation and justification. I shall ignore these well-known sources of scepticism, for there is a far more specific objection to the Classical HPS Model that is independent of contentious philosophical theses.

It is as a matter of well-attested fact extremely difficult to render comprehensible, whether by translation, gloss or exegesis, texts written in the context of theories (conceptual systems, ideologies, explanatory frameworks, world-pictures, paradigms, disciplinary matrices, research traditions, or what have you) alien to our own. How is the historian of science to proceed in this demanding task of interpretation?

No elaborate theory of historical interpretation is needed to show up the defects of the Classical HPS Model. We need consider only a few of the most elementary criteria of adequacy for a translation, exegesis or gloss of a scientific text or series of texts. For example: we should try to avoid ascribing to an author simultaneous adherence to beliefs of whose inconsistency he could not but have been aware; we should try to avoid ascriptions of belief which render his acquisition of particular theoretical beliefs inexplicable given his other theoretical beliefs, including beliefs about scientific method, and the factual information available to him through observation, experiment or the testimony of others; and we should try to avoid ascriptions of belief which make his actions, in particular his making of observations, carrying out of experiments and design of instruments, inexplicable given his likely purposes in performing those actions.It is a moot point whether the art of interpretation can be wholly digested into rules of this kind, or whether (as Gadamer has persuasively maintained) such rules can capture only the negative part of the art, the elimination of grossly fallacious or anachronistic interpretations, the positive and creative aspects of the art being such as to resist codification.[4] And even if the art of interpretation is in principle codifiable it remains doubtful whether such rules can ever determine unique interpretations, and, if they cannot, how far the objectivity of interpretation is vitiated. For our purposes, however, these hard questions can be set aside; it suffices to concede the uncontentious point that the art of interpretation is subject to constraints of this kind.

Once we admit the validity of such constraints the Classical HPS Model is seen to be misguided on several counts:

(1) As an interpreter, the historian is bound to consider an author's whole corpus of beliefs. Even if his primary aim is to assess the author's contributions to scientific knowledge, he cannot at the outset, as the Classical Model requires, simply pick out those beliefs that are by our standards both scientific (as opposed to, say, metaphysical, methodological or theological) and true, or approximately so.

(2) Interpretation does not, as the Classical Model implies consist of two stages, first description, then explanation. Rather the interpreter must seek translations and exegeses which make explicable past scientists' acquisition, maintenance and mutation of beliefs.

(3) Whilst the interpreter is bound to make certain assumptions about the forms of human rationality, he is not entitled, as the Classical Model requires, to assume that past scientists adhered to what we believe to be sound procedures of scientific enquiry. Rather he must try to make their first-order scientific beliefs explicable for us in the light of their strategies of enquiry and their second-order beliefs about the goals and proper procedures of scientific enquiry.

What expectations of each others' disciplines can historians and philosophers of science reasonably maintain in the light of these objections to the Classical HPS Model?

III

The philosopher's expectations are, in principle, the less radically undermined. For these objections do not in themselves cast doubt on the possibility of accurate accounts of the growth of scientific knowledge and of the strategies of enquiry and justification which have mediated theoretical innovations and their diffusion. So, in principle, the philosopher can hope to appeal to historians' chronicles of scientific progress both as a source of inspiration for and as a test of his accounts of the nature of science and scientific method. Nor do these objections cast practical doubt on the value of the history of science as a source of inspiration for the philosopher. I have in mind, for example, the use by Lakatos and by Hintikka and Remes of Pappus' account of geometrical analysis and synthesis as a source of inspiration for philosophical speculation on the role of heuristic in mathematics,[5] and the use of Galilean thought experiments by Kuhn as the basis for a philosophical hypothesis about the role of thought experiments in scientific discovery.[6] The objections do, however, raise grave doubts about the practical feasibility of using the history of science to test accounts of scientific method and rationality.

To test an account of heuristic method in this way a philosopher would need reliable information about the strategies of enquiry which have led individual scientists to theoretical innovations. Even in those rare cases in which a scientist has left a full record of his investigations, including detailed accounts of his actual observations and experiments, of his motives in carrying them out and of the conclusions he drew from them, there are formidable difficulties. Mastery of techniques of enquiry and problem-solving evidently constitutes for the most part practical not theoretical knowledge. In interpreting a past scientist's assertions about his own procedures and methods of enquiry we must allow for the likelihood of substantial misperception and distortion by then-current theories of scientific method, as well as the possibility (very evident in the case of Galileo) that his assertions were propagandist in intent.

For the philosopher interested not in heuristic method but in criteria for rational choice between theories, the prospects are yet worse. To test his account he would need full descriptions of the processes of diffusion of particular hypotheses and theories, including specifications of the reasons, both causal and justificatory, that have led to their eventual adoption or rejection. As Wisan emphasises in the case of Galilean science, we can scarcely expect to acquire such information. It is small wonder, then, that such philosophers of science as Lakatos who have taken the history of science to be an indispensable testing-ground for their canons for rational choice between theories have in practice succeeded only in showing that in certain cases the adoption of a theory that is, by our standards, better than its competitors, could have come about through adherence to those canons, not, as is surely required for a legitimate test, that it did in fact come about in that way.[7]

IV

Our objections to the Classical HPS Model legislate more decisively against certain of the uses to which historians have hoped to put the insights of philosophers. According to those objections, both the historian of science who tries to focus his investigations by appeal to modern criteria for the demarcation of the natural sciences from other disciplines, and the historian who seeks to explain scientific progress by appeal to some modern account of scientific rationality, are well set on a path of systematic misinterpretation. Is philosophy of science perhaps, as Wisan intimates, useless to the historian of science?

Consider a related question. Is conversancy with science indispensable to the historian of science? Clearly it is. For in assessing an interpretation we often have to consider the truth values of beliefs ascribed under that interpretation; for example, when one appeals to certain beliefs to explain a past scientist's action in designing a successful experiment or in constructing a successful instrument, the explanation will in general be plausible only if those beliefs are, under our interpretation, true or approximately so. More germane to our purposes, however, are the many indirect ways in which the skilled interpreter deploys his conversancy with our science. He does so when he works out what information a past scientist could have acquired using the techniques available to him; when he takes into account the likelihood of his having maintained, modified or rejected certain beliefs on the basis of observations he is known to have made; and when he speculates on the technological, conceptual, and mathematical difficulties he may have faced. Further, the interpreter needs such conversancy if he is to be aware of the full connotations of the scientific terminology available to him for purposes of translation and exegesis, an awareness which is needed if he is to achieve precision whilst avoiding the distortions of Whiggish hindsight.

According to our objections to the Classical HPS Model, the historian of science cannot afford to concentrate exclusively on past beliefs that are by our standards scientific, setting aside those that are methodological, epistemological or metaphysical. My suggestion now is that conversancy with the philosophy of science is necessary for the interpretation of such philosophical beliefs in ways very closely analogous to those in which conversancy with science is necessary for the interpretation of past scientific beliefs. Such conversancy is needed if the historian is to reconstruct the philosophical problems which exercised past scientists, whether he does so directly by appeal to their assertions about the nature of science, causation, explanation, probability, and so on, or indirectly by appeal to contemporary debates on those topics of which the scientist would have been aware. Such conversancy is needed if he is to see how their pronouncements about the nature of science and scientific method may have related to the actual procedures of enquiry and justification they adopted. And it is needed, crucially, if he is to achieve precision in translation and exegesis whilst avoiding the distortions likely to be imposed by the use of our technical philosophical terminology.

It is perhaps worth noting, in passing, a quite different reason why conversancy with philosophy is of value to the historian of science. Perhaps because in the history of philosophy (as in the history of political and legal theory) the attempt to foist linear progress and substantial past anticipation of present-day positions onto past systems of belief is more evidently doomed to failure, the historiography of philosophy, and especially of classical philosophy, provides paradigms of sensitive interpretation and reconstruction which historians of science may profitably emulate.[8]

But whilst there are ample grounds for holding that conversancy with the philosophy of science is indispensable for the adequate interpretation of past science, it is hard not to sympathize with Wisan's frustration. The historian who hopes, as Wisan does, to interpret in detail the strategies through which an individual scientist arrived at theoretical innovations is likely to be frustrated by the dearth of relevant recent work on heuristic method (at least in comparison with the plethora of recent work on assessment and comparison of theories).[9] More generally, the historian who attempts to integrate past scientists' scientific beliefs with their metaphysical and epistemological tenets might hope to be able to appeal to philosophically sophisticated histories of past views on, for example, scientific method, explanation, causation, probability and the weighing of evidence, and the classification of the arts and sciences. Though there are outstanding works in this genre (Hacking's *Emergence of Probability* springs to mind) there are also remarkable lacunae. Thus, whilst there are a number of scholarly studies of medievel classifications of knowledge, there is virtually nothing on this topic for the modern period.[10] Such lacunae are, I think, as unhealthy for the philosophy of science as they are frustrating to the historian of science.

To conclude on a more optimistic note: the historian of science who cannot find what he or she needs in the literature of the philosophy of science should surely be prepared to turn philosopher on his or her own account. As an *ad feminam* response to Wisan's pessimism about the role of philosophy of science in Galilean studies, I would cite two recent papers in which philosophical acumen is so deployed as to yield original and convincing interpretations of difficult Galilean texts. I have in mind Koertge's recent paper 'Galileo and the problem of accidents' and Wisan's own paper 'Galileo's scientific method: a re-examination'.[11]

University of Cambridge

NOTES

[1] W. Wisan, 'Galileo and the Emergence of a New Scientific Style', this volume.

[2] I. Lakatos, 'History of science and its rational reconstructions', in R. C. Buck and R. S. Cohen (eds.), *Boston Studies in the Philosophy of Science* **8** (1971), 9–135.

[3] E. Mach, *Die Mechanik in ihrer Entwicklung historisch-kritisch dargestellt*, Leipzig, 1883; Engl transl. by T. J. McCormack, *The Principles of Mechanics*, London, 1893.

[4] See, e.g., H.-G. Gadamer, 'The universality of the hermeneutical problem', (1966), in *Philosophical Hermeneutics* (transl. by D. E. Linge), Berkeley and Los Angeles, 1976, 3–17.

[5] I. Lakatos, 'The method of analysis-synthesis', in *Mathematics, Science and Epistemology* (ed. by J. Worrall and G. Currie), Cambridge, 1978, 70–103; K. J. J. Hintikka and U. Remes, *The Method of Analysis*, Dordrecht, 1974.

[6] T. S. Kuhn, 'A function for thought experiments' (1964); reprinted in *The Essential Tension*, Chicago, 1977, 240–265.

[7] The gulf between actual history and Lakatosian possible history is peculiarly evident in I. Lakatos and E. Zahar, 'Why did Copernicus' research programme supersede Ptolemy's?', in R. S. Westman (ed.), *The Copernican Achievement*, Berkeley, 1975, 354–383. This is an instructive case, for the impressive body of recent detailed historical studies of the diffusion and reception of Copernicus' innovations testifies to the sheer complexity, intractability and open-endedness of the task of elucidating the processes which eventuate in the widespread adoption of a theory; see, e.g., R. S. Westman (ed.), *op. cit.*, and J. Dobrzycki, *The Reception of Copernicus' Heliocentric Theory*, Dordrecht, 1972.

[8] Cf., the remarks of T. S. Kuhn, 'The history of science', *International Encyclopedia of the Social Sciences* **14** (1968), 74–83, on the influence of the historiography of philosophy on the historiography of science.

[9] There is, to be sure, a vast technical and philosophical literature in the heuristic fields of experimental design, statistical inference, decision theory, classification theory and artificial intelligence. But there is relatively little recent work on heuristic method in the natural sciences comparable with, say, Lakatos' investigations of heuristic method in mathematics in direct relevance to the historian concerned with past scientists' strategies of enquiry.

[10] The vital importance for the historian of science of understanding of past classifications of knowledge and their philosophical bases is well-argued, from very different points of view, by R. McRae, *The Problem of the Unity of the Sciences: Bacon to Kant*, Toronto, 1961, and by T. S. Kuhn, 'Mathematical versus experimental traditions in the development of physical science', *Journal of Interdisciplinary History* **7** (1976), 1–31.

[11] N. Koertge, *Journal of the History of Ideas* **38** (1977), 384–408; W. Wisan, in Butts and Pitt (eds.), *New Perspectives on Galileo*, Dordrecht, 1978, 1–57.

INDEX OF NAMES

INDEX OF SUBJECTS

SYNTHESE LIBRARY

Studies in Epistemology, Logic, Methodology,
and Philosophy of Science

1. J. M. Bochénski, *A Precis of Mathematical Logic*. 1959.
2. P. L. Guiraud, *Problèmes et méthodes de la statistique linguistique*. 1960.
3. Hans Freudenthal (ed.), *The Concept and the Role of the Model in Mathematics and Natural and Social Sciences*. 1961.
4. Evert W. Beth, *Formal Methods. An Introduction to Symbolic Logic and the Study of Effective Operations in Arithmetic and Logic*. 1962.
5. B. H. Kazemier and D. Vuysje (eds.), *Logic and Language. Studies Dedicated to Professor Rudolf Carnap on the Occasion of His Seventieth Birthday*. 1962.
6. Marx W. Wartofsky (ed.), *Proceedings of the Boston Colloquium for the Philosophy of Science 1961-1962*. Boston Studies in the Philosophy of Science, Volume I. 1963.
7. A. A. Zinov'ev, *Philosophical Problems of Many-Valued Logic*. 1963.
8. Georges Gurvitch, *The Spectrum of Social Time*. 1964.
9. Paul Lorenzen, *Formal Logic*. 1965.
10. Robert S. Cohen and Marx W. Wartofsky (eds.), *In Honor of Philipp Frank*. Boston Studies in the Philosophy of Science, Volume II. 1965.
11. Evert W. Beth, *Mathematical Thought. An Introduction to the Philosophy of Mathematics*. 1965.
12. Evert W. Beth and Jean Piaget, *Mathematical Epistemology and Psychology*. 1966.
13. Guido Küng, *Ontology and the Logistic Analysis of Language. An Enquiry into the Contemporary Views on Universals*. 1967.
14. Robert S. Cohen and Marx W. Wartofsky (eds.), *Proceedings of the Boston Colloquium for the Philosophy of Science 1964-1966. In Memory of Norwood Russell Hanson*. Boston Studies in the Philosophy of Science, Volume III. 1967.
15. C. D. Broad, *Induction, Probability, and Causation. Selected Papers*. 1968.
16. Günther Patzig, *Aristotle's Theory of the Syllogism. A Logical-Philosophical Study of Book A of the Prior Analytics*. 1968.
17. Nicholas Rescher, *Topics in Philosophical Logic*. 1968.
18. Robert S. Cohen and Marx W. Wartofsky (eds.), *Proceedings of the Boston Colloquium for the Philosophy of Science 1966-1968*. Boston Studies in the Philosophy of Science, Volume IV. 1969.

19. Robert S. Cohen and Marx W. Wartofsky (eds.), *Proceedings of the Boston Colloquium for the Philosophy of Science 1966-1968*. Boston Studies in the Philosophy of Science, Volume V. 1969.
20. J. W. Davis, D. J. Hockney, and W. K. Wilson (eds.), *Philosophical Logic*. 1969.
21. D. Davidson and J. Hintikka (eds.), *Words and Objections. Essays on the Work of W. V. Quine*. 1969.
22. Patrick Suppes, *Studies in the Methodology and Foundations of Science. Selected Papers from 1911 to 1969*. 1969.
23. Jaakko Hintikka, *Models for Modalities. Selected Essays*. 1969.
24. Nicholas Rescher *et al.* (eds.), *Essays in Honor of Carl G. Hempel. A Tribute on the Occasion of His Sixty-Fifth Birthday*. 1969.
25. P. V. Tavanec (ed.), *Problems of the Logic of Scientific Knowledge*. 1969.
26. Marshall Swain (ed.), *Induction, Acceptance, and Rational Belief*. 1970.
27. Robert S. Cohen and Raymond J. Seeger (eds.), *Ernst Mach: Physicist and Philosopher*. Boston Studies in the Philosophy of Science, Volume VI. 1970.
28. Jaakko Hintikka and Patrick Suppes, *Information and Inference*. 1970.
29. Karel Lambert, *Philosophical Problems in Logic. Some Recent Developments*. 1970.
30. Rolf A. Eberle, *Nominalistic Systems*. 1970.
31. Paul Weingartner and Gerhard Zecha (eds.), *Induction, Physics, and Ethics*. 1970.
32. Evert W. Beth, *Aspects of Modern Logic*. 1970.
33. Risto Hilpinen (ed.), *Deontic Logic: Introductory and Systematic Readings*. 1971.
34. Jean-Louis Krivine, *Introduction to Axiomatic Set Theory*. 1971.
35. Joseph D. Sneed, *The Logical Sstructure of Mathematical Physics*. 1971.
36. Carl R. Kordig, *The Justification of Scientific Change*. 1971.
37. Milic Capek, *Bergson and Modern Physics*. Boston Studies in the Philosophy of Science, Volume VII. 1971.
38. Norwood Russell Hanson, *What I Do Not Believe, and Other Essays* (ed. by Stephen Toulmin and Harry Woolf). 1971.
39. Roger C. Buck and Robert S. Cohen (eds.), *PSA 1970. In Memory of Rudolf Carnap*. Boston Studies in the Philosophy of Science, Volume VIII. 1971.
40. Donald Davidson and Gilbert Harman (eds.), *Semantics of Natural Language*. 1972.
41. Yehoshua Bar-Hillel (ed.), *Pragmatics of Natural Languages*. 1971.
42. Sören Stenlund, *Combinators, λ-Terms and Proof Theory*. 1972.
43. Martin Strauss, *Modern Physics and Its Philosophy. Selected Papers in the Logic, History, and Philosophy of Science*. 1972.
44. Mario Bunge, *Method, Model and Matter*. 1973.
45. Mario Bunge, *Philosophy of Physics*. 1973.
46. A. A. Zinov'ev, *Foundations of the Logical Theory of Scientific Knowledge (Complex Logic)*. (Revised and enlarged English edition with an appendix by G. A. Smirnov, E. A. Sidorenka, A. M. Fedina, and L. A. Bobrova.) Boston Studies in the Philosophy of Science, Volume IX. 1973.
47. Ladislav Tondl, *Scientific Procedures*. Boston Studies in the Philosophy of Science, Volume X. 1973.
48. Norwood Russell Hanson, *Constellations and Conjectures* (ed. by Willard C. Humphreys, Jr.). 1973.

49. K. J. J. Hintikka, J. M. E. Moravcsik, and P. Suppes (eds.), *Approaches to Natural Language.* 1973.
50. Mario Bunge (ed.), *Exact Philosophy – Problems, Tools, and Goals.* 1973.
51. Radu J. Bogdan and Ilkka Niiniluoto (eds.), *Logic, Language, and Probability.* 1973.
52. Glenn Pearce and Patrick Maynard (eds.), *Conceptual Change.* 1973.
53. Ilkka Niiniluoto and Raimo Tuomela, *Theoretical Concepts and Hypothetico-Inductive Inference.* 1973.
54. Roland Fraissé, *Course of Mathematical Logic* – Volume 1: *Relation and Logical Formula.* 1973.
55. Adolf Grünbaum, *Philosophical Problems of Space and Time.* (Second, enlarged edition.) Boston Studies in the Philosophy of Science, Volume XII. 1973.
56. Patrick Suppes (ed.), *Space, Time, and Geometry.* 1973.
57. Hans Kelsen, *Essays in Legal and Moral Philosophy* (selected and introduced by Ota Weinberger). 1973.
58. R. J. Seeger and Robert S. Cohen (eds.), *Philosophical Foundations of Science.* Boston Studies in the Philosophy of Science, Volume XI. 1974.
59. Robert S. Cohen and Marx W. Wartofsky (eds.), *Logical and Epistemological Studies in Contemporary Physics.* Boston Studies in the Philosophy of Science, Volume XIII. 1973.
60. Robert S. Cohen and Marx W. Wartofsky (eds.), *Methodological and Historical Essays in the Natural and Social Sciences. Proceedings of the Boston Colloquium for the Philosophy of Science 1969-1972.* Boston Studies in the Philosophy of Science, Volume XIV. 1974.
61. Robert S. Cohen, J. J. Stachel, and Marx W. Wartofsky (eds.), *For Dirk Struik. Scientific, Historical and Political Essays in Honor of Dirk J. Struik.* Boston Studies in the Philosophy of Science, Volume XV. 1974.
62. Kazimierz Ajdukiewicz, *Pragmatic Logic* (transl. from the Polish by Olgierd Wojtasiewicz). 1974.
63. Sören Stenlund (ed.), *Logical Theory and Semantic Analysis. Essays Dedicated to Stig Kanger on His Fiftieth Birthday.* 1974.
64. Kenneth F. Schaffner and Robert S. Cohen (eds.), *Proceedings of the 1972 Biennial Meeting, Philosophy of Science Association.* Boston Studies in the Philosophy of Science, Volume XX. 1974.
65. Henry E. Kyburg, Jr., *The Logical Foundations of Statistical Inference.* 1974.
66. Marjorie Grene, *The Understanding of Nature. Essays in the Philosophy of Biology.* Boston Studies in the Philosophy of Science, Volume XXIII. 1974.
67. Jan M. Broekman, *Structuralism: Moscow, Prague, Paris.* 1974.
68. Norman Geschwind, *Selected Papers on Language and the Brain.* Boston Studies in the Philosophy of Science, Volume XVI. 1974.
69. Roland Fraissé, *Course of Mathematical Logic* – Volume 2: *Model Theory.* 1974.
70. Andrzej Grzegorczyk, *An Outline of Mathematical Logic. Fundamental Results and Notions Explained with All Details.* 1974.
71. Franz von Kutschera, *Philosophy of Language.* 1975.
72. Juha Manninen and Raimo Tuomela (eds.), *Essays on Explanation and Understanding. Studies in the Foundations of Humanities and Social Sciences.* 1976.

73. Jaakko Hintikka (ed.), *Rudolf Carnap, Logical Empiricist. Materials and Perspectives.* 1975.
74. Milic Capek (ed.), *The Concepts of Space and Time. Their Structure and Their Development.* Boston Studies in the Philosophy of Science, Volume XXII. 1976.
75. Jaakko Hintikka and Unto Remes, *The Method of Analysis. Its Geometrical Origin and Its General Significance.* Boston Studies in the Philosophy of Science, Volume XXV. 1974.
76. John Emery Murdoch and Edith Dudley Sylla, *The Cultural Context of Medieval Learning.* Boston Studies in the Philosophy of Science, Volume XXVI. 1975.
77. Stefan Amsterdamski, *Between Experience and Metaphysics. Philosophical Problems of the Evolution of Science.* Boston Studies in the Philosophy of Science, Volume XXXV. 1975.
78. Patrick Suppes (ed.), *Logic and Probability in Quantum Mechanics.* 1976.
79. Hermann von Helmholtz: *Epistemological Writings. The Paul Hertz/Moritz Schlick Centenary Edition of 1921 with Notes and Commentary by the Editors.* (Newly translated by Malcolm F. Lowe. Edited, with an Introduction and Bibliography, by Robert S. Cohen and Yehuda Elkana.) Boston Studies in the Philosophy of Science, Volume XXXVII. 1977.
80. Joseph Agassi, *Science in Flux.* Boston Studies in the Philosophy of Science, Volume XXVIII. 1975.
81. Sandra G. Harding (ed.), *Can Theories Be Refuted? Essays on the Duhem-Quine Thesis.* 1976.
82. Stefan Nowak, *Methodology of Sociological Research. General Problems.* 1977.
83. Jean Piaget, Jean-Blaise Grize, Alina Szeminska, and Vinh Bang, *Epistemology and Psychology of Functions.* 1977.
84. Marjorie Grene and Everett Mendelsohn (eds.), *Topics in the Philosophy of Biology.* Boston Studies in the Philosophy of Science, Volume XXVII. 1976.
85. E. Fischbein, *The Intuitive Sources of Probabilistic Thinking in Children.* 1975.
86. Ernest W. Adams, *The Logic of Conditionals. An Application of Probability to Deductive Logic.* 1975.
87. Marian Przelecki and Ryszard Wójcicki (eds.), *Twenty-Five Years of Logical Methodology in Poland.* 1977.
88. J. Topolski, *The Methodology of History.* 1976.
89. A. Kasher (ed.), *Language in Focus: Foundations, Methods and Systems. Essays Dedicated to Yehoshua Bar-Hillel.* Boston Studies in the Philosophy of Science, Volume XLIII. 1976.
90. Jaakko Hintikka, *The Intentions of Intentionality and Other New Models for Modalities.* 1975.
91. Wolfgang Stegmüller, *Collected Papers on Epistemology, Philosophy of Science and History of Philosophy.* 2 Volumes. 1977.
92. Dov M. Gabbay, *Investigations in Modal and Tense Logics with Applications to Problems in Philosophy and Linguistics.* 1976.
93. Radu J. Bogdan, *Local Induction.* 1976.
94. Stefan Nowak, *Understanding and Prediction. Essays in the Methodology of Social and Behavioral Theories.* 1976.
95. Peter Mittelstaedt, *Philosophical Problems of Modern Physics.* Boston Studies in the Philosophy of Science, Volume XVIII. 1976.

96. Gerald Holton and William Blanpied (eds.), *Science and Its Public: The Changing Relationship*. Boston Studies in the Philosophy of Science, Volume XXXIII. 1976.
97. Myles Brand and Douglas Walton (eds.), *Action Theory*. 1976.
98. Paul Gochet, *Outline of a Nominalist Theory of Proposition. An Essay in the Theory of Meaning*. 1980. (Forthcoming.)
99. R. S. Cohen, P. K. Feyerabend, and M. W. Wartofsky (eds.), *Essays in Memory of Imre Lakatos*. Boston Studies in the Philosophy of Science, Volume XXXIX. 1976.
100. R. S. Cohen and J. J. Stachel (eds.), *Selected Papers of Léon Rosenfeld*. Boston Studies in the Philosophy of Science, Volume XXI. 1978.
101. R. S. Cohen, C. A. Hooker, A. C. Michalos, and J. W. van Evra (eds.), *PSA 1974: Proceedings of the 1974 Biennial Meeting of the Philosophy of Science Association*. Boston Studies in the Philosophy of Science, Volume XXXII. 1976.
102. Yehuda Fried and Joseph Agassi, *Paranoia: A Study in Diagnosis*. Boston Studies in the Philosophy of Science, Volume L. 1976.
103. Marian Przelecki, Klemens Szaniawski, and Ryszard Wójcicki (eds.), *Formal Methods in the Methodology of Empirical Sciences*. 1976.
104. John M. Vickers, *Belief and Probability*. 1976.
105. Kurt H. Wolff, *Surrender and Catch: Experience and Inquiry Today*. Boston Studies in the Philosophy of Science, Volume LI. 1976.
106. Karel Kosík, *Dialectics of the Concrete*. Boston Studies in the Philosophy of Science, Volume LII. 1976.
107. Nelson Goodman, *The Structure of Appearance*. (Third edition.) Boston Studies in the Philosophy of Science, Volume LIII. 1977.
108. Jerzy Giedymin (ed.), *Kazimierz Ajdukiewicz: The Scientific World-Perspective and Other Essays, 1931-1963*. 1978.
109. Robert L. Causey, *Unity of Science*. 1977.
110. Richard E. Grandy, *Advanced Logic for Applications*. 1977.
111. Robert P. McArthur, *Tense Logic*. 1976.
112. Lars Lindahl, *Position and Change. A Study in Law and Logic*. 1977.
113. Raimo Tuomela, *Dispositions*. 1978.
114 Herbert A. Simon, *Models of Discovery and Other Topics in the Methods of Science*. Boston Studies in the Philosophy of Science, Volume LIV. 1977.
115. Roger D. Rosenkrantz, *Inference, Method and Decision*. 1977.
116. Raimo Tuomela, *Human Action and Its Explanation. A Study on the Philosophical Foundations of Psychology*. 1977.
117. Morris Lazerowitz, *The Language of Philosophy. Freud and Wittgenstein*. Boston Studies in the Philosophy of Science, Volume LV. 1977.
118. Stanislaw Leśniewski, *Collected Works* (ed. by S. J. Surma, J. T. J. Srzednicki, and D. I. Barnett, with an annotated bibliography by V. Frederick Rickey). 1980. (Forthcoming.)
119. Jerzy Pelc, *Semiotics in Poland, 1894-1969*. 1978.
120. Ingmar Pörn, *Action Theory and Social Science. Some Formal Models*. 1977.
121. Joseph Margolis, *Persons and Minds. The Prospects of Nonreductive Materialism*. Boston Studies in the Philosophy of Science, Volume LVII. 1977.
122. Jaakko Hintikka, Ilkka Niiniluoto, and Esa Saarinen (eds.), *Essays on Mathematical and Philosophical Logic*. 1978.
123. Theo A. F. Kuipers, *Studies in Inductive Probability and Rational Expectation*. 1978.

124. Esa Saarinen, Risto Hilpinen, Ilkka Niiniluoto, and Merrill Provence Hintikka (eds.), *Essays in Honour of Jaakko Hintikka on the Occasion of His Fiftieth Birthday.* 1978.

125 Gerard Radnitzky and Gunnar Andersson (eds.), *Progress and Rationality in Science.* Boston Studies in the Philosophy of Science, Volume LVIII. 1978.

126. Peter Mittelstaedt, *Quantum Logic.* 1978.

127. Kenneth A. Bowen, *Model Theory for Modal Logic. Kripke Models for Modal Predicate Calculi.* 1978.

128. Howard Alexander Bursen, *Dismantling the Memory Machine. A Philosophical Investigation of Machine Theories of Memory.* 1978.

129. Marx W. Wartofsky, *Models: Representation and the Scientific Understanding.* Boston Studies in the Philosophy of Science, Volume XLVIII. 1979.

130. Don Ihde, *Technics and Praxis. A Philosophy of Technology.* Boston Studies in the Philosophy of Science, Volume XXIV. 1978.

131. Jerzy J. Wiatr (ed.), *Polish Essays in the Methodology of the Social Sciences.* Boston Studies in the Philosophy of Science, Volume XXIX. 1979.

132. Wesley C. Salmon (ed.), *Hans Reichenbach: Logical Empiricist.* 1979.

133. Peter Bieri, Rolf-P. Horstmann, and Lorenz Krüger (eds.), *Transcendental Arguments in Science. Essays in Epistemology.* 1979.

134. Mihailo Marković and Gajo Petrović (eds.), *Praxis. Yugoslav Essays in the Philosophy and Methodology of the Social Sciences.* Boston Studies in the Philosophy of Science, Volume XXXVI. 1979.

135. Ryszard Wójcicki, *Topics in the Formal Methodology of Empirical Sciences.* 1979.

136. Gerard Radnitzky and Gunnar Andersson (eds.), *The Structure and Development of Science.* Boston Studies in the Philosophy of Science, Volume LIX. 1979.

137. Judson Chambers Webb, *Mechanism, Mentalism, and Metamathematics. An Essay on Finitism.* 1980. (Forthcoming.)

138. D. F. Gustafson and B. L. Tapscott (eds.), *Body, Mind, and Method. Essays in Honor of Virgil C. Aldrich.* 1979.

139. Leszek Nowak, *The Structure of Idealization. Towards a Systematic Interpretation of the Marxian Idea of Science.* 1979.

140. Chaim Perelman, *The New Rhetoric and the Humanities. Essays on Rhetoric and Its Applications.* 1979.

141. Wlodzimierz Rabinowicz, *Universalizability. A Study in Morals and Metaphysics.* 1979.

GPSR Compliance

The European Union's (EU) General Product Safety Regulation (GPSR) is a set of rules that requires consumer products to be safe and our obligations to ensure this.

If you have any concerns about our products, you can contact us on ProductSafety@springernature.com

In case Publisher is established outside the EU, the EU authorized representative is:

Springer Nature Customer Service Center GmbH
Europaplatz 3
69115 Heidelberg, Germany

Batch number: 10761599

Printed by Printforce, the Netherlands